Contributions to Management Science

For further volumes:
www.springer.com/series/1505

Francesco Ferrucci

Pro-active Dynamic Vehicle Routing

Real-Time Control and Request-Forecasting Approaches to Improve Customer Service

Physica-Verlag
A Springer Company

Francesco Ferrucci
WINFOR–Business Computing
and Operations Research
University of Wuppertal
Wuppertal, Germany

Additional material to this book can be downloaded from http://extras.springer.com.

ISSN 1431-1941 Contributions to Management Science
ISBN 978-3-642-33471-9 ISBN 978-3-642-33472-6 (eBook)
DOI 10.1007/978-3-642-33472-6
Springer Heidelberg New York Dordrecht London

Library of Congress Control Number: 2013932351

Keywords: Perishable goods, Dynamic vehicle routing, Real-time control, Request forecasting, Stochastic knowledge, Past request information, Tabu Search, Request data quality

Printed on acid-free paper

Physica-Verlag is a brand of Springer
Springer is part of Springer Science+Business Media (www.springer.com)

You can see far because you are standing on the shoulders of giants.

(Inspired by Bernard of Chartres)

I dedicate this book to my mother and my grandmother

Foreword

Obtaining timely transportation processes is increasingly becoming a crucial factor in modern supply chains. By implementing nearly bufferless processes, production systems strongly depend on a timely delivery of required intermediate products or materials. Since possible delays may result in costly outages, implementing efficient transportation processes is crucial. Additionally, ensuring the loyalty of end-consumers requires fulfilling of their rising expectations on a fast order delivery. This applies particularly to scenarios where these orders comprise goods that are perishable and/or are already subsequent deliveries. Besides the existing urgency, various sources of dynamism, which are present in real-world distribution processes, make a timely and efficient implementation difficult. Consequently, the development of suitable adaptive real-time control approaches is a vital and current research topic in transportation science. Here, pure deterministic approaches are distinguished from approaches that additionally integrate stochastic knowledge about the probability that future requests or events will occur. Since deterministic approaches only adapt existing tour plans according to already known requests, they are frequently denoted as reactive real-time concepts. In contrast, the exploitation of existing stochastic information about the arrival of future requests additionally enables a pro-active guidance of vehicles to request-likely areas. The PhD thesis written by Francesco Ferrucci provides significant contributions to this vital research field. It was mainly motivated by real-world subsequent delivery processes of newspapers. Due to the high perishability of the transported goods, these distribution processes have to be executed under considerable urgency. Starting from the application of a pure deterministic real-time approach that already enables an efficient execution of dynamic transportation processes, this research work analyzes whether and in particular under which circumstances the integration of stochastic knowledge about expected future request arrivals leads to further improvements. For this purpose, a general cluster scheme for exploiting existing past request information in an offline step is proposed. Subsequently, the resulting clusters are transformed into so-called dummy customers with specific time window constraints. These artificial requests are integrated into the tour plans by applying the deterministic real-time approach in order to allow for more efficient and robust tour plans. In contrast to earlier approaches that assume the availability of distributions which can be defined in a closed form, the utilized stochastic knowledge is systematically attained by the offline application of a sophisticated cluster analysis that integrates specific quality restrictions. Therefore, the proposed method can be applied to a larger number of real-world applications.

In order to analyze the performance of the proposed concept, extensive computational experiments are conducted. The results of these experiments strongly indicate that the integration of the generated dummy customers is promising when a minimum temporal as well as spatial diversity of the past request data is given. This is impressively shown by several specifically designed evaluations. On top of that, besides the degree of structural distortion, the degree of structural diversity (dosd) is introduced as a new measure that allows an offline prediction as to whether the integration of stochastic knowledge is useful for a given past request data set, or not. Specifically, a strong correlation between the dosd and the improvements that can be yielded is statistically proven. Furthermore, the PhD thesis provides a detailed analysis of the impact of allowed vehicle en-route diversions. In contrast to the literature, where vehicle en-route diversion is either completely allowed or forbidden, this research work analyzes intermediate scenarios in detail. Here, due to the increased stress level of the drivers that may lead to additional costs, the allowance of an en-route diversion requires a minimum cost reduction. On the basis of extensive computational results, this important trade-off is analyzed in detail for the first time.

Having known Francesco for a long time and intensively worked with him on sophisticated real-time control concepts for vehicle routing, it is not at all surprising to me that his PhD thesis is one in which I have taken great pleasure and interest to read. He has always been very ambitious and zealous in trying to understand research problems thoroughly and has proven to possess the necessary endurance and scientific understanding to finally obtain the needed insights into the underlying problem structure. Moreover, he reprograms the source codes in order to integrate new ideas immediately. Therefore, after discussing new concepts and possible workarounds late in the evening, you may receive an email early in the morning informing you about their performance in recent computational tests. All in all, I can summarize these fruitful years by just saying that it was such a pleasure for me to be able to supervise him as a member of my team. I am convinced beyond doubt that we will keep in touch and cooperate in further practical and scientific projects in future.

Wuppertal, Germany
July 2012

Professor Stefan Bock

Preface

In many business relationships, unexpectedly occurring events lead to unfavorable situations which cause a high inconvenience or high costs for affected customers. Two examples of such situations are the unsuccessful delivery of newspapers to subscribers and the breakdown of important machines during production processes. In such situations, affected customers demand urgent assistance from a service company which resolves the situation by delivering required goods which can be material or immaterial, e.g., newspapers or repair services. In order to maintain customer loyalty, the minimization of customer inconvenience which increases with the required time for delivering the demanded goods (denoted as response time) is increasingly becoming the most crucial objective in such processes. Due to this urgent demand, these goods are considered as *perishable goods* in this book. In real-world applications such as the mentioned subsequent delivery of newspapers or in repairman companies operating under Service Level Agreements (SLAs), a quick service of customer requests which dynamically arrive during the execution of the transportation process is the highest priority. The considered applications are denoted as the Real-time Distribution Of Perishable Goods (RDOPG).

In this book, three contributions are made to the field of dynamic vehicle routing. First, based on results that have been achieved by recent marketing research, an appropriate dynamic routing model for RDOPG applications is formulated. In this model, customer inconvenience arising along with the response time achieved for occurring customer requests is considered and its minimization is pursued as the main objective. Specifically, in this book, we evaluate the impact of a linear and a quadratic evolution of customer inconvenience depending on response time combined with a maximum allowed response time on the attained solution quality. Moreover, a comprehensive introduction to tour planning is given and existing classifications for routing problems in the literature are presented. Based on these classifications, a new approach for classifying routing problems according to RDOPG-specific attributes is proposed. The second contribution represents the main part of this book. It evaluates the benefits that can be attained in the solution quality through the integration of stochastic knowledge about expected future customer requests in RDOPG applications. In order to evaluate these benefits, a new deterministic real-time control approach which centrally coordinates the transportation process during its execution is proposed. This real-time control approach considers the concurrency of tour plan execution and tour plan adaptation. The transportation process is adapted during its execution by solving individual problem instances utilizing a proposed stage-based Tabu Search metaheuristic. Moreover, the mathematical model used for representing the individual problem instances is introduced. The proposed deterministic real-time control approach is then extended to a pro-active real-time control approach which additionally makes use of stochastic knowledge about the spatial and temporal occurrence of expected future customer requests. The utilized stochastic knowledge is solely generated from past request information by a proposed forecasting approach and no prescribed probability distributions are utilized. Different quality criteria are considered in the proposed forecasting approach in order to ensure a minimum defined quality of the generated stochastic knowledge. In order to implement the pro-active real-time control approach, ways of handling and integrating the stochastic knowledge into the solution process by using dummy customers are proposed. According to the considered objective of minimizing customer inconvenience, the stochastic knowledge is used to guide vehicles to request-likely areas before real requests arrive there in order to service them quicker and thereby attain a higher solution quality.

Comprehensive computational experiments are carried out on real-world request data from a subsequent delivery of newspapers process (S^{REAL}) as well as on generated test scenarios (S^{GEN}). In all experiments, the effects of a variety of parameter settings are evaluated in order to analyze the effects on the solution quality. In order to identify scenarios in which the application of the proposed pro-active real-time control approach achieves superior results compared to the proposed deterministic real-time control approach, we introduce *structural quality* for classifying the value of stochastic knowledge in a given routing problem scenario according to the available request data. Specifically, we derive the structural quality of a request data set from two newly proposed criteria denoted as *structural distortion* and *structural diversity*. In addition to defining both criteria in detail, we also propose methods for determining the structural distortion and the structural diversity of a given request data set. Besides evaluating a proposed approach for determining structural distortion on request data sets with different levels of structural distortion, we introduce a two-dimensional structural diversity classification which is used to generate request data sets of S^{GEN} with significant differences in structural quality. Computational results show that in scenarios with high levels of structural diversity, the proposed pro-active real-time control approach is able to achieve significantly improved results compared to its proposed deterministic counterpart. Specifically, depending on the evolution of customer inconvenience, improvements in the solution quality of up to 26 % (linear2X customer inconvenience) and 47 % (quadratic customer inconvenience) above the ones attained by the deterministic real-time control approach can be achieved. Thanks to the proposed classification of structural diversity, differences in results obtained among the tested scenarios become explainable. On the S^{REAL} request data, additional improvements of up to 7 % could be achieved by using the pro-active real-time control approach. In order to explain the differences in the additionally attainable improvements on S^{GEN} and S^{REAL}, we performed a visual analysis on both types of request data. In this analysis, further characteristics of scenarios which support an efficient application of the pro-active real-time control approach could be identified. Moreover, we propose a measure denoted as the degree of structural diversity (*dosd*) for determining the structural quality of given, arbitrary request data sets. We show the effectiveness of the *dosd* by analyzing correlations existing between the *dosd* and the improvements attained by the pro-active real-time control approach on S^{GEN} and S^{REAL}. The results show that there exists a strong linear correlation between the proposed *dosd* and the attainable improvements. Therefore, the proposed structural diversity and the *dosd* provide good starting points for assessing the structural quality of given request data sets which supports an efficient application of stochastic knowledge in real-time control approaches under different scenarios. Moreover, the *dosd* also provides management with decision support in efficiently customizing the proposed stochastic knowledge generation approach in given scenarios using results achieved in the computational experiments.

The third contribution of this book is a method for considering driver inconvenience aspects which arise from vehicle en-route diversion activities. Specifically, a method is proposed which makes it possible to restrict the number of vehicle en-route diversion activities performed by the real-time control approach by restricting the allowed vehicle en-route diversion activities to those which improve the solution quality by at least a defined threshold value. Computational experiments show the efficiency of the proposed method. Specifically, by utilizing appropriate threshold parameter values, driver inconvenience can be reduced by significantly lowering vehicle en-route diversion activities and yet still achieving significant improvements compared to completely forbidding vehicle en-route diversion activities. As also shown in the second contribution, the pro-active real-time control approach provides a significantly higher solution quality in request data scenarios comprising a high structural quality compared to the deterministic real-time control approach. Even when all vehicle en-route diversion activities are completely forbidden, the pro-active real-time control approach still achieves a significantly better solution quality compared to the deterministic real-time control approach with all vehicle en-route diversion activities allowed.

Wuppertal, Germany
July 2012

Francesco Ferrucci

Acknowledgments

During my very exciting time as a research assistant from October 2006 to July 2012 at the Department of Business Computing and Operations Research (WINFOR) at the University of Wuppertal in Germany, I gathered many terrific experiences in science, practical research, and life. Motivated by the challenges in a real-world application that we got to know from a company, I prepared my PhD thesis originally entitled *"Development and evaluation of pro-active real-time control and forecasting approaches for the distribution of perishable goods utilizing past request information to improve quality of customer service and reduce driver inconvenience"* on which this book is based.

My thesis was supervised by Professor Stefan Bock whom I have personally known for more than 7 years now. While attending Stefan's lectures during my study time, I quickly realized that we both eagerly share the same enthusiasm for developing and evaluating sophisticated algorithms and solution methods for real-world problems. While writing my diploma thesis under his supervision, Stefan also supported me with many creative ideas that greatly inspired me. Having known Stefan as someone who is also always keen at discussing ideas and analyzing results in detail, I decided that it would be a great idea to write my PhD thesis under his supervision. During this time, I had the opportunity of gathering a lot of new expertise as a teaching assistant, research assistant, project assistant, and project coordinator. I would like to thank Stefan for insightfully supervising my thesis, always having time for exchanging research ideas and also for being both a good lecturer and friend. Thank you very much, Stefan, for all the possibilities you gave me during my time writing my thesis. I have enjoyed working with you from the very first day.

One of the highlights during my PhD time was my three months' visit to Montreal in Canada. In addition to experiencing the beauty of Canada in winter time, I also had the extraordinary possibility of jointly working on my thesis with Professor Michel Gendreau, one of the world's most famous professors in transportation science at the CIRRELT—the leading research institute for logistics and transportation. By turning my thesis into an international research project, I had a very enriching and fruitful collaboration with Michel. After my research time in Montreal, I was allowed to give talks about our joint work with Michel at the international conferences of INFORMS in San Diego, INFORMS ALIO in Buenos Aires, EURO in Lisbon, and OR in Munich. Thanks, Michel, for all of your ideas and fruitful advice. It has been an honor for me to work with you. Furthermore, I thank Verena Schmid who I first met in Canada where we had the unforgettable ability to discover great places in Montreal together. She is always a valuable conversational partner for scientific and non-scientific discussions.

I would also like to thank all of my colleagues at WINFOR for creating a friendly as well as very productive work environment. The department has always been a place with a relaxed atmosphere while working together with highly skilled colleagues who are also experts at table top soccer and other great activities. Thank you, Volker Arendt, Rudolf Bauer, Benjamin Dahmen, Daniel Dittmar, Paul Göpfert, Simon Görtz, Torben Kuschel, Christian Rathjen, and all others for great and helpful discussions and the wonderful time. I would also like to thank Simon Görtz for his valuable ideas we discussed in our scientific work group which have been greatly helpful for my research work and thesis. In the same spirit, my thanks go to my Master's students Britta Kortmann, Sandra Klewer, and Jens Keune for providing high-quality results which were very useful for my thesis. I also thank Professor Krumke who provided us with access to his VRP solver. Special thanks also go to DDS Digital Data Services GmbH in Karlsruhe, Germany for providing me with the real road network data for my research.

I am greatly indebted to Maria Muwanga for her great ideas and for her dedicated support in polishing the English of my thesis turning it into a shiny piece of gold. I would also like to thank my friends who additionally proofread my thesis and also provided insightful remarks.

Finally, I would like to wholeheartedly thank my family and my friends. All their love, support, kindness, and encouragement have helped me a lot in preparing my thesis and mean more to me than words can say.

Wuppertal, Germany Francesco Ferrucci
July 2012

Contents

List of Figures

List of Tables

List of Algorithms

Abbreviations

Abbreviation	Description	Defined on page
ADFSVRPSTW	The Asymmetric Dynamic Forecasted Stochastic Vehicle Routing Problem with Soft Time Windows	85
ADP	Approximate Dynamic Programming	123
ADTSPTW	The A Priori Dynamic Traveling Salesman Problem with Time Windows	114
ADW strategy	Advanced Dynamic Waiting strategy in Mitrovic-Minic and Laporte (2004)	105
ASAP	The After Salesman Problem	106
B2B	Business To Business	1
B2C	Business To Customer	1
B&B	Branch-and-Bound	28
B&C	Branch-and-Cut	28
B&C&P	Branch-and-Cut-and-Price	28
B&P	Branch-and-Price	28
BRH	Branch-and-Regret Heuristic in Hvattum et al. (2007)	134
BUSIEST strategy	BUSIEST strategy in Larsen (2000)	115
CCC	Capacity-Cut Constraints	21
CCP	Chance Constrained Program	53
CRC-32	Cyclic Redundancy Check with a checksum size of 32 bits	197
CRN	Common Random Numbers	209
CVRP	The Capacitated Vehicle Routing Problem	21
DARP	The Dial-A-Ride Problem	24
DDARP	The Dynamic Dial-A-Ride Problem	93
DF strategy	Drive-First strategy in Mitrovic-Minic and Laporte (2004)	104

Abbreviation	Description	Defined on page
DFSVRP	The Dynamic Forecasted Stochastic Vehicle Routing Problem	56
DPDP	The Dynamic Pickup and Delivery Problem	93
DPDPSTW	The Dynamic Pickup and Delivery Problem with Soft Time Windows	103
DSHH	Dynamic Scenario Hedging Heuristic in Hvattum et al. (2006)	133
DSVRP	The Dynamic Stochastic Vehicle Routing Problem	56
DTRP	The Dynamic Traveling Repairman Problem	109
DTSP	The Dynamic Traveling Salesman Problem	96
DVA	The Dynamic Vehicle Allocation Problem	119
DVRP	The Dynamic Vehicle Routing Problem	56
DVRPSTW	The Dynamic Vehicle Routing Problem with Soft Time Windows	85
DVRPTW	The Dynamic Vehicle Routing Problem with Time Windows	96
DW strategy	Dynamic Waiting strategy in Mitrovic-Minic and Laporte (2004)	105
EDGE	Enhanced Data Rates for GSM Evolution	62
FCD	Floating Car Data	58
FCFS	First-come, first-served	107
FCFS policy	First-come, first-served policy in Bertsimas and van Ryzin (1991)	110
FCFS-mSQM policy	First-come, first-served multiple-vehicle Stochastic Queue Median policy in Bertsimas and van Ryzin (1993a)	111
FCFS-SQM policy	First-come, first-served Stochastic Queue Median policy in Bertsimas and van Ryzin (1991)	110
FCM	Fuzzy Clustering Method	125
FES	Future Event Set	207
FIFO	First In-First Out	35
GIS	Geographic Information System	63
GPDP	The General Pickup and Delivery Problem	22
GPRS	General Packet Radio Service	62
GPS	Global Positioning System	61
GSM	Groupe Spécial Mobile, Global System for Mobile Communications	62
HAPC	Hybrid Adaptive Predictive Control	125
HI-REQ strategy	HI-REQ strategy in Larsen (2000)	115
HSDPA	High Speed Downlink Packet Access	62
HSUPA	High Speed Uplink Packet Access	62
HVRP	The Heterogeneous Vehicle Routing Problem	35

Abbreviation	Description	Defined on page
JiT	Just in Time concepts	5
LNS	The Large Neighborhood Search neighborhood operator	194
LTL	Less-Than-truckLoad	7
MCVRP	The Multiple Compartment Vehicle Routing Problem	34
MDH	Myopic Dynamic Heuristic in Hvattum et al. (2006) and Hvattum et al. (2007)	133
MDP	Markov Decision Process	55
MDVRP	The Multi-Depot Vehicle Routing Problem	33
MIP	Mixed Integer Programming	19
MPA	Multiple Plan Approach in, e.g., Bent and van Hentenryck (2004b)	102
MPA^c	The Multiple Plan Approach with consensus ranking function in, e.g., Bent and van Hentenryck (2004b)	103
MPA^d	The Multiple Plan Approach with distance ranking function in, e.g., Bent and van Hentenryck (2004b)	103
MPS	Maximum Position Shift	95
MREL	The MultiRelocate neighborhood operator	194
MSA	Multiple Scenario Approach in, e.g., Bent and van Hentenryck (2004b)	129
MSA^c	The Multiple Scenario Approach with consensus ranking function in, e.g., Bent and van Hentenryck (2004b)	130
MSA^d	The Multiple Scenario Approach with distance ranking function in, e.g., Bent and van Hentenryck (2004b)	130
$MSA\text{-}LC^c$	The Multiple Scenario Approach with least commitment and consensus ranking function in, e.g., Bent and van Hentenryck (2003)	130
MST	Minimum Spanning Tree	29
MTF	Many-To-Few	24
MTM	Many-To-Many	24
MTO	Many-To-One	24
MTVRSP	The Multiple Trip Vehicle Routing and Scheduling Problem	35
NEAREST strategy	NEAREST strategy in Larsen (2000)	115
NN policy	Nearest Neighbor policy in Bertsimas and van Ryzin (1991)	110
NO REPOS strategy	NO REPOS strategy in Larsen (2000)	115
NP-hard	Non-deterministic polynomial time hard	17
OR	Operations Research	12
OSCO	Online Stochastic Combinatorial Optimization	128
OVRP	The Open Vehicle Routing Problem	34

Abbreviation	Description	Defined on page
Pμ policy	Pμ policy in the approach of Bertsimas and van Ryzin (1993a)	112
PART policy	Partitioning policy in Bertsimas and van Ryzin (1991)	110
PDP	The Pickup and Delivery Problem	24
PDTRP	The Partially Dynamic Traveling Repairman Problem	114
PROACTIVE10	The pro-active real-time control approach applied in a time-limit fashion for 10 seconds when dynamic events have occurred in the previous anticipation horizon	234
PVRP	The Probabilistic Vehicle Routing Problem	54
RA policy	Randomized Assignment policy in Bertsimas and van Ryzin (1993a)	111
RB strategy	Request Buffering strategy in Pureza and Laporte (2008)	106
RDOPG	Real-Time Distribution Of Perishable Goods	2
REL	The Relocate neighborhood operator	192
ROLLING20	The deterministic real-time control approach applied in a rolling fashion for 20 seconds in each anticipation horizon	220
SEO	Single event optimization	55
SFC policy	Space Filling Curve policy in Bertsimas and van Ryzin (1991)	110
SLA	Service Level Agreement	6
SPR	Stochastic Program with Recourse	53
SPS	Standard Positioning Service	61
SPSP	Single-Point Shortest-Path	158
SQMP	The Stochastic Queue Median Problem	110
SSSP	Single-Source Shortest-Path	158
$SSSP_R$	Single-Source Shortest-Path Reverse	158
STSP	The Symmetric Traveling Salesman Problem	29
SVRP	The Stochastic Vehicle Routing Problem	51
TIMELIMIT10	The deterministic real-time control approach applied in a time-limit fashion for 10 seconds when dynamic events have occurred in the previous anticipation horizon	220
TSP	The Traveling Salesman Problem	19
UMTS	Universal Mobile Telecommunications System	62
VRP	The Vehicle Routing Problem	20
VRPB	The Vehicle Routing Problem with Backhauls	33
VRPSC	The Vehicle Routing Problem with Stochastic Customers	53
VRPSCD	The Vehicle Routing Problem with Stochastic Customers and Stochastic Demands	54
VRPSD	The Vehicle Routing Problem with Stochastic Demands	53

Abbreviation	Description	Defined on page
VRPSPD	The Vehicle Routing Problem with Simultaneous Pickup and Delivery	33
VRPSST	The Vehicle Routing Problem with Stochastic Service Times	54
VRPSTT	The Vehicle Routing Problem with Stochastic Travel Times	54
VRPSTW	The Vehicle Routing Problem with Soft Time Windows	33
VRPTD	The Vehicle Routing Problem with Time Deadlines	33
VRPTW	The Vehicle Routing Problem with Time Windows	33
VRSP	The Vehicle Routing and Scheduling Problem	34
VWTH	Vehicular Waiting Time Heuristic in de Oliveira et al. (2008)	105
WDP	The Waiting Drivers Problem	116
WF strategy	Wait-First strategy in Mitrovic-Minic and Laporte (2004)	104
WTI	The Within Tour Insertion neighborhood operator	191
XBT	The Exchange Between Tours neighborhood operator	195
ZEROTIME	The deterministic real-time control approach applied in a zero time fashion when dynamic events have occurred in the previous anticipation horizon	220

General RDOPG Symbols

Symbol	Description	First used on page
∞	Infinity	21
A	The set of arcs of digraph G	157
$\tilde{A}$	The arcs of the road network $\tilde{R}$	156
a_i	Region i used in the S^{GEN} generation process	211
C	The set of valid clusters used in the proposed forecasting approach	177
C^{max}	Upper bound on the number of selectable clusters	178
C_n^{latest}	The set of the n clusters of C which start latest	231
c_{is}	The binary parameter indicating if segment s belongs to cluster c_i	178
c_i^{end}	The end time of cluster c_i	179
c_i^{endTL}	The temporal level at which cluster c_i ends	179
c_i^{start}	The start time of cluster c_i	179
c_i^{startTL}	The temporal level at which cluster c_i starts	178
c_{temp}	Temporal cluster examined during the cluster generation process	173
c_{temp}^{avgTT}	The average travel time from node $n_{c_{temp}}$ to all requests observed within the segments of c_{temp} over the last n^{f} working days	175
D	Representation of the depot node in a tour plan	189
D^{p}	The geographical position of the depot represented by the node in the road network $\tilde{R}$ at which the depot is located	157
D_τ	The set of pending requests at τ	153
$DC^{\lambda\text{rem}}$	Dummy customer removal threshold	181

Symbol	Description	First used on page
DC^{maxAvgTT}	The predefined maximum average travel time from $n_{c_{temp}}$ to all past requests observed within c_{temp} over the last n^{f} working days	175
DC^{mse}	The maximum spatial extension of a valid cluster	173
DC^{mte}	The maximum temporal extension of a valid cluster	173
$DC^{\text{min}\lambda}$	The minimum rate parameter required for a valid cluster	173
DC^{minPQ}	The minimum demanded Poisson quality of a valid cluster	227
DC^{radiusTT}	Maximum predefined travel time radius of $n'_{c_{temp}}$	175
DC^{se}	The spatial extension of all segments s of the set S	172
DC^{te}	The temporal extension of all segments s of the set S	172
d_{ij}	The travel distance between locations i and j	22
dod	The Degree of Dynamism	65
$dosd$	The Degree of Structural Diversity	249
E	Representation of the dummy end tour plan element E	189
$edod$	The Effective Degree of Dynamism	66
$edod_{\text{tw}}$	The Effective Degree of Dynamism with Time Windows	66
G	The digraph derived from the road network $\tilde{R}$	157
H	The solution approach (often a heuristic approach) evaluated in the value of information	74
K	The set of available vehicles, i.e., the available fleet	81
$\mathcal{K}$	The number of available vehicles, i.e., the fleet size	81
$\lambda(c_{temp})$	The rate parameter of cluster c_{temp}	174
$\lambda(a_i, t_j)$	Request arrival rate parameter in region i in time slice j	212
$\lambda(c_i, \tau^+)$	The remaining rate parameter of cluster c_i at time point τ^+	179
$\lambda(p_s)$	The rate parameter of the time-space Poisson process of segment s	172
$\lambda(t_j)$	Total request arrival rate parameter in time slice j	212
LNS^{execMax}	Maximum number of executions of LNS performed within one iteration	195
LNS^{execMin}	Minimum number of executions of LNS performed within one iteration	195
LNS^{reqRemF}	Maximum request removal factor applied in LNS	195
LNS^{reqSelR}	The ratio of requests considered within an execution of LNS	195
M	Big number	178
m	The length of each time period used to determine the *dosd*	250
$MREL^{\text{execMax}}$	Maximum number of executions of MREL performed within one iteration	194
$MREL^{\text{execMin}}$	Minimum number of executions of MREL performed within one iteration	194

Symbol	Description	First used on page
$MREL^{\text{reqMax}}$	Maximum number of removed requests within an execution of MREL	194
$MREL^{\text{reqMin}}$	Minimum number of removed requests within an execution of MREL	194
$MREL^{\text{reqSelR}}$	The ratio of requests considered within an execution of MREL	194
N	The set of nodes of digraph G	157
$\tilde{N}$	The nodes of the road network $\tilde{R}$	156
$n_{c_{temp}}$	The node in the road network $\tilde{R}$ at which the dummy customer is placed in cluster c_{temp}	175
n^{f}	The number of past working days utilized in the segment-based forecasting approach	171
P_{τ}^{bt}	The best theoretical tour plan at τ	152
$P_{\tau}^{\text{r}*}$	The best tour plan of the set $\tilde{P}_{\tau}$	154
P_{τ}^{r}	The relevant tour plan at τ	152
P_{τ}^{t}	The theoretical tour plan at τ	152
$P_{\tau+}^{\text{bt}}$	The best tour plan found during the invocation of the solution method	155
$P_{\tau+}^{\text{r}}$	The tour plan $P_{\tau}^{\text{r}*}$ adapted to its anticipated state	155
$P_{\tau+}^{\text{t}*}$	The tour plan $\tilde{P}_{\tau}^{\text{t}}$ adapted to its anticipated state in accordance with $P_{\tau+}^{\text{r}}$	155
$P_{\tau+}^{\text{t}}$	The theoretical tour plan examined last during the invocation of the solution method	155
$p_{S'}^{*}$	Compound Poisson process comprising the set of segments S'	173
p_s	The time-space Poisson process of segment s	172
$p(a_i, t_j)$	Request arrival probability in region i in time slice j used in the S^{GEN} generation process	212
$PS_{\lambda}(x)$	The probability mass function of the Poisson distribution with rate parameter λ and value x	176
$\tilde{P}_{\tau}^{\text{bt}}$	The adapted best theoretical plan P_{τ}^{bt} with integrated requests of set D_{τ}	154
$\tilde{P}_{\tau}^{\text{r}}$	The adapted relevant plan P_{τ}^{r} with integrated requests of set D_{τ}	154
$\tilde{P}_{\tau}^{\text{t}}$	The adapted theoretical plan P_{τ}^{t} with integrated requests of set D_{τ}	154
$\tilde{P}_{\tau}$	The set of adapted tour plans at τ	154
R	The set of customer requests	81
$\mathbb{R}$	The set of real numbers	157
R^*	The set of requests selected for reinsertion by a neighborhood operator	188
R^2	Coefficient of determination	252

Symbol	Description	First used on page
R^{mrt}	The maximum response time allowed for servicing requests without high penalty costs R^{pen}	82
R^{pen}	High penalty costs occurring in case requests are serviced with a response time larger than R^{mrt}	82
R^{st}	The service time assumed to be required for all requests	82
R^{w}	A list containing all currently unserviced requests in order of non-increasing objective function contribution values	194
R_τ	The set of requests known at time point τ	151
R_τ^{U}	The set of known unserviced requests at time point τ	157
$\mathcal{R}_\tau^{\text{U}}$	The number of unserviced requests at time point τ	157
$\mathcal{R}$	The number of customer requests	81
RD	Level of RegionDiversity present in a request data set of S^{GEN}	212
r	Pearson product-moment correlation coefficient	251
r_i^{a}	The arrival time of request i in the system	151
r_i^{e}	The start time of the time window of request i	159
$r_i^{\text{e}}(\tau^+)$	The dynamically updated start of the time window of dummy customer i at τ^+	181
r_i^{l}	The end time of the time window of request i	159
r_i^{p}	The geographical position of request i in the road network, i.e., the node at which it is located	159
r_i^{s}	The service time of request i	159
$r_i^{\text{s}}(\tau^+)$	The dynamically updated service time of dummy customer i at τ^+	181
r_i^{w}	The weight factor of request i in the objective function	159
$r_i^{\text{w}}(\tau^+)$	The dynamically updated weight factor of dummy customer i at τ^+	181
$\tilde{R}$	The utilized road network	156
S	The set of segments in the segment-based forecasting approach	172
S_C	Segments which are assigned to clusters of the set C	178
S^{GEN}	The designed request data class evaluated in the computational experiments	210
$S^{\text{GEN}_{\text{all}}}$	The request data set which results from 30 systematically chosen days from S^{GEN}	220
S^{MIXa}	Generated request data set with high structural diversity	227
S^{MIXa2x}	Generated request data set with high structural diversity	227
S^{MIXb}	Generated request data set with high structural diversity	227
S^{MIXc}	Generated request data set with high structural diversity	227
S^{REAL}	The real-world request data class evaluated in the computational experiments	210

Symbol	Description	First used on page
S^{REAL^*}	The request data set derived from S^{REAL}	241
t^{a}	The length of the anticipation horizon	152
t_i^{res}	Response time of request i	161
t_j	Time slice j used in the S^{GEN} generation process	212
t_j^{e}	End time of time slice j	212
t_j^{s}	Start time of time slice j	212
t_τ^{calc}	The time period used for anticipating the static problem instance generated at τ	152
t_{ij}	The travel time required to travel from location i to location j in the digraph G	157
τ^+	The start time of the subsequent anticipation horizon	152
τ^-	The start time of the previous anticipation horizon	153
τ	The current time during the execution of the transportation process	151
$t_{c_i}^{\mathrm{rem}}$	The time point at which the corresponding dummy customer of cluster i is removed from the tour plan	181
TD	Level of TimeDiversity present in a request data set of S^{GEN}	212
$\Theta(x)$	A function defined as $\Theta(x) = 1$ if $x > 0$ and 0 otherwise	37
$\tilde{t}$	A function describing the travel time for each arc $\tilde{a} \in \tilde{A}$ of the road network $\tilde{R}$	157
$V_H(I)$	The value of information of solution approach H on instance I	74
v_k	Tuple representing vehicle k	159
v_k^+	The start location of vehicle k	157
v_k^-	The end location of vehicle k	157
v_k^{p}	The geographical position of vehicle k represented by the node that the vehicle reaches next in the road network $\tilde{R}$	159
$v_k^{\delta t}$	The availability time of vehicle k at its start location v_k^+ relative to the time point at which the anticipated system state is generated	157
X	The set of binary decision variables	160
x_i	Binary decision variable indicating if cluster $i \in C$ is selected	178
x_{kij}^{d}	Binary decision variable indicating if vehicle k is driving from location i to location j	160
x_{ki}^{s}	Binary decision variable indicating if vehicle k services request i	160
Y	The set of decision variables depending on the set X	160
y_i^{a}	The arrival time of the assigned vehicle at request i	160
y_i^{s}	The start time of service at request i	160

Symbol	Description	First used on page
$z_H(I)$	The objective function value of the solution generated by solution approach H for problem instance I in the value of information	74
$z_H(I^{O})$	The objective function value of the solution which has been generated by solution approach H on the corresponding offline instance of I in the value of information	74

Symbols in Chap. 2

Symbol	Description	First used on page
A	The set of arcs in a directed graph	18
A_i	The arrival time of the servicing vehicle at the location of request i in a GPDP	24
C	The loading capacity of the homogeneous vehicles in a CVRP	21
$c_1(P^a)$	The costs which occur in an SVRP according to the information known before the execution of the transportation process	53
$c_2(P^a)$	The costs which are expected to occur in an SVRP during the execution of the transportation process	53
c_{ij}	The costs for traveling from location i to location j	18
CR_A	The competitive ratio of solution approach A	73
D_i	The departure time of the servicing vehicle at location i in a GPDP	23
d_i	The demand of request i in a CVRP	21
E	The set of edges in an undirected graph	18
e_i	The start time of the time window of request i in a GPDP	24
$f(x)$	The general objective function of a GPDP	23
G	The directed or undirected graph	18
I	Instance I of the considered dynamic routing problem in the competitive analysis	73
K	The set of available vehicles in a routing problem	20
K^+	The unification of all vehicle start positions in a GPDP	22
K^-	The unification of all vehicle end positions in a GPDP	22
$\mathcal{K}$	The number of available vehicles in a routing problem	20

Symbol	Description	First used on page
k^+	The start location of vehicle k in a GPDP	22
k^-	The end location of vehicle k in a GPDP	22
L	The unification of the pickup and delivery locations in a GPDP	22
l_i	The end time of the time window of request i in a GPDP	24
M	The set of all theoretically possible instances of the considered dynamic problem in the competitive ratio	73
n_{all}	The number of all requests in a routing problem, i.e., $n_{\text{all}} = n_{\text{stat}} + n_{\text{dyn}}$	65
n_{dyn}	The number of dynamically arriving requests during the execution of the transportation process	65
n_{stat}	The number of known requests during the initial tour plan generation process	65
P^{a}	The a priori tour plan in an SVRP	53
p_i	Occurrence probability of request i in a VRPSC	53
Q_k	The individual capacity of vehicle k in a GPDP	22
$\bar{q}_i$	The weight of the total cargo which needs to be transported for servicing request i in the GPDP	22
q_j	The weight which is assigned to each location j in a GPDP	22
R	The set of customer requests	19
R_i^+	The set of pickup locations of request i in a GPDP	22
R_i^-	The set of delivery locations of request i in a GPDP	22
$\mathcal{R}$	The number of customer requests in a routing problem	19
$r(S)$	The minimum number of vehicles required to service all requests of set S	21
r_i	The request $i \in R$	16
r_i^{a}	The point in time when customer request i arrives in the system	37
r_i^{e}	The start time of the time window of request i	32
r_i^{l}	The end time of the time window of request i	32
r_i^{lt}	Length of the lateness that has occurred at request i	36
r_i^{re}	The available reaction time of request i	66
S	Subset of all requests denoted by R	21
T	The time during the considered planning horizon up to which new requests can arrive in the system used in the degree of dynamism and its variants	65
T'	Set of time intervals representing time-dependent travel times	35
t_{ij}	The travel time required to travel from location i to location j	23

Symbol	Description	First used on page
$t_{ij}(\tau)$	A function denoting the travel time between two relevant locations i and j depending on the current time τ	35
V	The set of vertices (or nodes) in a graph	18
v_k	The vehicle $k \in K$	16
W	The unification of all vehicle start and end positions in a GPDP	22
X	Set of binary decision variables	19
x_{ij}	Binary decision variable indicating if location j is directly visited after location i, used in, e.g., the TSP	19
x_{ijk}	Binary decision variable that is 1 if and only if vehicle $k \in K$ directly travels from location i to location j $(i, j \in V)$ in a three-index vehicle flow formulation	21
Y	The set of binary variables	21
y_i	The current load of the servicing vehicle when arriving at location i in a GPDP	23
y_i^{S}	The point in time when the servicing vehicle arrives at request i	33
y_{ik}	The binary decision variable which is 1 if and only if vehicle $k \in K$ visits location $i \in V$ in a CVRP	21
z^*	The objective function value of the solution generated by an optimal solution approach when applied to the corresponding offline instance I^{O} in the competitive analysis	73
$z_A(I)$	The objective function value of the solution generated by solution approach A applied to instance I in the competitive analysis	73
z_{ik}	A binary decision variable indicating whether request i is serviced by vehicle k in a GPDP	23

Symbols in Chap. 4

Symbol	Description	First used on page
A_i, B_i	The start time and the end time of the inner time window of request i in the presented objective functions of Fagerholt (2000) and Fagerholt (2001)	138
a_i, b_i	The start time and the end time of the time window of request i in the presented objective functions of Sexton and Bodin (1985), Sexton and Choi (1986), and Dumas et al. (1990)	138
$D/G/m$	Queuing system with deterministic request arrivals, general service time distribution, and m servers (cf. Kendall 1953)	112
δt	The length of the system state anticipation at time point t in Ichoua et al. (2000)	143
δt_k	Amount of time that vehicle k should wait at its current position in the approach of Ichoua (2001)	121
E_i, L_i	The start time and the end time of the outer time window of request i in the presented objective functions of Fagerholt (2000), Fagerholt (2001)	138
$G/G/m$	Queuing system with general independently distributed request interarrival times, general service time distribution, and m servers (cf. Kendall 1953 and Tijms 2003, p. 341 et seq.)	111
H	The considered planning horizon in Bent and van Hentenryck (2004c)	128
$\overline{H}$	The end of the planning horizon in Bent and van Hentenryck (2004c)	128
$\underline{H}$	The beginning of the planning horizon in Bent and van Hentenryck (2004c)	128

Symbol	Description	First used on page
λ_i	Individual intensity parameter for request arrivals in subregion i in Larsen (2000)	114
λ	Intensity parameter denoting the request arrival rate in Bertsimas and van Ryzin (1991)	110
$M/G/1$	Queuing system with a Poisson (Markovian) input, i.e., request arrivals, general service time distribution, and one server (cf. Kendall 1953 and Tijms 2003, p. 341 et seq.)	111
p_i^{MAX}	Penalty costs of request i in the objective functions of Fagerholt (2000) and Fagerholt (2001)	138
r	The chosen request out of the set R to be serviced next in Bent and van Hentenryck (2004c)	128
S	The set of sample scenarios	130
s	A sample scenario $s \in S$	130
t	The current time during the execution of the transportation process	121
w	The objective function value of the solution of the transportation process in Bent and van Hentenryck (2004c)	129
$w(r)$	The gain of request r according to the considered objective function	128

Chapter 1
Introduction

In many real-world business to customer (B2C) and business to business (B2B) applications, unexpectedly occurring events may lead to unfavorable situations that cause inconvenience or high costs for affected customers (or companies; in what follows, we will use the term customers). For example, considering the case of a newspaper subscription, a newspaper that has not been delivered in the morning creates a high amount of inconvenience for the subscriber since reading newspapers is a common habit in the daily lives of many people. Another example can be found in manufacturing companies. If an important machine breaks down during the production process, high outage costs occur. In order to provide a rough idea about the dimension of a production outage, consider two of the major automobile brands BMW (BMW Group) and Mercedes-Benz (Daimler Group) with a total number of produced cars of 1,236,989 (BMW) and 1,312,456 (Mercedes-Benz) in 2010. One hour of a production outage in all manufacturing facilities of these brands causes on average a loss of 140 or 150 produced cars, or, according to their revenue, a financial loss of over 6 million euros (cf. BMW Group 2011, Daimler Group 2011).

Due to the occurring inconvenience and costs in such situations, affected customers want to have these situations quickly resolved. Therefore, customers demand assistance from a specific *service company* which then resolves the situation by delivering required *goods* with a high urgency. These goods may be material or immaterial, e.g., spare parts, services, or a combination of both. It is assumed that for the mentioned exceptional situations, a contract with a specific service company which is responsible for quickly resolving the situation has been negotiated in advance. The key aspect in these situations is that, due to the high urgency, service companies are forced to deliver the required goods to the affected customer as quickly as possible. Specifically, since the occurrence of the unpleasant situation has already caused a high amount of inconvenience to the customer, a delayed delivery of these goods causes additional significant customer inconvenience. Therefore, this additional arising customer inconvenience decreases the value of the requested goods with increasing time required for their delivery. Hence, we will refer to these goods as *perishable goods* throughout this book. Their flow is always directed from the service company to the customer so that the triggered service processes can be

F. Ferrucci, *Pro-active Dynamic Vehicle Routing*, Contributions to Management Science,
DOI 10.1007/978-3-642-33472-6_1, © Springer-Verlag Berlin Heidelberg 2013

referred to as *ad-hoc distribution processes*. According to these characteristics, the real-world processes considered in this book are denoted as *Real-Time Distribution of Perishable Goods* (RDOPG) applications.

In the marketing literature, a significant amount of research on customer satisfaction has been conducted. According to Tax et al. (1998), customer satisfaction relates to how a company deals with a customer complaint. Moreover, Smith et al. (1999) state that in case of a failure, a company has the possibility of either restoring customer satisfaction and recovering loyalty or risk customer dissatisfaction. Specifically, failing to deliver the expected service quality may result in customer complaints, reduced customer satisfaction, and even in loss of customers. Parasuraman (1987) states that in "a market crowded by competitors claiming to be concerned about customer satisfaction, the real winners will be the ones with a truly customer-oriented corporate culture". Therefore, an adequate handling of customer complaints about service quality is crucial for service companies in order to maintain customer loyalty (cf. Brady and Cronin Jr. 2001, Pollack 2009). Specifically, for the service companies considered in RDOPG applications, providing a high service quality is considered as the essential factor for successfully maintaining and expanding their market positions by achieving a high customer loyalty (cf. Parasuraman et al. 1985, Zeithaml et al. 1996). Hence, it is the topmost priority of the considered service companies to fulfill or exceed the service quality expected by their customers in order to generate high customer satisfaction and to establish long-term customer connectivity.

The considered RDOPG applications are based on a variant of the Dynamic Vehicle Routing Problem (cf., e.g., Gendreau and Potvin 1998) since new *customer requests* or in short requests dynamically arrive at the considered service company during the planning horizon. A relevant characteristic is that customer inconvenience starts at the very moment when the corresponding customer makes a new request at the service company. This is different from most research on dynamic routing problems in the literature which focuses on applications where it is sufficient to service requests at *any time* within assigned time windows. In contrast, in RDOPG applications, a quick delivery of demanded goods has the highest priority. Due to the fact that customers become more inconvenienced the longer they have to wait for the requested goods, the considered objective is to *minimize the total customer inconvenience* which evolves with the request *response time*, i.e., the time period between the arrival of the customer request and start of service at the customer request location (cf. Larsen et al. 2007).

In order to attain low customer inconvenience values and thus a high service quality, the development and evaluation of a new *pro-active real-time control approach* for RDOPG applications is the main focus of this book. Specifically, this approach is used for coordinating the service companies' daily operative distribution processes which are also called *transportation processes* in what follows. The coordination of these transportation processes consists of efficiently utilizing the available service workers in real-time. Since it is assumed that each service worker uses a vehicle, in what follows, the terms service workers and vehicles will be synonymously used. With regard to the considered objective, an efficient vehicle utilization consists of coordinating the vehicles so that customer inconvenience is minimized. Since customer inconvenience depends on the achieved request response time as mentioned above, using methods that reduce the attained response time of requests is important. Therefore, in the proposed pro-active real-time control approach, the available vehicles are pro-actively coordinated. Specifically, they are guided in advance to areas in which future requests are expected within a short amount of time so that a faster service of these newly arriving requests and hence a higher solution quality can be attained. A challenging aspect here is obtaining such information about expected request arrivals since this is often not directly available in many real-world applications. In order to achieve this aim, new methods for automatically generating information about expected future requests from past request information are proposed in this book. Since such information contains uncertainty, it is denoted as *stochastic knowledge* in what follows. This stochastic knowledge is directly utilized in the pro-active real-time control approach during the coordination of the transportation process.

In this chapter, a *motivation* is given in which real-life applications belonging to the considered RDOPG applications are presented and relevant background information is provided. This is followed by a *categorization of the considered problem* whose tasks are also classified according to the different layers of transportation problems. Furthermore, the *aims and contributions of the book* as well as an *outline of the book* are presented.

1.1 Motivation

In B2C and B2B markets, customers expect high service quality from companies, especially in urgent ad-hoc after-sales services (cf. Parasuraman et al. 1991). The demand for such services is triggered by a failure in prior service or a failure in a system for which a service contract has been established. Since customers mainly rate service quality according to the customer inconvenience that they experience, which depends on the resulting response times, these increased requirements force service companies to deliver requested perishable goods under high time pressure. Specifically, the longer an affected customer waits, the more inconvenience will occur. On the other hand, quick assistance opens up possibilities for service companies to improve their image and expand their market position.

In what follows, three real-world examples which motivate the research on RDOPG applications considered in this book are given. In B2C markets, a typical application is the *subsequent delivery of newspapers*. This process occurs when a subscriber has not received his newspaper in the morning due to a delivery error (e.g., the delivery person forgot to deliver the newspaper), a delivery failure (e.g., drop-out of the corresponding delivery person), or a thievery. In each of these cases, it is of particular importance to the newspaper publisher to compensate the subscriber. This compensation can either be fulfilled by a voucher or a subsequent delivery. Since a quick subsequent delivery allows market differentiation for the

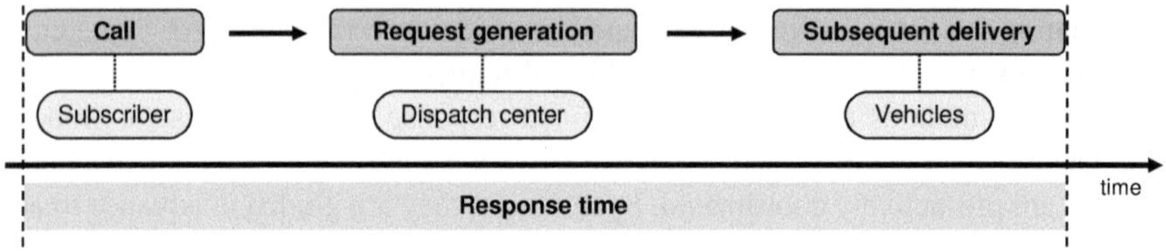

Fig. 1.1 Example RDOPG application: The subsequent delivery of newspapers

publishing house compared to providing a voucher, the subsequent delivery process considered in this book is of high practical relevance. Newspapers can be considered as perishable goods since customers demand a quick subsequent delivery of their newspaper since they are used to reading it at the same time every day. Specifically, many people are used to and want to read their newspaper during a particular time period during the day, e.g., while having breakfast or while traveling to work by public means. In the evening, the news is already outdated and therefore the newspaper is rather uninteresting at this time of day. Furthermore, emerging low-cost and convenient alternatives to printed newspapers such as cheaper or even free newspaper websites of competitors give customers more market options and a better position to enforce their demands. According to a recent study, these alternatives are increasingly used by U.S. consumers (cf. The Media Audit 2011). This exerts pressure on publishing houses to satisfy their customers in order to maintain customer connectivity which is important for two reasons. First, a high number of subscribers results in high revenues from selling newspapers. Second, a large subscriber base provides a good negotiating position to the publishing house and enables it to make good bargains for advertising space. This is especially important due to the emerging market share of Internet advertisements (see Table A.2 in Appendix A) which may lead to a decline in revenue of print advertisements as well as revenue from newspaper inserts (cf. Edmonds 2010).

In Fig. 1.1, the individual activities in the subsequent newspaper delivery process are illustrated. First, when a newspaper is missing, a subscriber calls the designated call and dispatch center. Here, a new request is generated which is subsequently serviced by designated vehicles. In order to facilitate this process, Bieding et al. (2009) propose the application of a system for the electronic communication of required information as well as utilizing online routing algorithms for coordinating the subsequent newspaper delivery process. Moreover, Ferrucci and Bock (2008) have shown that the utilization of an appropriate solution approach for generating the vehicles' delivery tours has a high potential of increasing the solution quality in such applications. A key aspect of this RDOPG application is that, when a vehicle has reached the customer's location, the *service time* required to service the customer, i.e., by handing over the newspaper, can be assumed to be quite short (about 60 seconds) as well as homogeneous for all customer requests. This allows a reliable estimation of the time that a vehicle will require at a customer request. Furthermore, capacity constraints are also not an issue since customers demand a newspaper to be subsequently delivered from a limited set of newspapers. Moreover, since newspapers are light, a sufficient number required for subsequent delivery activities during the day can be loaded onto the vehicles at the beginning of the day. Therefore, each vehicle can service any pending customer request. What makes the problem complex is that newspaper subscribers are usually geographically dispersed so that new requests can theoretically arise anywhere in the city. Another example of RDOPG applications in B2C environments is quick road assistance in case of a car failure. This application also belongs to the distribution of perishable goods since customers want to receive quick help in case of problems with their cars. The main difference from the previously described RDOPG application is that complex technical problems may occur due to high-tech automobiles so that the service time has a significantly higher degree of uncertainty.

In B2B markets one application of high economic relevance arises in the area of repairman companies which provide specialized repair services for severe failures occurring in production companies (cf. van de Klundert and Wormer 2010). Such repair services have gained a much higher importance recently since manufacturers have significantly changed their production habits over the last years in order to deal with increased complexities in production processes and in order to reduce costs (cf. Bock 2004, pp. 37–41):

- **Decrease of the production depth:** Due to technological advances, the complexity of produced goods is steadily increasing. Manufacturers try to cope with this increased complexity of their production processes by focusing on their core competencies so that the in-house production depth can be decreased. Another reason for reducing the production depth arises if required input goods can be cost-efficiently procured. Possibilities of cheap procurement have arisen in recent years due to a reduction in transportation costs. Specifically, since the 1990s, the European transportation market has dramatically changed so that foreign companies performing truck-based transportation of goods at lower costs are permitted to enter foreign markets (cf. Stumpf 1998, p. 19 et seq., Bock 2004, p. 30 et seq.). As shown in Fig. A.1 (see Appendix A), their market share has steadily increased leading to reductions in transportation costs.
- **Just in Time concepts (JiT):** In order to reduce storage costs for input goods, manufacturers instruct component suppliers to deliver input goods just at the time point when they are required (cf. Faria et al. 2010).

As mentioned above, these developments lead to leaner in-house production processes and cost-savings. However, they also make production processes more prone to errors. Due to the mentioned reductions in the production depth of individual manufacturers, a significantly increasing demand for transportation services for delivering required input goods arises (see Table A.1 in Appendix A). Furthermore, the mentioned increased complexity of many products due to technological advances has led to another development:

- **Special requirements and globalization:** Raw materials and expertise knowledge required for producing specific goods are only available in a limited number of countries. However, since these goods are required in a variety of production processes all over the world, strategic supply chain alliances formed by different companies according to input goods required in production processes are gaining more importance (cf. Ghiani et al. 2004, pp. 16–17). "As a result of globalization, more emphasis must be put on the efficient design and management of supply chains, sometimes at the world level" (Ghiani et al. 2004, p. 17).

If some input goods cannot be delivered as planned, manufacturers may encounter problems in keeping their production process running. This situation becomes even more critical when using JiT in which such bottlenecks arise much faster due to reduced storage capacities of required input goods. Furthermore, since many companies are connected to each other and their production processes have strong interdependencies, major disturbances throughout the supply chain may occur in case of failures in a single company in the supply chain (cf. Rice and Caniato 2003).

Since broken-down production lines cause significant financial losses every minute as mentioned above in the case of automotive companies, companies in supply chains often make individual contracts which specify expected service levels and define penalties in case the service level is not met by the supplying company (cf. Faria et al. 2010). Hence, in case of production failures, quick support from repairman companies is mandatory. Therefore, repairman companies themselves are also often bound to individual contracts with their customers. This type of contract is referred to as Service Level Agreement (SLA) and defines the type of problems which are covered as well as terms and conditions, e.g., a maximum allowed response time to a request of the corresponding customer. If a repairman company does not service a customer request within this maximum response time, high penalty costs occur. Moreover, it is advantageous for repairman companies to service customer requests before the end of this time period for two reasons. First, customers will experience better service quality so that the repairman company can attain a better image. Second, since resuming a production process earlier than stated in the SLA imposes less financial damage to the manufacturer, cost savings can be shared between both parties. In addition to unknown customer requests, the time period lengths required for service at different customer request locations often contain a high degree of uncertainty in repairman applications. Hence, the estimation of these time period lengths poses an additional challenge. Furthermore, for manufacturing companies, the point in time when a broken-down production is repaired is important. Hence, SLA contracts can also specify a maximum tolerated *downtime* (cf. Bertsimas and van Ryzin 1991) consisting of the sum of response time and required service time. However, since the service time of requests significantly depends on the considered RDOPG application and cannot be influenced by optimizing the transportation process, in this book we focus on customer inconvenience and solution quality aspects which depend on the achieved request response time.

Economic Need for Action As can be derived from the mentioned real-world applications, the considered RDOPG applications are of high practical relevance. In these applications, a quick reaction towards dynamically arriving customer demands is the topmost priority so that customer loyalty is maintained and customer inconvenience is kept at a low level. Since operational costs incurred by travel activities of vehicles can be regarded as negligible compared to costs incurred by a late response to customer requests, it is advantageous for companies to quickly service pending customer requests even if this requires additional travel activities. Hence, as mentioned above, it is advantageous to use a pro-active real-time control approach so that a quicker response to customer requests is attained thereby yielding high customer satisfaction and quality of service.

1.2 Categorization of the Considered Problem

The considered problem of coordinating the operative transportation processes in RDOPG applications is part of the *service operations management* in service companies (Johnston and Clark 2008, p. 3):

> Service operations management is concerned with delivering service to the customers or users of the service. It involves understanding the needs of our customers, managing the processes that deliver the services, ensuring our objectives are met, while also paying attention to the continual improvement of our services. As such, operations management is a central organisational function and one that is critical to organisational success.

Hence, "Service operations management is the term that is used to cover the activities, decisions and responsibilities of operations managers in service organisations. These managers are often called operations managers but many other titles are used, such as [...], fleet managers in transport companies, [...] and so on" (Johnston and Clark 2008, p. 4).

Furthermore, the considered RDOPG applications exhibit some characteristics of *short-haul freight transportation* processes (cf. Ghiani et al. 2004, p. 247 et seq.). Specifically, the requested goods are distributed to customers in a relatively small service area without transshipment. Since the used vehicles can carry more than one good at the same time, the resulting transportation problem is referred to as a less-than-truckload (LTL) problem (cf. Ghiani et al. 2004, p. 12). Hence, in such transportation problems it is likely that several vehicles are close to each other so that sophisticated methods for assigning and sequencing requests on vehicles can be advantageously applied. In contrast to classic short-haul freight transportation processes, there are differences in the applied objective function and the type of goods to be delivered in the considered RDOPG applications.

In the literature, short-haul transportation processes consist of different activities divided into three different *layers of planning* which can also be applied to the considered RDOPG applications. Besides different activities, each of these layers has a different time horizon (cf. Ghiani et al. 2004, p. 247 et seq.):

- **Strategic layer:** In this layer, long-term decisions are made. Generally, in transportation processes, this comprises choosing appropriate depot locations, defining operation zones as well as choosing transportation services that are offered. In

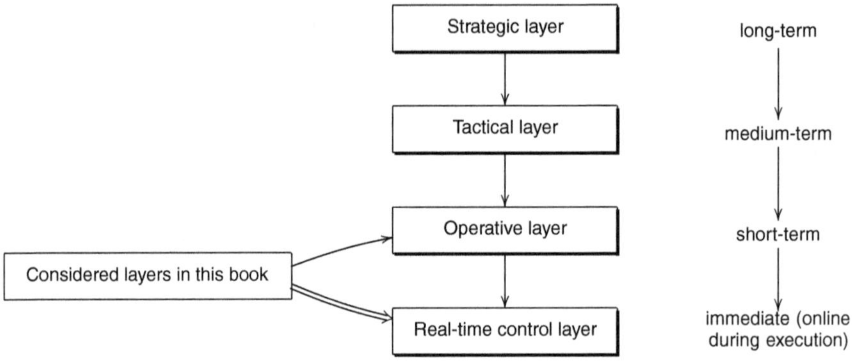

Fig. 1.2 Layers of the transportation process including the real-time control layer

RDOPG applications, besides establishing appropriate depots, this also includes choosing the branches of industry for which services are offered, e.g., repair services for production systems of automobile manufacturers or subsequent delivery of newspapers for publishing houses.

- **Tactical layer:** This layer specifies decisions that are applied for a medium-term time horizon. In RDOPG applications for example, this involves determining the available fleet size.
- **Operative layer:** In this layer, short-term tasks are performed. For example, requests which are already known before the execution of the transportation process are assigned and sequenced on the available vehicles. In so doing, the tour plan of the next planning horizon, usually the next working day, is generated (cf. Stumpf 1998, p. 52 et seq.).

As can be seen from these three layers, the described planning activities end when the vehicles leave the depot in order to service assigned requests. However, in dynamic routing problems such as the RDOPG applications considered in this book, this is not sufficient since the majority of requests to be serviced only occur *during* the execution of the transportation process.

In order to overcome this limitation, Bock (2004, p. 63) extends these three layers by introducing a fourth layer denoted as **Real-time control layer**. This layer deals with the coordination of the transportation process in real-time parallel to its execution. Figure 1.2 illustrates the four described layers and their time range. As can be seen in this figure, the real-time control approaches which are developed in this book are applied in the operative layer and real-time control layer since an initial solution is calculated before the execution of the transportation process which is then adapted during its execution. As mentioned above, since in the considered RDOPG applications the majority of customer requests only arrive during the execution of the transportation process, the activities of the real-time control layer are the main focus of this book.

1.3 Aims and Contributions of the Book

The described RDOPG applications cover various real-world problems. Motivated by the challenges faced in the considered real-world problems, a new pro-active real-time control approach is developed and evaluated in this book. Moreover, since real-world request data of the subsequent delivery of newspapers process is provided from a publishing house, this application is used as the representative RDOPG application in this book.

In this section, first the aims of this book are presented. Specifically, the *central research questions to be answered in this book* are described. Afterwards, the *contributions of this book* are described.

1.3.1 Central Research Questions to Be Answered in This Book

During the evaluation of the proposed pro-active real-time control approach, we investigate the following three questions:

Question 1: Does the integration of information about expected future request arrivals into a deterministic real-time control approach, thus forming a pro-active real-time control approach, allow for a significant improvement of the solution quality by reducing customer inconvenience?

Question 2: If question 1 yields a positive result, under which criteria does this apply? For example, what impact does the structural quality of existing request data and the resulting stochastic knowledge which is generated on the basis of this request data have on the solution quality? Furthermore, which characteristics are suitable for measuring the structural quality of available request data?

Question 3: What impact does vehicle en-route diversion have on the solution quality? Since this increases driver inconvenience, is there a conflict between customer inconvenience and driver inconvenience and does a useful trade-off between both objectives exist?

1.3.2 Contributions of This Book

The proposed models, approaches, and methods for pro-actively managing and controlling the operative transportation processes of RDOPG applications can be characterized by the following contributions:[1]

[1] Some of the contributions of this thesis have also been published in concise form in Ferrucci et al. (2013).

- **Maximization of customer loyalty:** We integrate results from marketing research dealing with customer loyalty. Specifically, customer inconvenience is operationalized in the objective function linearly and quadratically. Besides, an additional penalty occurs for requests that exceed a maximum allowed response time. This combines the aims of servicing customer requests quickly and maintaining negotiated conditions, e.g., those defined by SLA contracts.
- **Real-time control of time-critical distribution of perishable goods:** Since dynamically arriving requests have to be immediately serviced due to the considered objective function, the tour plan execution process needs to be efficiently controlled in real-time. Therefore, a sophisticated deterministic real-time control approach is proposed. It is utilized in order to conduct the dynamic transportation process by solving individual problem instances. Due to the perishability of the goods to be delivered, this real-time control approach faces a high degree of dynamism and urgency (cf. Larsen et al. 2002, Ichoua et al. 2007). Furthermore, the proposed approach accounts for the concurrency of tour plan execution and tour plan adaptation during the transportation process. In order to classify the considered RDOPG applications with regard to other routing problems known in the literature, a new classification based on suitable classifications in the literature is proposed.
- **Universally applicable forecasting approach for the generation and integration of stochastic knowledge:** In this book, forecasting methods which are able to automatically extract stochastic knowledge about expected future requests solely from past request information are developed. Specifically, a customizable segment-based clustering approach which identifies and selects suitable, well-defined request-likely areas by analyzing request arrivals of past working days is proposed. Since underlying request distributions of the available past request information are unknown, the clustering approach comprises methods which ensure that only request-likely areas complying with defined quality criteria are selected. Additionally, methods for integrating the generated stochastic knowledge in form of dummy customers into the proposed deterministic real-time control approach are developed. This also includes the definition of handling dummy customers during the execution of the transportation process. In doing so, the deterministic real-time control approach is extended to the proposed pro-active real-time control approach.
- **Integration of vehicle en-route diversion in a real road network:** In order to increase the practicability of the proposed real-time control approaches, vehicles move on a fully-detailed urban road network. In order to allow vehicle en-route diversion in the road network, the implemented discrete event-based simulator maps vehicle travel activities in detail.
- **Tabu Search with a variable neighborhood structure:** In order to quickly solve individual problem instances during the execution of the transportation process, a Tabu Search metaheuristic with a variable neighborhood structure is proposed. Depending on solutions explored in previous iterations, intensifying or diversifying neighborhood operators are applied. In order to efficiently and precisely determine the tabu-active state of solutions, an appropriate tabu-active attribute is developed.

- **Classification of structural request data quality:** In order to support the identification of scenarios in which the application of the proposed pro-active real-time control approach is advantageous, we investigate the *structural quality* of request data sets. For this purpose, we derive the structural quality of a request data set from two newly proposed criteria denoted as *structural distortion* and *structural diversity*. In addition to defining both criteria in detail, we also propose methods for determining the structural distortion and the structural diversity of a given request data set. By evaluating designed request data sets with different structural qualities, we show the efficiency of an approach proposed for distinguishing between request data sets which have a low structural distortion and those that have a high structural distortion. Note that this is important since a low structural distortion means that request arrivals on the next day follow specific patterns identified by the forecasting method based on the past request information. This is required in order to enable reliable predictions of expected future requests which is a prerequisite for an efficient utilization of pro-active real-time control approaches. Furthermore, by examining request data sets with low structural distortion, we identify structural diversity as a crucial criterion for classifying the value of the attainable stochastic knowledge thereof. In computational experiments, the impact of different levels of structural diversity on the attainable improvement is assessed in detail. Specifically, for each tested request data set, the solution quality improvements attainable by using the pro-active real-time control approach are evaluated by comparing its results with the results attained by its deterministic counterpart. Furthermore, besides a qualitative assessment of the structural diversity of request data sets, a quantitative approach denoted as the *degree of structural diversity* (*dosd*) is proposed. For request data sets evaluated in the main computational study, we compare the attained improvements with the values calculated by the proposed *dosd*. The results show the effectiveness of the *dosd* in estimating improvements that can be attained by utilizing the proposed pro-active real-time control approach compared to its deterministic version. Therefore, the proposed criteria provide a good starting point for assessing the structural quality of request data sets in order to efficiently estimate the attainable improvements by using stochastic knowledge in pro-active real-time control approaches. Furthermore, these criteria allow for an efficient customization of the proposed pro-active real-time control approach depending on the existing scenario.
- **Evaluation of vehicle en-route diversion effects and related driver inconvenience:** Consistent with results of other research work on dynamic routing problems in the literature, we show that allowing vehicle en-route diversion has a significantly positive effect on the attained solution quality. However, allowing vehicle en-route diversion activities increases the demands on the drivers and thereby raising driver inconvenience. Since the positive effects on solution quality make the complete prohibition of vehicle en-route diversion unattractive, an approach for limiting the number of vehicle en-route diversion activities is proposed and resulting effects on the solution quality are investigated. Reasonable trade-off parameter settings with regard to the contradicting aims of minimizing driver inconvenience and customer inconvenience are presented.

In this book, the above-mentioned contributions are achieved by developing, implementing, and evaluating real-time control and forecasting approaches which use various methodologies from the area of Operations Research (OR). Hillier and Lieberman (2008) describe OR as follows:

> As its name implies, operations research involves "research on operations." Thus, operations research is applied to problems that concern how to conduct and coordinate the *operations* (i.e., the *activities*) within an organization. (...) The *research* part of the name means that operations research uses an approach that resembles the way research is conducted in established scientific fields. To a considerable extent, the *scientific method* is used to investigate the problem of concern. In fact, the term *management science* [MS] sometimes is used as a synonym for operations research. (Hillier and Lieberman 2008, p. 2)

According to the authors, the following steps are required to be undertaken in an OR project:

> In particular, the process begins by carefully observing and formulating the problem, including gathering all relevant data. The next step is to construct a scientific (typically mathematical) model that attempts to abstract the essence of the real problem. It is then hypothesized that this model is a sufficiently precise representation of the essential features of the situation that the conclusions (solutions) obtained from the model are also valid for the real problem. Next, suitable experiments are conducted to test this hypothesis, modify it as needed, and eventually verify some form of the hypothesis. (Hillier and Lieberman 2008, p. 2)

Another important aspect is that "OR is also concerned with the practical management of the organization. Therefore, to be successful, OR must also provide positive, understandable conclusions to the decision maker(s) when they are needed" (Hillier and Lieberman 2008, p. 3). This is important since executives seem not to trust systems that make decisions on a black box basis. According to Stenfors et al. (2007), a survey in Finland's 500 largest companies which assessed the utilized tools when making major decisions revealed that only 10 % of the used tools belong to OR/MS tools. Nevertheless, executives see benefits in using OR tools in order to make operational processes more efficient. According to Stenfors et al. (2007), "Tools which support creativity are needed". In order to attain a higher acceptance of the proposed methods in practical applications, the process of generating stochastic knowledge before the execution of the transportation process as well as the vehicle activities performed during the execution of the transportation process are visualized in the graphical user interfaces of the developed software tools. However, since the focus of this book is to evaluate the attainable improvements by utilizing stochastic knowledge in real-time control approaches, the implemented visualizations are not considered any further in this book. Moreover, skilled users can interactively influence the stochastic knowledge generation process by defining specific values for different parameters.

Further information about the field of OR can be found among others in Hillier and Lieberman (2008) and Domschke and Drexl (2011).

1.4 Outline of the Book

This book is organized as follows. After this chapter, a comprehensive introduction to tour planning approaches is given in Chap. 2. Here, classic static routing problems as well as its variants known in the literature are described. Furthermore, differences between static and dynamic routing problems are presented. Moreover, classifications of routing problems published in the literature are presented and, based on these classifications, a unified approach for classifying routing problems according to RDOPG-specific attributes is proposed. As mentioned above, since RDOPG applications are based on dynamic routing problems, a major part of this chapter considers characteristics of dynamic routing problems.

In Chap. 3, a problem description of the considered RDOPG applications is presented in detail. According to this problem description, in the literature review in Chap. 4, research on dynamic routing problems which covers aspects relevant for RDOPG applications is examined. The major focus is on describing approaches which provide an increased flexibility for integrating newly arriving requests during the execution of the transportation process.

Chapter 5 introduces the proposed deterministic real-time control approach which is used to control the transportation process during its execution. Furthermore, the mathematical model utilized for modeling individual problem instances arising during the execution of the transportation process is described. Afterwards, the proposed real-time control approach is classified according to a general classification scheme for real-time control approaches.

In order to generate stochastic knowledge out of past request information, a new segment-based clustering approach is proposed in Chap. 6. Specifically, we elaborate on how our forecasting approach generates and subsequently integrates stochastic knowledge into the deterministic real-time control approach thereby creating the proposed pro-active real-time control approach. Furthermore, the handling of the generated stochastic knowledge in the real-time control approach during the execution of the transportation process is described.

In Chap. 7, the developed solution method which is used for solving the individual problem instances arising in the real-time control approaches is presented. This is a stage-based Tabu Search metaheuristic which utilizes different neighborhood operators, depending on solutions visited in previous iterations. The neighborhood operators as well as the tabu-active attribute utilized in this solution approach are presented in detail. Furthermore, the stage-based scheme used for selecting a neighborhood operator in each iteration is described.

In Chap. 8, the results of the computational experiments are extensively presented. Test instances based on real-world and designed request data sets comprising defined structural quality characteristics are presented. Specifically, in order to evaluate the proposed pro-active real-time control approach under different levels of structural request data quality, the criteria structural distortion and structural diversity are described in detail. Using these criteria, request data sets with different levels of structural distortion as well as different levels of structural diversity are designed. Results from different preliminary computational studies are presented. In the main computational experiments, the performance of the pro-active real-time control approach is evaluated by comparing its attained results with the results achieved by its deterministic counterpart. Moreover, new approaches for assessing the structural quality of given request data sets with regard to their suitability in our proposed pro-active real-time control approach are presented. After identifying suitable request data sets, measures for determining achievable improvements by using the pro-active real-time control approach are proposed. Furthermore, a new approach for limiting vehicle en-route diversion activities is introduced. Finally, answers to the three central research questions are given.

Chapter 9 concludes this book by summarizing the main results and discussing avenues for further research that are opened by the contributions of this book.

Chapter 2
Introduction to Tour Planning: Vehicle Routing and Related Problems

In the literature, tour planning problems (or routing problems) are a vital research area. As illustrated in Fig. 2.1, the number of journal publications on Vehicle Routing Problems (VRPs, see Sect. 2.3.2) has steadily increased over the years. This increased scientific interest can be explained by several factors. For example, progress in computational resources has opened new possibilities for modeling more complex routing problems such as dynamic routing problems (see Sect. 2.5.3.2). Furthermore, new arising real-world applications provide inspiration for developing new approaches for coordinating complex transportation processes.

In this chapter, a general introduction to tour planning problems as well as aspects of specific types of routing problems are given. Note that this chapter mainly deals with routing problems which comprise characteristics of short-haul routing problems, i.e., routing problems in which transportation activities are performed in a bounded service area without transshipment during a planning horizon which usually has a length of up to one working day (see Sect. 1.2 and cf. Ghiani et al. 2004, p. 247 et seq.). First, the *general task and definitions* of tour planning problems are introduced. Afterwards, the *representation of requests and general types of routing problems*, i.e., node-based and edge-based routing problems, are described. Since the considered RDOPG applications can be modeled as variants of node-based routing problems, different *node-based routing problems* known in the literature and characteristics of these, including solution complexity issues and existing solution methods, are presented. Furthermore, since the considered RDOPG applications can be modeled as a variant of the well-known Vehicle Routing Problem, *extensions to the Vehicle Routing Problem* which can be found in the literature, e.g., specific problem extensions or other objective functions, are presented. Afterwards, different types of *information revelation in routing problems* are discussed by describing appropriate classifications known in the literature. Moreover, a new classification with regard to the type of information revelation and coordination of the transportation process is proposed in order to describe specific attributes which are characteristic of RDOPG applications.

Since RDOPG applications utilize a centralized coordination of the transportation process, characteristics of *dynamic routing problems which utilize a centralized*

F. Ferrucci, *Pro-active Dynamic Vehicle Routing*, Contributions to Management Science, DOI 10.1007/978-3-642-33472-6_2, © Springer-Verlag Berlin Heidelberg 2013

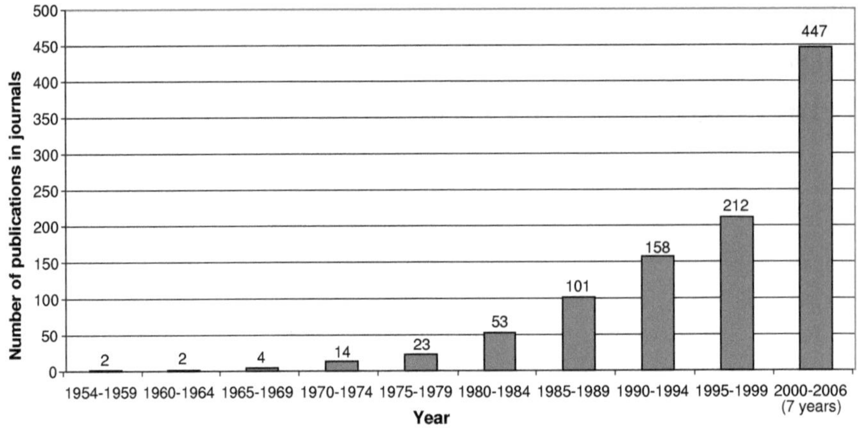

Fig. 2.1 Number of journal publications with regard to VRPs (cf. Eksioglu et al. 2009)

coordination are presented in the following section. In addition to typical objectives considered in dynamic routing problems and a description of different types of possible dynamic events, a measure known in the literature for determining the difficulty of solving dynamic routing problems is presented. Furthermore, dynamic real-world routing applications are described and classified according to the presented measure. Moreover, two measures for evaluating the performance of solution approaches for dynamic routing problems are presented. Finally, a *general classification scheme for routing problems* known in the literature is described and extended according to characteristics which are specific for RDOPG applications.

2.1 General Task and Definitions

In general, the tour planning process can be defined as follows: A set of requests needs to be assigned to a set of resources, for example vehicles that execute the transportation process. Since resources always represent vehicles in the RDOPG applications considered in this book, the terms vehicle and resource are used synonymously. For each vehicle, the assigned requests are required to be ordered in a specific sequence in which the requests will be serviced by the vehicle afterwards. Performing the assignment and sequencing tasks results in a *tour* for each vehicle. Both of these tasks can either be consecutively or simultaneously performed. The step in a tour which consists of traveling from one request to the next is called a *route*. The set of all vehicle tours forms the *tour plan* which is also denoted as the *solution* of the considered routing problem. Note that problem-specific constraints, e.g., time window restrictions of requests, need to be fulfilled by the tour plan in order to obtain a feasible solution. Figure 2.2 illustrates the described tour planning process using five pending requests and two vehicles which start and end their tour at a depot D.

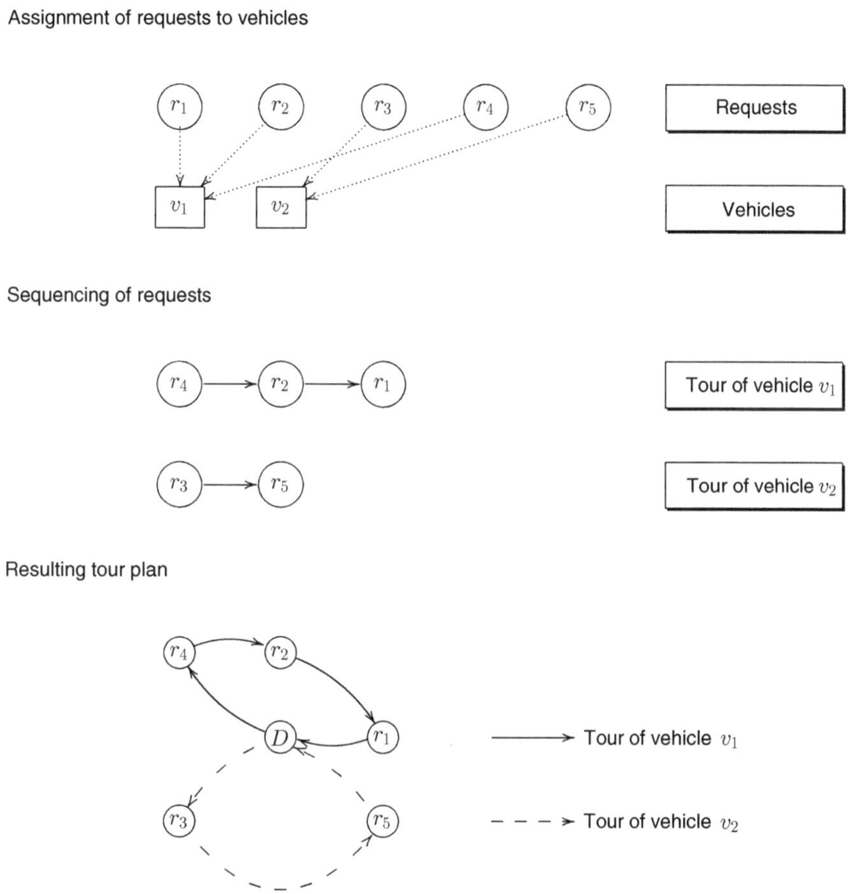

Fig. 2.2 Visualization of the tasks in a routing problem

The generation of a tour plan is connected to a specific *aim* which is pursued, e.g., minimizing the total travel distance. Mathematically, this aim is defined by the *objective function* which assigns a specific *objective function value* to each solution. By using this objective function value, different solutions can be compared and ordered according to their *solution quality*. The goal is to find the *optimal solution*, i.e., the one with the best objective function value among all feasible solutions. In what follows, we assume that the aim is always to minimize a given objective function value, i.e., an optimal solution has the minimum objective function value among all feasible solutions. Since routing problems usually belong to the class of strongly non-deterministic polynomial time hard (NP-hard) combinatorial optimization problems (cf. Lenstra and Rinnooy Kan 1981), it is often not possible to determine an optimal solution in reasonable time due to complexity reasons. Hence, the aim is to determine a good solution whose objective function value is as close as possible to the objective function value of an optimal solution. Details on the complexity of routing problems are described in Sect. 2.3.4. Furthermore, different objectives which are pursued in routing problems are presented in Sect. 2.4.12.

2.2 Representation of Requests and General Types of Routing Problems

In routing problems, requests are represented by *relevant locations* and are modeled in a *graph*. This graph consists of a set of vertices (or nodes) defined by V and of a set of arcs A or edges E. Specifically, if the graph is directed, it is defined by $G = (V, A)$. If it is undirected, it is defined by $G = (V, E)$. In the directed case, direction-dependent travel costs can be assigned so that the costs for traveling from location $i \in V$ to location $j \in V$ represented by c_{ij} can be different from c_{ji} whereas in the undirected case, it holds that $c_{ij} = c_{ji}\ \forall i, j \in V$.

Routing problems can be distinguished according to the type of requests that need to be serviced. Specifically, a distinction is made between *arc-based* routing problems and *node-based* routing problems. In arc-based routing problems, customer requests are represented by arcs or edges, depending on whether the graph is directed or undirected. A characteristic of these types of problems is that the vehicle is at a different position before and after servicing a request. In the Chinese Postman Problem introduced by Kwan (1962), each road is assumed to be undirected so that each edge of G needs to be visited at least once. Moreover, if G also contains directed roads, the problem is called the Mixed Chinese Postman Problem. Another variant is the Rural Postman Problem in which only a given subset of edges $E' \subset E$ needs to be serviced. For more information on arc-based routing problems, see Eiselt et al. (1995) and Golden et al. (2008, p. 37 et seq.). In contrast to arc-based routing problems, node-based routing problems arise if customers are located at vertices $v \in V$ so that vehicles remain at the same position while servicing a customer request. In addition to pure node-based and pure arc-based routing problems, mixed routing problems where requests are located at arcs as well as nodes exist.

Since the RDOPG applications considered in this book belong to the class of node-based routing problems, we consider node-based routing problems in what follows.

2.3 Node-Based Routing Problems

In this section, classic node-based routing problems are introduced in their asymmetric variant so that different costs for both travel directions can be modeled. First, the *Traveling Salesman Problem (TSP)* and the *Vehicle Routing Problem (VRP and CVRP)* are described. Afterwards, routing problems in which goods or persons have to be transported between customer locations are presented. These problems consist of the *General Pickup and Delivery Problem (GPDP) and related variants* known as the Pickup and Delivery Problem and the Dial-A-Ride Problem. Furthermore, the *complexity of node-based routing problems* as well as *solution methods for node-based routing problems* are addressed.

2.3.1 The Traveling Salesman Problem (TSP)

The first routing problem which was considered in the literature is the Traveling Salesman Problem (TSP, cf. Dantzig et al. 1954). In the TSP, one salesman needs to service a given set of $\mathcal{R}$ customer requests $R = \{1, \ldots, \mathcal{R}\}$, also denoted as cities. In this routing problem, the set of requests is equal to the set of relevant locations V in the digraph $G = (V, A)$ which are defined by $V = \{1, \ldots, \mathcal{R}\}$. The salesman is originally located at one of the customer requests and needs to service the requests by visiting each customer location exactly once and finally returning to its origin. Traveling from i to j, $(i, j) \in A = V \times V$ produces costs denoted by c_{ij}. For example, these costs can represent travel distances (see Sect. 2.4.12). The sought optimal solution is a cost-minimal tour plan that represents the sequence in which the requests are to be visited. Specifically, the solution is given by an assignment of the binary variables $X = \{x_{ij}\}$ $(i, j \in V,\ i \neq j)$ which are 1 if the salesman directly travels from location i to j and 0 otherwise. The TSP is constituted by the following mathematical model (cf. Laporte 1992b):

$$\min z = \sum_{i \in V} \sum_{j \in V} c_{ij} \cdot x_{ij} \quad \text{s.t.} \tag{2.1}$$

$$\forall i \in V: \quad \sum_{j \in V} x_{ij} = 1 \tag{2.2}$$

$$\forall i \in V: \quad \sum_{j \in V} x_{ji} = 1 \tag{2.3}$$

$$\forall S \subset V,\ 2 \leq |S| \leq \mathcal{R} - 2: \quad \sum_{i \in S} \sum_{j \in \{V \setminus S\}} x_{ij} \geq 1 \tag{2.4}$$

The objective function value of a solution is determined by the objective function (2.1). Constraints (2.2) and (2.3) ensure that each location is visited and left exactly once. Constraints (2.4) eliminate subtours in the solution (see Fig. 2.3). Note that this subtour formulation results in $2^{\mathcal{R}} - 2\mathcal{R} - 2$ subtour elimination constraints which cannot be reasonably handled by using Mixed Integer Programming (MIP) solvers even for small problem instances (cf. Laporte 1992b). Besides this subtour elimination formulation, other subtour elimination constraint formulations are possible. For example, Miller et al. (1960) propose a formulation which only contains $O(\mathcal{R}^2)$ restrictions but at the expense of additional variables. Moreover, it may produce weaker lower bounds than the previous formulation (cf. Desrochers and Laporte 1991). Other subtour elimination formulations can be found for example in Laporte (1992b) and in Domschke and Scholl (2010, p. 100).

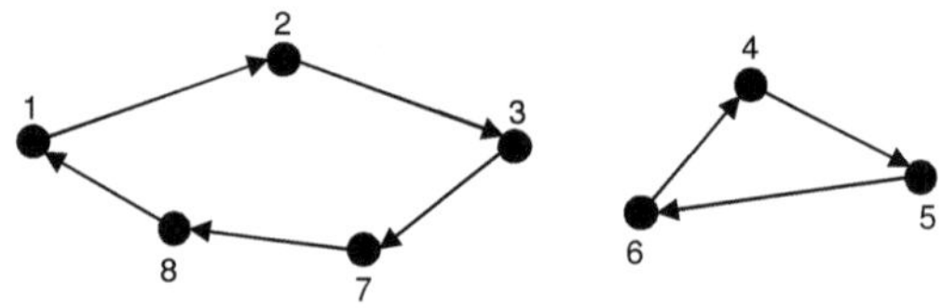

Fig. 2.3 Infeasible TSP solution due to a subtour in the solution

2.3.2 The Vehicle Routing Problem (VRP and CVRP)

The Vehicle Routing Problem (VRP) was introduced by Dantzig and Ramser (1959). In its basic version, a set of requests has to be serviced by a fleet of homogeneous vehicles. Each vehicle begins and ends its tour at the same location called the *depot*. The VRP is used to model two different types of real-world business cases:

1. Goods that are loaded onto the vehicles at the depot are delivered to the customers.
2. Goods that are located at customer locations are collected and transported to the depot.

The uncapacitated VRP in which capacity constraints of the vehicles can be neglected is equal to a multiple-salesman variant of the previously introduced TSP (m-TSP, cf. Laporte and Nobert 1980 and Bektas 2006). In the VRP, $R = \{1, \ldots, \mathcal{R}\}$ defines the set of $\mathcal{R}$ customer requests. The set of nodes is defined by $V = R \cup \{0\}$, i.e., customer requests and the depot denoted by node 0. Furthermore, the set K represents the set of $\mathcal{K}$ uncapacitated vehicles $K = \{1, \ldots, \mathcal{K}\}$. As is the case in the TSP, travel costs on the arcs are defined by c_{ij} with $i, j \in V$. Since each request location is visited exactly once by a vehicle and since the utilized vehicles are homogeneous, it is not required to specify which tour belongs to which vehicle in the solution. Hence, it is possible to model the VRP using only a *two-index vehicle flow formulation* that uses $O(\mathcal{R}^2)$ binary variables as is the case in the TSP (cf. Toth and Vigo 2002, p. 30). In this formulation, the solution consists of a tour plan represented by a set of binary variables $X = \{x_{ij}\}$ $(i, j \in V,\ i \neq j)$ equal to 1 if and only if a vehicle directly travels from location i to location j. The VRP can then be defined by the following mathematical model (cf. Laporte and Nobert 1980 and Bektas 2006):

$$\min z = \sum_{i \in V} \sum_{j \in V} c_{ij} \cdot x_{ij} \quad \text{s.t.} \tag{2.5}$$

$$\forall i \in R: \quad \sum_{j \in \{V \setminus i\}} x_{ij} = 1 \tag{2.6}$$

$$\forall i \in R: \quad \sum_{j \in \{V \setminus i\}} x_{ji} = 1 \tag{2.7}$$

$$\sum_{i \in R} x_{0i} = \mathcal{K} \tag{2.8}$$

$$\sum_{i \in R} x_{i0} = \mathcal{K} \tag{2.9}$$

$$\forall S \subset R,\ 2 \le |S| \le \mathcal{R} - 2\colon \quad \sum_{i \in S} \sum_{j \in \{V \setminus S\}} x_{ij} \ge 1 \tag{2.10}$$

The objective function is given by (2.5). Constraints (2.6) and (2.7) ensure that each customer request is visited and left exactly once. These constraints also determine that exactly one vehicle services each request. Constraints (2.8) and (2.9) make sure that each vehicle leaves the depot and returns to it exactly once. Finally, constraints (2.10) eliminate subtours in the vehicle tours.

The Capacitated VRP The Capacitated Vehicle Routing Problem (CVRP) is closely related to the uncapacitated VRP. In addition, each request $i \in R$ has a demand d_i and all vehicles have the same loading capacity denoted by C. The CVRP can be modeled using the previously introduced formulation of the VRP and replacing constraints (2.10) by

$$\forall S \subset R,\ 2 \le |S| \le \mathcal{R} - 2\colon \quad \sum_{i \in S} \sum_{j \in \{V \setminus S\}} x_{ij} \ge r(S) \tag{2.11}$$

Constraints (2.11) are known as *capacity-cut constraints* (CCCs) and they ensure both the elimination of subtours as well as vehicle capacity limits. The value $r(S)$ defines the minimum number of vehicles required to service all requests of set S. It can be calculated by solving a bin-packing problem for determining the minimum number of required bins, each with capacity C, for loading the items of set S (cf. Toth and Vigo 2002, p. 7) or by using a trivial lower bound given by $r(S) = \lceil \frac{\sum_{i \in S} d_i}{C} \rceil$. Note that setting $C = \infty$ leads to constraints (2.10).

Despite requiring only the same amount of binary variables as in a TSP, the described two-indexed formulation has certain drawbacks. Specifically, since no tour-specific attributes can be integrated, e.g., individual vehicle arrival times at customer requests, it does not allow to model more complex variants of the VRP. In order to overcome this limitation, a *three-index vehicle flow formulation* is utilized (cf. Toth and Vigo 2002, p. 15). In this formulation, let $X = \{x_{ijk}\}$ be the set of binary variables that are 1 if and only if vehicle $k \in K$ directly travels from location i to location j $(i, j \in V,\ i \neq j)$. Furthermore, each binary variable of the set $Y = \{y_{ik}\}$ is 1 if and only if vehicle $k \in K$ visits location $i \in V$. Using this notation, the CVRP can be modeled as follows:

$$\min z = \sum_{i \in V} \sum_{j \in V} \sum_{k \in K} c_{ij} \cdot x_{ijk} \quad \text{s.t.} \tag{2.12}$$

$$\forall i \in R\colon \quad \sum_{k \in K} y_{ik} = 1 \tag{2.13}$$

$$\sum_{k \in K} y_{0k} = \mathcal{K} \tag{2.14}$$

$$\forall k \in K, i \in V: \quad \sum_{j \in V} x_{ijk} = \sum_{j \in V} x_{jik} = y_{ik} \tag{2.15}$$

$$\forall k \in K: \quad \sum_{i \in V} d_i \cdot y_{ik} \leq C \tag{2.16}$$

$$\forall S \subseteq R, h \in S, k \in K: \quad \sum_{i \in S} \sum_{j \notin S} x_{ijk} \geq y_{hk} \tag{2.17}$$

Besides the objective function defined in (2.12), constraints (2.13) ensure that each request is serviced by exactly one vehicle while constraints (2.14) make sure that the depot belongs to the tour of each vehicle. Constraints (2.15) ensure the connectedness of the vehicle tours and by constraints (2.16) vehicle capacity restrictions are taken into account. Subtours are eliminated by constraints (2.17).

2.3.3 The General Pickup and Delivery Problem (GPDP) and Related Variants

In this section, we describe the General Pickup and Delivery Problem as well as two variants denoted as the Pickup and Delivery Problem and the Dial-A-Ride Problem. In contrast to the previously introduced routing problems, these problems consider applications in which goods or passengers are picked up at specific locations and need to be transported to other locations on the same tour. Hence, the vehicles are empty at the start and at the end of their tour.

2.3.3.1 The General Pickup and Delivery Problem

The General Pickup and Delivery Problem (GPDP) was described by Savelsbergh and Sol (1995). In the GPDP, each customer request $i \in R$ consists of a total cargo of weight $\bar{q}_i$ which needs to be transported from a set of pickup locations R_i^+ to a set of delivery locations R_i^- which are individual for each customer request. Each location $j \in \{R_i^+ \cup R_i^-\}$ has a weight q_j assigned. Positive weights assigned to pickup locations indicate the weight of goods that have to be picked up whereas negative weights at delivery locations denote the quantity of goods that need to be delivered. Moreover, it holds that $\bar{q}_i = \sum_{j \in R_i^+} q_j = -\sum_{j \in R_i^-} q_j$. The set $R^+ = \bigcup_{i \in R} R_i^+$ comprises all pickup locations and set $R^- = \bigcup_{i \in R} R_i^-$ is composed of all delivery locations of all requests. Additionally, set $L = R^+ \cup R^-$ unifies all request locations. Requests are serviced by a *heterogeneous* vehicle fleet $K = \{1, \ldots, \mathcal{K}\}$. Specifically, each vehicle $k \in K$ has an individual capacity Q_k, as well as an individual start location k^+ and end location k^-. Furthermore, we define $K^+ = \bigcup_{k \in K} k^+$, $K^- = \bigcup_{k \in K} k^-$, and $W = K^+ \cup K^-$. Hence, in the corresponding digraph G, the set of nodes V is defined as $V = L \cup W$. For all $i, j \in V$, d_{ij} denotes the travel distance, t_{ij} the travel time, and c_{ij} the costs for traveling from location i to j. Note that service times required at request locations can be easily integrated into the travel times, hence they are not considered explicitly.

The model of the GPDP is formulated using four types of variables. Each binary decision variable z_{ik} $(i \in R, k \in K)$ is equal to 1 if request i is assigned to vehicle k, 0 otherwise. Each binary decision variable x_{ijk} $((i, j) \in (L \times L) \cup \{(k^+, j)|k \in K, j \in L\} \cup \{(j, k^-)|j \in L, k \in K\},\ i \neq j)$ is 1 if vehicle k directly travels from location i to j, 0 otherwise. Each non-negative variable D_i $(i \in V)$ specifies the departure time at location i. Finally, each non-negative variable y_i $(i \in V)$ denotes the current load of the corresponding vehicle when it arrives at location i. The GPDP can be modeled as follows:

$$\min z = f(x) \quad \text{s.t.} \tag{2.18}$$

$$\forall i \in R\colon \quad \sum_{k \in K} z_{ik} = 1 \tag{2.19}$$

$$\forall i \in R, l \in \left\{R_i^+ \cup R_i^-\right\}, k \in K\colon \quad \sum_{j \in V} x_{ljk} = \sum_{j \in V} x_{jlk} = z_{ik} \tag{2.20}$$

$$\forall k \in K\colon \quad \sum_{j \in \{L \cup \{k^-\}\}} x_{k^+jk} = 1 \tag{2.21}$$

$$\forall k \in K\colon \quad \sum_{i \in \{L \cup \{k^+\}\}} x_{ik^-k} = 1 \tag{2.22}$$

$$\forall k \in K\colon \quad D_{k^+} = 0 \tag{2.23}$$

$$\forall i \in R, p \in R_i^+, q \in R_i^-\colon \quad D_p \leq D_q \tag{2.24}$$

$$\forall i, j \in V, k \in K\colon \quad x_{ijk} = 1 \Rightarrow D_i + t_{ij} \leq D_j \tag{2.25}$$

$$\forall k \in K\colon \quad y_{k^+} = 0 \tag{2.26}$$

$$\forall i \in R, l \in \left\{R_i^+ \cup R_i^-\right\}\colon \quad y_l \leq \sum_{k \in K} Q_k \cdot z_{ik} \tag{2.27}$$

$$\forall i, j \in V, k \in K\colon \quad x_{ijk} = 1 \Rightarrow y_j = y_i + q_i \tag{2.28}$$

The objective function is defined by (2.18). The authors state that depending on the considered application, different functions can be used for $f(x)$, for example travel distance, travel time, or customer inconvenience. For details, we refer to Savelsbergh and Sol (1995) and Sect. 2.4.12. Constraints (2.19) make sure that each customer request is assigned to exactly one vehicle. Constraints (2.20) ensure that a vehicle only visits and leaves customer request locations which are belonging to requests assigned to the vehicle exactly once. Constraints (2.21) and (2.22) define that each vehicle starts and ends its tour at the specified location. Constraints (2.23) enforce that all vehicles start their tour at time point 0. Constraints (2.24) form the request precedence constraints. Specifically, all pickup locations of a request have to be serviced before its delivery locations are allowed to be serviced. Constraints (2.25) set the correct departure times at visited locations. Note that the less or equal formulation allows the integration of time windows defined by $[e_i, l_i],\ i \in L$ and $D_i = \max(A_i, e_i)$ with A_i denoting the arrival time of the servicing vehicle at request location i. Hence, vehicles are allowed to wait at request locations if they arrive there before e_i. Constraints (2.26), (2.27), and (2.28) together form capacity constraints.

In what follows, two special cases of the GPDP are presented. Note that the previously described TSP and VRP also represent special cases of the GPDP.

2.3.3.2 The Pickup and Delivery Problem (PDP)

The Pickup and Delivery Problem (PDP) is a special case of the GPDP. Specifically, in the PDP, each request $i \in R$ has exactly one pickup and one delivery location and only one depot exists at which all vehicles start and end their tour. The PDP can be formulated by using the model of the GPDP and setting its parameters to $|W| = 1$ and $\forall i \in R$: $|R_i^+| = |R_i^-| = 1$ so that all vehicles have the same start and end position and each request consists of exactly one pickup and one delivery location.

2.3.3.3 The Dial-A-Ride Problem (DARP)

In Dial-A-Ride Problems (DARP, cf., e.g., Cordeau and Laporte 2003, 2007), applications which deal with the transportation of passengers are considered. The DARP can be formulated by using the mathematical model of the GPDP and defining its parameters as $|W| = 1$, $\forall i \in R$: $|R_i^+| = |R_i^-| = 1$, and $\forall i \in R$: $\bar{q}_i = 1$ since each request is assumed to consist of exactly one passenger. Common real-world applications of the DARP are taxi cab related problems or scheduled transportation of patients to medical services. Since in the DARP people are transported instead of goods, the objective function often includes specific passenger convenience attributes. For example, the ride time of individual passengers can be restricted or longer ride times penalized. According to Larsen (2000), the DARP exists in three different versions:

- **Many-to-One (MTO):** In this variant of the DARP, only one destination to which passengers are transported is considered. This case represents for example a real-life application where different customers from a small residential area are collected and afterwards transported to the same destination.
- **Many-to-Few (MTF):** In this case, the number of destinations is still relatively small compared to the number of origins. In real-life applications, this case represents for example picking up customers from a residential area and transporting them to different stops at a large shopping mall.
- **Many-to-Many (MTM):** This variant of the DARP can be considered as a taxi cab system in which multiple individual passengers with different origins and destinations have to be serviced.

More information on the DARP and its dynamic variants can be found in Sects. 2.6.5.1 and 2.6.5.3.

2.3.4 Complexity of Node-Based Routing Problems

All the previously described node-based routing problems represent *combinatorial optimization problems*. Specifically, since the corresponding tasks of assigning and sequencing requests on vehicles can only be performed in a finite number of different *combinations*, the number of existing solutions is also finite. Hence, one method for determining an optimal solution consists of just evaluating all feasible solutions. However, as we illustrate in this section, the computational time required to solve routing problems by using complete enumeration is often too high. Combinatorial problems can be distinguished by their complexity in order to estimate the required computational time for solving them. According to Garey and Johnson (1979, p. 6), a function $f(n)$ is $O(g(n))$ if and only if there exists a constant c so that it holds that $|f(n)| \leq c \cdot |g(n)|$ for all values of $n \geq 0$ with n denoting the length of the input, e.g., the number of customer requests in a routing problem. According to the authors and this notation, an algorithm is called a *polynomial time algorithm* if its computational time complexity function is $O(p(n))$ for some polynomial function p. The authors state that all algorithms with no polynomial function bounding their time complexity are called *exponential time algorithms*. Hence, the group of exponential time algorithms covers all algorithms that have at least an exponential time complexity (cf. Gass 2003, p. 304). Using this classification, it can be decided whether an algorithm can be used to solve the considered problem with reasonable computational effort. Specifically, in Table 2.1, the computational time complexity for algorithms with a polynomial, exponential, and factorial time complexity are illustrated. In this table, it is assumed that each computational operation consumes the same amount of time which is defined as one microsecond, i.e., 10^{-6} seconds. As can be derived from this table, complete enumeration cannot be used to solve a given problem which belongs to the exponential classification in reasonable time if larger problem instances are considered.

In order to link the results illustrated above to the time complexity of algorithms for solving routing problems, we consider the required computational time for finding an optimal solution of a TSP instance with n customer requests. Since the number of feasible solutions amounts to $(n-1)!$, the number of feasible solutions grows in a factorial manner with n. Therefore, even TSP instances with a small size cannot be practically solved by enumerating all possible solutions in a reasonable amount of time (see again Table 2.1). Although more sophisticated solution methods have been developed for solving TSPs, the TSP still belongs to the class of strongly NP-hard combinatorial optimization problems (cf., e.g., Garey and Johnson 1979, pp. 211–214, Laporte 1992b, and Korte and Vygen 2012, p. 405). In computational complexity theory, the problem class P comprises all problems which can be solved in polynomial time using a deterministic Turing machine while those of

Table 2.1 Comparison of several polynomial and exponential time complexity functions (cf. Garey and Johnson 1979, p. 7)

Time complexity function	Problem size n					
	10	20	30	40	50	60
n	0.00001 seconds	0.00002 seconds	0.00003 seconds	0.00004 seconds	0.00005 seconds	0.00006 seconds
n^2	0.0001 seconds	0.0004 seconds	0.0009 seconds	0.0016 seconds	0.0025 seconds	0.0036 seconds
n^3	0.001 seconds	0.008 seconds	0.027 seconds	0.064 seconds	0.125 seconds	0.216 seconds
n^5	0.1 seconds	3.2 seconds	24.3 seconds	1.7 minutes	5.2 minutes	13.0 minutes
2^n	0.001 seconds	1.0 seconds	17.9 minutes	12.7 days	35.7 years	366 centuries
3^n	0.059 seconds	58 minutes	6.5 years	3,855 centuries	$2 \cdot 10^8$ centuries	$1.3 \cdot 10^{13}$ centuries
$n!$	3.6288 seconds	771 centuries	$8.4 \cdot 10^{16}$ centuries	$2.6 \cdot 10^{32}$ centuries	$9.6 \cdot 10^{48}$ centuries	$2.6 \cdot 10^{66}$ centuries

the problem class NP can only be solved in polynomial time by a non-deterministic Turing machine (cf., e.g., Garey and Johnson 1978; 1979, pp. 27 and 31). Moreover, a problem is denoted as NP-hard if all problems of the problem class NP are *reducible* to the considered problem. According to Lenstra and Rinnooy Kan (1979), a problem A is defined as reducible to problem B if for each instance of problem A, an instance of problem B can be generated in polynomial time so that solving the instance of problem B will also solve the instance of problem A. Hence, the reducibility of problem A to problem B implies that problem A represents a special case of problem B. Therefore, problem B is at least as hard as problem A. Furthermore, a problem is NP-complete if it is NP-hard and also belongs to problem class NP. For more information on computational complexity theory, see, e.g., Garey and Johnson (1978; 1979, pp. 23–33) and Lenstra and Rinnooy Kan (1979).

An NP-hard combinatorial optimization problem is called strongly (or unary) NP-hard if it remains NP-hard even when all input numbers are encoded in unary, cf., e.g., Garey and Johnson (1978; 1979, p. 120) and Lenstra and Rinnooy Kan (1979). Otherwise, NP-hard combinatorial problems are called weakly (or binary) NP-hard if a pseudo-polynomial time algorithm for solving the problem is known: "A pseudo-polynomial time algorithm can be useful even when there is no natural bound on the input numbers we expect. It will display 'exponential behavior' only when confronted with instances containing 'exponentially large' numbers, and instances of this sort might be rare for the application we are interested in. If so, this type of algorithm might serve our purposes almost as well as a polynomial time algorithm" (Garey and Johnson 1979, p. 91).

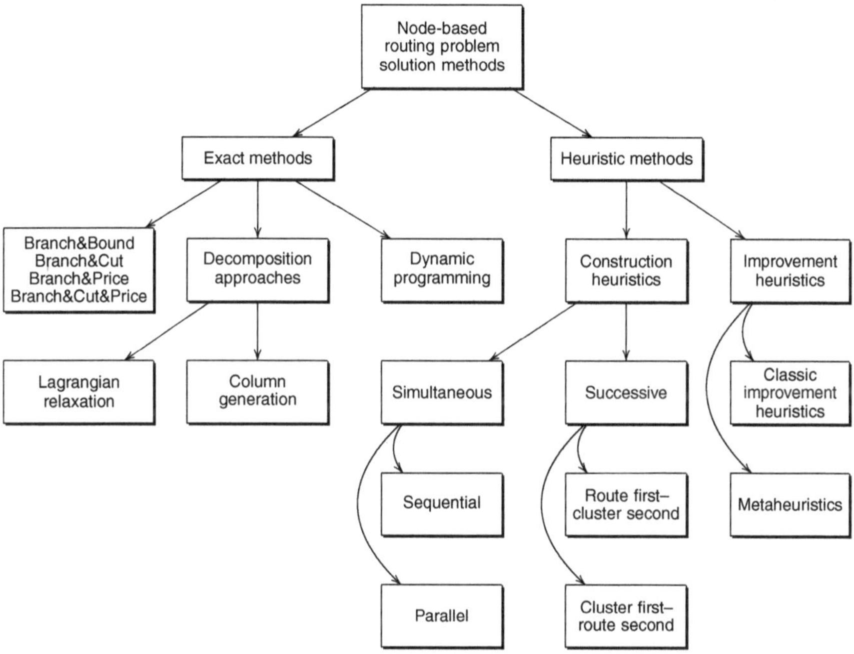

Fig. 2.4 Classification of different solution methods for node-based routing problems (in accordance with Richter 2005, p. 42 and Aberle 2009, p. 543)

Since the TSP is a strongly NP-hard combinatorial optimization problem as mentioned above, all currently known algorithms for solving TSPs to optimality have at least an exponential computational time complexity in the worst case. Note that since all of the previously described routing problems, such as the VRP and the GPDP, are extensions of the TSP, these problems are also strongly NP-hard combinatorial problems. Nevertheless, by utilizing specific characteristics of these routing problems or by accepting close-to-optimum solutions, various solution techniques which are applicable for solving practical-sized problem instances in reasonable time have been developed for the considered node-based routing problems. An overview of various solution techniques is presented in the next section.

2.3.5 Solution Methods for Node-Based Routing Problems

In order to solve node-based routing problems, various exact and heuristic solution methods have been developed (cf., e.g., Laporte 1992a, Toth and Vigo 2002, Richter 2005, Cordeau et al. 2007b, and El-Sherbeny 2010). Figure 2.4 gives an overview of several techniques developed for solving node-based routing problems exactly and heuristically.

2.3.5.1 Exact Solution Methods

Despite their exponential computational time complexity in the worst case, exact solution methods determine an optimal solution of a routing problem in finite time. Using problem-specific characteristics, they can be used to derive an optimal solution for certain routing problems within reasonable time in the average case. Moreover, they usually provide information about the solution quality of the currently best known solution during the solution process by calculating the current *gap*. This gap denotes the maximum percentage deviation from the optimal objective function value. Specifically, this gap is determined by calculating the deviation of the currently best known solution's objective function value from the value of a calculated lower bound of the considered problem (cf. Toth and Vigo 2002, p. 159).

One class of exact solution methods are *Branch-and-Bound* (B& B) and related approaches which were first introduced by Land and Doig (1960) and extended by Dakin (1965) for solving MIP problems using *tree-search* approaches. In general, first a relaxed version of the MIP problem, e.g., its linear relaxation, is solved in the *root node* of the *search tree*. If all variables in the solution of the relaxed problem are integer, the procedure stops. Otherwise, an integer variable x_k is selected whose value x_{k0} is continuous in the current relaxed solution. Using x_k, two new problems are constructed in the *branching step* by adding the restriction $x_k \leq \lfloor x_{k0} \rfloor$ and $x_k \geq \lceil x_{k0} \rceil$, respectively, to the corresponding relaxed problem, thus creating two new nodes, i.e., subproblems, in the search tree. By solving these relaxed versions of the original problem again, lower bounds for the optimal objective function value of the original problem existing in the corresponding subtree of the current node can be calculated. In doing so, it is known with certainty that no better solution can be found in the subtree belonging to this node if the lower bound of the current node is higher than the objective function value of the currently best known solution of the original problem. Hence, this subtree can be removed from further consideration—this activity is called *bounding*. The aim of B&B approaches is to find an optimal solution by examining as few nodes as possible in the search tree by calculating tight lower bounds in each node so that subtrees of the search tree which do not lead to an optimal solution can be early bounded (cf. Gass 2003, p. 313). One of the first B&B related approaches for the TSP is the approach of Little et al. (1963). In Branch-and-Cut (B& C) approaches, additional cutting planes are subsequently integrated in order to strengthen the relaxed problem solved in each node in order to calculate better lower bounds. In Branch-and-Price (B&P, cf. Barnhart et al. 1998), column generation is applied in each node with the same aim of achieving better lower bounds. For example, Savelsbergh and Sol (1998) propose a B&P approach for solving a real-life GPDP. Furthermore, both techniques can be combined to Branch-and-Cut-and-Price (B&C&P) approaches (cf., e.g., Fukasawa et al. 2006).

Lagrangian relaxation and column generation based approaches belong to the class of decomposition approaches. *Lagrangian relaxation* (cf. Fisher 2004) is often used as a sub-procedure in order to determine good lower bounds for B&B and related approaches. Specifically, by removing certain restrictions from the original problem, a relaxed problem which can be solved significantly faster than the original problem is created. Due to the removed restrictions, an optimal solution of the relaxed problem is likely to be infeasible for the original problem. Nevertheless, it represents a lower bound of the optimal solution's objective function value of the original problem. In order to reduce such infeasibilities, a violation of the removed restrictions is penalized in the objective function of the relaxed problem. The challenge then is to find good Lagrange multipliers by which these penalties are integrated into the objective function of the relaxed problem so that infeasibilities with regard to the original problem are reduced as much as possible. In so doing, the calculated lower bounds are tightened, i.e., they are closer to the objective function value of an optimal solution of the original problem thereby gaining a higher quality. In order to find such good Lagrange multipliers, subgradient methods are frequently applied. One approach for computing tight lower bounds for the Symmetric Traveling Salesman Problem (STSP) using Lagrangian relaxation is proposed by Held and Karp (1970). In this approach, the problem is relaxed to a Minimum Spanning Tree (MST) problem which can be solved in polynomial time (cf. Sedgewick 1983, p. 413). Since the solution of an MST often disregards tour connectivity characteristics of a TSP solution, this Lagrangian relaxation-based approach is used as a sub-procedure of a B&B solution approach proposed in Held and Karp (1971).

Column generation (cf. Desaulniers et al. 2005) is another solution technique for reducing the average computational effort which is required to compute an optimal solution of combinatorial optimization problems. Starting with a small subset of decision variables, duality costs of the generated optimal solution of this reduced problem are examined in order to add further promising decision variables which lead to better solutions. Column generation has been successfully applied to VRPs, e.g., by Krumke et al. (2002) and Westphal and Krumke (2008). In these approaches, binary decision variables are used to select tours out of a given set of available tours. Specifically, set partitioning is utilized in order to construct feasible tour plans out of the currently available tours. New tours, i.e., columns, are subsequently generated by extending the set of available tours with tours comprising more customer requests. Since in the practical application considered in both approaches optimal solutions often consist of tours with a low number of requests, optimal solutions are often achieved within acceptable time limits. Note that this is one example of efficiently utilizing the given problem structure in developing practically applicable solution methods. More information about the mentioned approaches can be found in Sect. 4.6 of the literature review.

Dynamic programming (cf. Bellman 1954, 2003) can be applied to combinatorial problems which can be modeled as *multi-stage decision processes*. The current state of such a process is determined by a set of quantities denoted as *state variables* (cf. Bellman 1954, p. 503). The current stage is advanced by making *decisions* which transform the state variables. According to Bellman (1954, p. 503), a "sequence of decisions [is] called a policy, and a policy which is most advantageous according to some preassigned criterion [is] called an optimal policy" which represents an optimal solution of the considered problem. Moreover, the author states the principle of optimality as follows, "An optimal policy has the property that whatever the initial state and initial decisions are, the remaining decisions must constitute an optimal policy with regard to the state resulting from the first decisions" (Bellman 1954, p. 504). According to Lew and Mauch, this also means that optimal policies have optimal subpolicies, "That this observation is valid follows from the observation that, if a policy has a subpolicy that is not optimal, then replacement of the subpolicy by an optimal subpolicy would improve the original policy" (Lew and Mauch 2007, p. 5). Hence, the optimal policy is determined by making optimal decisions in the current stage by using the optimal decision of previous stages with regard to the considered objective function. This procedure is also known in the literature as the *optimal substructure* property. In each stage, the size of the considered subproblems is increased so that an optimal solution of the original problem is finally calculated. One issue is that in NP-hard routing problems "the cardinality of the state space is usually exponential in the problem size" (Desrochers et al. 1988). If problem-specific characteristics can be used to limit the set of feasible subproblems early, i.e., removing states which cannot be extended to a feasible solution of the original problem, these characteristics can be utilized to reduce the number of states required to be solved for calculating an optimal solution of the original problem. For instance, the approach of Dumas et al. (1995) for solving a TSP with hard time windows uses different feasibility tests in order to detect infeasible states due to time window violations. At each stage, the set of currently considered requests is increased until finally an optimal policy for all requests of the original problem can be derived using the calculated optimal policies for the previously considered subproblems. If tight time windows are present in a problem instance, the approach is able to significantly reduce the search space and hence also the computational time required to solve the problem instance to optimality.

2.3.5.2 Heuristic Solution Methods

In contrast to exact solution approaches, heuristic solution approaches aim at finding good solutions quickly, often without providing information about the gap to optimality. Heuristic solution methods for routing problems can be divided into *construction heuristics* and *improvement heuristics*. Construction heuristics generate a tour plan from scratch while improvement heuristics improve a given solution.

Construction heuristics can be classified into simultaneous and successive approaches. In *simultaneous* approaches, the assignment and sequencing of requests is performed at the same time. Such approaches can be further distinguished according to how the tours are constructed. Specifically, if the tours are constructed one after another, a construction heuristic is called *sequential*. If the vehicles' tours are constructed at the same time so that a pending request can be inserted in different tours, the construction heuristic is called *parallel*. Apart from simultaneous construction heuristics, *successive* construction heuristics exist in which the assignment and sequencing tasks are consecutively performed. For example, if a *route first–cluster second* heuristic is applied to a VRP, a TSP is solved first resulting in a so-called giant tour which determines the sequence of all requests. In a second step, restrictions such as capacity constraints are used to divide the giant tour into individual vehicle tours. In *cluster first–route second* heuristics, first the assignment of requests to vehicles is performed by considering existing constraints. Afterwards, a TSP is solved for each vehicle tour. Further information and examples of construction heuristics can be found among others in Bräysy and Gendreau (2005a, pp. 6–10).

Besides construction heuristics, improvement heuristics are used to modify an existing solution in order to improve its solution quality. Improvement heuristics can be categorized into *classic improvement heuristics* and *metaheuristics*. What both types of heuristics have in common is that a given current solution is modified using one or more *neighborhood operators*. The set of solutions that can currently be generated by using a specific neighborhood operator is called the *neighborhood* of the current solution. After evaluation of the neighborhood, one solution out of the neighborhood is selected as the new current solution. Using this solution as the new current solution, the described neighborhood search process starts all over again.

Classic improvement heuristics (cf. Bräysy and Gendreau 2005a) modify a given tour plan until a local optimum is reached, i.e., until there exists no solution in the neighborhood of the current solution with a better objective function value than the current solution. Besides classic improvement heuristics, various metaheuristic concepts for routing problems have been developed. The basic idea is to embed one or more problem-dependent neighborhood operators into a problem-independent framework which controls the application of these neighborhood operators. The purpose of these frameworks is to overcome local optima in order to find better solutions in other areas of the solution space. In Hoos and Stützle (2005), a large variety of known metaheuristics is described where the most popular are: Variable Neighborhood Descent (VND), Simulated Annealing (SA), Tabu Search (TS), Greedy Randomized Adaptive Search Procedure (GRASP), Ant Colony Optimization (ACO), and Memetic Algorithms (MAs). Furthermore, metaheuristic solution approaches for routing problems are also based on Genetic Algorithms (GAs, cf. Mitchell 1998) and Variable Neighborhood Search (VNS, cf. Hansen and Mladenovic 2001). Another sophisticated solution technique is Large Neighborhood Search (LNS) which has for example been applied to routing problems by Bent and van Hentenryck (2004a). Ropke and Pisinger (2006) utilize LNS in a modified version called Adaptive Large Neighborhood Search (ALNS). Further information on metaheuristics for routing problems can be found in Toth and Vigo (2002), Cordeau et al. (2005), Bräysy and Gendreau (2005b), Gendreau et al. (2006), and Gendreau et al. (2008). The Tabu Search metaheuristic developed and utilized in the real-time control approaches will be presented in Chap. 7 of this book.

The solution of a routing problem can be regarded as a graph in which customer requests and the depot represent the nodes of the graph and in which the arcs represent the routes, i.e., the vehicle travel activities between the relevant locations. In doing so, neighborhood operators which are applied in improvement heuristics can be subdivided into *arc-exchanging* and *node-exchanging* operators. In arc-exchanging operators, two or more arcs which, depending on the operator, belong to the same vehicle tour or to different tours, are exchanged so that one or more customer requests are indirectly moved to a different position in the tour plan. In contrast, node-exchanging operators directly move customer requests to another position in the same tour or in another tour. The following neighborhood operators are frequently applied in improvement heuristics for VRPs:

- **2-opt and 3-opt operator:** These operators introduced by Lin (1965) exchange two or three arcs within the same tour.
- **Or-opt operator:** This operator proposed by Or (1976) exchanges three arcs in the same tour in order to move a sequence of customer requests to another position on the same tour while preserving the sequence in which the exchanged customers are serviced.
- **2-opt* operator:** This operator by Potvin and Rousseau (1995) exchanges the latter part of two different tours while preserving the sequence of the requests in the exchanged tour parts.
- **Relocate, exchange, and cross operator:** Savelsbergh (1992) introduces these three node-exchanging operators. The relocate operator moves one customer to another tour while the exchange operator swaps two customers in two different tours preserving their positions in the tours. The cross operator works similar to 2-opt*.
- **CROSS-exchange operator:** Badeau et al. (1997) propose the CROSS-exchange operator. It considers two tours in which a part of each tour is exchanged. Both exchanged tour parts can have a different number of customers. Since one part can also consist of zero customers, it is possible to move one or more customers from one tour to another thereby generalizing other neighborhood operators such as the previously mentioned relocate, exchange, Or-Opt, and 2-opt* operator.

The neighborhood operators which are utilized in the proposed Tabu Search metaheuristic will be presented in Sect. 7.2.

2.4 Extensions to the Vehicle Routing Problem

As mentioned above, since RDOPG applications can be modeled as a variant of the VRP, the remainder of this chapter mainly deals with characteristics and aspects of VRPs. In this section, several extensions and special cases of the classic VRP and other objective functions are presented. For further variants and explanations we refer to Toth and Vigo (2002, p. 177), Baldacci et al. (2010), and El-Sherbeny (2010).

2.4.1 Time Windows

In VRPs with time windows, a time window $[r_i^{\mathrm{e}}, r_i^{\mathrm{l}}]$ is assigned to each customer request $i \in R$. We denote the point in time when the servicing vehicle arrives at request i by y_i^s. In general, two types of time windows exist. In case of *hard time windows*, service of a request is required to start before the end of the time window, i.e., $\forall i \in R$: $y_i^s \leq r_i^l$. Tour plans which do not fulfill these hard time window restrictions are infeasible. If a vehicle arrives at the location of request i before r_i^e, it is usually allowed to wait there. This problem variant is called the Vehicle Routing Problem with Time Windows (VRPTW, cf. Toth and Vigo 2002, p. 8). The second variant is the Vehicle Routing Problem with *Soft Time Windows* (VRPSTW, cf. Toth and Vigo 2002, p. 179). In the VRPSTW, service of requests is allowed to be performed after the end of the time window but at the expense of additional *penalty costs*. These penalty costs may either be *fixed*, i.e., occurring for each late serviced request regardless of the length of the lateness, *variable*, i.e., depending on the length of the lateness, or a combination of both. More information on fixed and variable penalty costs are given in Sect. 2.4.12.

Another variant is the Vehicle Routing Problem with Multiple Time Windows in which each request i can be serviced during different time windows (cf. Toth and Vigo 2002, p. 179). A further variant of the VRPTW is the Vehicle Routing Problem with Time Deadlines (VRPTD, cf. Thangiah et al. 1993), where each request i can be serviced at any point in time up to a given latest time point denoted by r_i^l. This variant can be modeled using a VRPTW and setting $r_i^e = 0$ for all requests $i \in R$.

2.4.2 Backhauls

In the Vehicle Routing Problem with Backhauls (VRPB), both mentioned real-world business cases of VRPs dealing with delivering and collecting goods at customer locations are combined. For this purpose, the set of requests is split into the ones requiring *linehauling* (delivering goods) and those that require *backhauling* (collecting goods). In VRPB problems, usually all linehaul customer requests must be serviced before the backhaul requests (cf. Toth and Vigo 2002, p. 9, Parragh et al. 2008).

2.4.3 Simultaneous Pickup and Delivery

The Vehicle Routing Problem with Simultaneous Pickup and Delivery (VRPSPD) is an extension of the VRPB. Specifically, in this problem, backhaul requests can already be collected by a vehicle before it has delivered all linehaul requests (cf. Toth and Vigo 2002, p. 10).

2.4.4 Multi-Depot Problems

In the Multi-Depot Vehicle Routing Problem (MDVRP, cf. Golden et al. 2008, p. 164), dispatchable vehicles are located at different depots. There may be special restrictions for assigning requests to vehicles. Specifically, certain requests may require service by a vehicle from a specific depot, for example if goods to be delivered are only available at certain depots.

2.4.5 Open Routing Problems

In the Open Vehicle Routing Problem (OVRP), vehicles do not need to return to their depot after completing their assigned tour. This problem is for example considered in the work of Tarantilis et al. (2004).

2.4.6 Vehicle Scheduling

While in VRPs decisions only comprise the assignment and sequencing of requests to vehicles, in Vehicle Routing and Scheduling Problems (VRSP) scheduling decisions of vehicles are additionally considered (cf. Bodin and Golden 1981, Solomon 1987, and Mitrovic-Minic et al. 2004). Specifically, vehicles leave the depot and customer locations only at specified time points. Vehicle scheduling decisions are useful for different purposes. For example, in order to consider law-regulated driving time restrictions (cf. Brandão and Mercer 1997 and Sect. 2.4.10), vehicles can be instructed to wait for a certain amount of time at a specified request location. Furthermore, this allows the implementation of waiting strategies (see Sect. 4.3). Note that the integration of scheduling significantly increases the decision space so that the problem complexity is increased.

2.4.7 Multiple Compartments

In the VRP variant denoted as the Multiple Compartment Vehicle Routing Problem (MCVRP, cf. Muyldermans and Pang 2010), vehicles have different storage areas for goods. This is important for transporting goods on the same vehicle which must not be mixed, e.g., different liquids.

2.4.8 Multiple Trips

In Multiple Trip Vehicle Routing Problems, vehicles can return to the depot during the planning horizon and set for another trip afterwards. Such intermediate returns to the depot are useful if distribution problems are considered in which the load of all requests assigned to a vehicle's tour exceeds its capacity. Hence, the vehicle can replenish its capacity and continue to service further requests. Furthermore, vehicle scheduling decisions may become relevant, e.g., for determining when the vehicle sets for its next trip. For example, Brandão and Mercer (1997) consider a Multiple Trip Vehicle Routing and Scheduling Problem (MTVRSP) in which legal time breaks and maximum legal driving times per day are additionally considered.

2.4.9 Time-Dependent Travel Times

VRPs with time-dependent travel times are denoted as Time-Dependent Vehicle Routing Problems (TDVRP). A characteristic of these problems is that the same route requires a different amount of time depending on the current time of the day (cf. Malandraki and Daskin 1992, Ichoua et al. 2003). Specifically, the travel time between two relevant locations $i, j \in V$ is a function $t_{ij}(\tau)$ which depends on the current time τ at which the travel activity starts at location i. In order to model time-dependent travel times, the time during the day is often divided into m time intervals $T' = \{T_1, T_2, \ldots, T_m\}$. Within each of these intervals $t \in T'$, the travel times are assumed to be constant. Note that when modeling time-dependent travel times it is reasonable to consider the first in-first out (FIFO) property, i.e., a later departure at location i does not lead to an earlier arrival at location j (cf. Ichoua et al. 2003 and Fleischmann et al. 2004a). Depending on the type of problem, time-dependent travel times can be known in advance with certainty, with some uncertainty, or even completely unknown (see also Sect. 2.5.2.1). Further approaches which consider time-dependent travel times can be found in the survey of Flatberg et al. (2005). If time-dependent travel times are present, scheduling decisions can become important in order to reduce travel times.

2.4.10 Legal Driving Time Regulations

In order to comply with legal regulations, different VRP approaches consider legal driving time regulations (cf. Brandão and Mercer 1997 and Ferrucci 2006, pp. 38–42). Depending on the country, these regulations consist of a maximum driving time per work shift or legal time breaks required after a certain driving time has been completed.

2.4.11 Heterogeneous Fleet

In the Heterogeneous Vehicle Routing Problem (HVRP), the available vehicles are divided into different classes. Differences between vehicle classes consist for example of different fixed and variable operational costs as well as individual loading capacities (cf. Gendreau et al. 1999b and Tarantilis and Kiranoudis 2007) and different travel speeds (cf. Ferrucci 2006, pp. 36–37).

2.4.12 *Objective Functions in Vehicle Routing Problems*

As mentioned in Sect. 2.3.3, different objective functions are pursued in VRPs according to the considered application. In what follows, we present different types of objective functions. Note that these individual objectives can also be combined by utilizing a weighted sum of different objective functions.

2.4.12.1 Minimization of Travel-Dependent Parameters

Besides the minimization of operative transportation costs denoted by c_{ij} as has been mentioned for example in the description of the TSP and the VRP, the minimization of travel distances d_{ij} or travel times t_{ij} is often considered $(i, j \in V)$.

2.4.12.2 Minimization of the Number of Utilized Vehicles

This objective function aims at minimizing the number of vehicles which are utilized to service the requests. This aim is often combined with the minimization of the total travel distance (cf., e.g., Homberger and Gehring 2005) or the total travel time (cf., e.g., Hvattum et al. 2006).

2.4.12.3 Minimization of the Sum of Tour Durations

This objective pursues the minimization of the total time utilized by all vehicles in order to service the requests (cf. Savelsbergh 1992). If time windows are present, vehicle scheduling aspects become relevant.

2.4.12.4 Minimization of the Completion Time

This objective function aims at minimizing the time point at which the last vehicle returns to the depot after all requests have been serviced. Note that if in the previously described objective all vehicles have to depart from the depot directly at the beginning of the execution of the transportation process, this and the previous objective become equal (cf. Savelsbergh 1992).

2.4.12.5 Minimization of Lateness Costs

In VRPSTWs, costs occurring for late serviced requests can be classified into fixed and variable lateness costs. We denote the length of the lateness that has occurred at request i by $r_i^{\mathrm{lt}} = \max(0, y_i^{\mathrm{s}} - r_i^{\mathrm{l}})$ where, as mentioned above, y_i^{s} describes the start time of service at request i. Furthermore, we define $F(x)$ as a function representing the amount of the penalty costs that occur at a lateness of length x. By using the objective function $\min z = \sum_{i \in R} F(r_i^{\mathrm{lt}})$, fixed penalty costs of amount P can be modeled by defining $F(x) = P \cdot \Theta(x)$ with $\Theta(x) = 1$ if $x > 0$ and 0 otherwise. Accordingly, variable penalty costs can be modeled by defining $F(x)$ as a monotonously increasing function, e.g., a linear or a quadratic function. As mentioned above, both types of penalty costs can also be combined.

2.4.12.6 Minimization of the Number of Unserviced Customers

This objective function is related to the previously presented one, but instead of soft time windows, hard time windows exist. Hence, late service of a customer is not allowed. Instead, if it holds that $y_i^{\mathrm{s}} > r_i^{\mathrm{l}}$, request i is removed from the system and is counted as an unserviced customer. Hence, in order to minimize the number of unserviced customers, it is the primary objective to service as many customers as possible within their time windows. In general, this objective can be modeled by fixed penalty costs which have been introduced in the previous section. However, it is advisable to provide more information to the solution method about how urgent a customer is to the solution method, i.e., how far away customer request i is still from its last allowed time point for start of service denoted by r_i^{l}. The utilization of such additional information has also been addressed in van de Klundert and Wormer (2010). In their work, the authors state that just using fixed penalty costs in order to fulfill maximum request response times defined by SLAs (which is related to the considered objective in this book, see also Sect. 1.1) is not advisable. By integrating appropriate additional information into the objective function, the solution method can precisely discriminate between different solutions in which all customer requests are feasibly serviced within their time windows. Therefore, the solution method is able to identify solutions in which requests are currently scheduled for service with larger buffer times before the end of their time windows. This is also advantageous in the considered RDOPG applications since it increases the possibilities for integrating future arriving requests into the tour plan within the maximum allowed response time so that no penalty costs occur (see also Sect. 5.3.2.4).

2.4.12.7 Minimization of Customer Inconvenience and Request Response Time

As mentioned in Chap. 1, the considered RDOPG applications aim at servicing requests so that customer inconvenience is minimized and a maximum request response time whose violation causes penalty costs is maintained if possible. While a maximum allowed request response time can be represented by using fixed lateness costs, customer inconvenience can be modeled by using variable lateness costs (see Sect. 2.4.12.5) and defining $r_i^{\mathrm{e}} = r_i^{\mathrm{l}} = r_i^{\mathrm{a}}$ with r_i^{a} denoting the point in time when customer request i arrives in the system. Depending on the utilized customer inconvenience function in the objective function $\min z = \sum_{i \in R} F(r_i^{\text{lt}})$, different objectives can be achieved. Specifically, if a linear customer inconvenience function for $F(x)$ is assumed, the pursued aim is equal to the minimization of the average request response time. In contrast, by utilizing an over-linear customer inconvenience function such as a quadratic function for $F(x)$, more fairness with regard to the request response times of individual requests can be attained since the longer a customer waits, the quicker the inconvenience grows. More information on the evaluated customer inconvenience functions and the utilized objective function in the considered RDOPG applications will be presented in Sect. 5.3.2.4. Different effects on the solution quality resulting from the application of these customer inconvenience functions will be analyzed in the computational experiments in Chap. 8.

2.5 Information Revelation in Routing Problems

In all of the previously described routing problems, the solution is obtained by solving the corresponding model on basis of *problem data* which is given for the parameters of the model. In what follows, we define that this data is derived from the *relevant information* of the corresponding routing problem. Depending on the considered routing problem, the problem data is either constant or changes during the transportation process of the considered routing problem. Moreover, we define that the problem data changes during the execution of the transportation process if during its execution different parameter values are present in the model. Depending on the type of relevant information, such changes in the problem data may be known in advance with certainty or can contain uncertainty. Moreover, if changes in the relevant information are not known with certainty, stochastic knowledge about expected changes may be available. This is explained in detail in Sect. 2.5.3.

Basically, the problem data of a routing problem consists of the following types of information:

- **Customer requests:** Information about customer requests comprises the request location, the demand, start and end times of time windows, and fixed as well as variable penalty costs.
- **Vehicles:** In order to efficiently assign the requests to vehicles according to the given objective, information about the available fleet size and properties of the vehicles is required.
- **Travel times and travel distances:** This type of information is provided by the digraph G for all connections between the considered relevant locations (see Sect. 2.2).

In this section, we describe differences in the *revelation of relevant information in vehicle routing problems*, i.e., how and when changes in the relevant information occur in the considered routing problem during the execution of its transportation process. Furthermore, we present *approaches in the literature for characterizing relevant information and distinguishing between static and dynamic routing problems*.

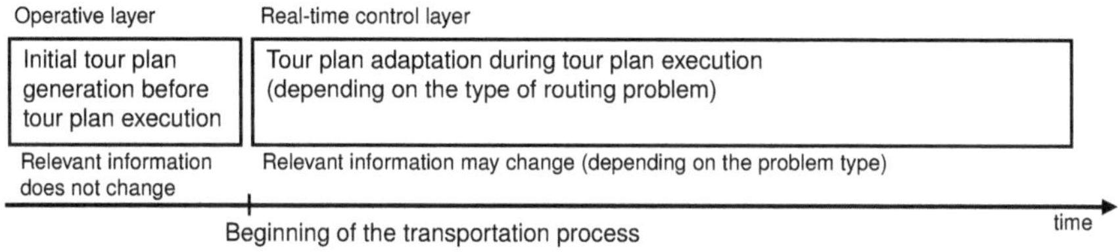

Fig. 2.5 General process of tour plan generation and adaptation before and during the execution of the transportation process

Based on the results of these approaches, *a unified approach for classifying routing problems with regard to characteristics and processing of relevant information*, which are both important for RDOPG applications, in routing problems is proposed. Using this approach, different types of VRPs known in the literature are presented in detail.

2.5.1 Revelation of Relevant Information in Vehicle Routing Problems

With regard to information revelation in VRPs, a common practice is to distinguish between the relevant information which is known *before* the execution of the transportation process and the relevant information which only becomes known *during* the execution of the transportation process (see Fig. 2.5).

Before the execution of the transportation process, an initial tour plan is generated based on the problem data given by the relevant information which is known at that time point. With regard to the layers of transportation processes (see Sect. 1.2), this initial tour plan generation corresponds to the operative layer. It is assumed that the time available for generating the initial tour plan is not restricted and that the relevant information does not change during this generation. After the initial tour plan generation is complete, the execution of the transportation process starts. Specifically, the vehicles leave the depot in order to service customer requests according to the initial tour plan. An important factor is that, depending on the considered routing problem, the existing relevant information can change during the execution of the transportation process. If such changes are caused by events whose time of occurrence during the execution of the transportation process cannot be precisely predicted in advance, we denote them as *dynamic events*—one example of a dynamic event is the arrival of a new request. Otherwise we state that *incomplete relevant information* is present. Specifically, although the relevant information does not change by itself, some parts of it are unknown during the initial tour plan generation process and are only revealed later. For example, the actual customer demand is only disclosed when the vehicle reaches the corresponding customer location. Routing problems with changing relevant information and appropriate tour plan adaptation techniques are described in detail in Sect. 2.5.3.2. In the next section, we deal with characteristics of relevant information as well as with differences between static and dynamic routing problems.

2.5.2 Approaches in the Literature for Characterizing Relevant Information and Distinguishing Between Static and Dynamic Routing Problems

In what follows, differences in routing problems according to characteristics of relevant information are presented. Moreover, approaches published in the literature for distinguishing between *static* and *dynamic* routing problems are described.

2.5.2.1 Characteristics of Relevant Information

In Psaraftis (1995), the author introduces four different criteria for classifying relevant information in routing problems:

- **Evolution of information:** In static routing problems, the relevant information remains constant while in dynamic routing problems the relevant information changes during the execution of the transportation process. More information about differences in static and dynamic routing problems is given in the next section.
- **Quality of information (about future events):** Information available about future events is of one of four types; future events can either be *known with certainty (deterministic)*, *known with uncertainty (forecasted)*, *follow prescribed probability distributions (probabilistic information)*, or *unknown (no information)*.
- **Availability of information:** Changes which occur in the relevant information during the transportation process can either be *locally* or *globally* available. One example of a local information availability is if a driver is informed of the actual amount of goods that the current customer requires only when the corresponding customer location is reached and there is no possibility of informing other vehicles or a central dispatching center about this change. In contrast, in routing problems where such changes are globally available, they are recorded in a central system.
- **Processing of information:** If a *centralized processing* of relevant information is applied, all changes in the relevant information are recorded and handled by a central dispatching center. In a *decentralized processing*, changes in the relevant information are autonomously handled by the vehicles.

2.5.2.2 Static and Dynamic Routing Problems

In this section, we describe different approaches proposed in the literature which classify static and dynamic routing problems. In what follows, we present the twelve aspects by Psaraftis (1988) in which static and dynamic routing problems can be distinguished from each other:

1. **Time dimension is essential:** In a static routing problem the current time dimension during the execution of the transportation process is not important since, as stated above, static routing problems only consider the generation of the initial tour plan before the execution of the transportation process and the relevant information does not change during its execution. In dynamic routing problems, time is essential for several reasons. First, since changes in the relevant information often occur unexpectedly during the execution of the transportation process, depending on the current time, vehicles have a different position and status which needs to be considered in order to efficiently adapt the tour plan according to occurring changes in the relevant information. Furthermore, during the adaptation of the tour plan according to the new system situation, time elapses in the transportation process. Hence, further changes in the system situation that occur during this adaptation need to be carefully considered (see also Sects. 2.5.3.2 and 4.7.4).
2. **Problem may be open-ended:** In contrast to static routing problems, in some dynamic routing problems the length of the planning horizon is not determined in advance.
3. **Future information may be imprecise or unknown:** As mentioned above, in static routing problems all relevant information is known in advance with certainty. In dynamic routing problems, future events are only known with uncertainty or are not known at all before the execution of the transportation process.
4. **Near-term events are more important:** In static applications where all relevant information is known with certainty, no drawbacks exist in directly committing vehicles to decisions which are only taking place in the far future. For example, a vehicle can be sent to an isolated customer request without negative consequences, even if it needs to wait there for a long time until this request can be serviced. In contrast, in dynamic routing problems, postponing far future decisions can be advantageous in order to make better decisions later when more relevant information is available, for example due to additional requests which have arrived intermediately.
5. **Information update mechanisms are essential:** Since in static problems the relevant information does not change, no tour plan adaptations are required during the transportation process. In dynamic routing problems, it is necessary to integrate occurring relevant information updates into the real-time control approach. Specifically, ongoing processes, e.g., requests currently being serviced or vehicle movements being performed, have to be considered in tour plan adaptations. Moreover, in order to inform vehicles about new decisions made during an adaptation, the existence of a bidirectional communication between the vehicles and the dispatching center is important.
6. **Resequencing and reassignment decisions may be warranted:** In dynamic routing problems, changes in the relevant information may render the currently executed tour plan infeasible so that it needs to be adapted. During this adaptation, existing requests may be required to be resequenced or reassigned to

other vehicles. Moreover, newly arrived requests are required to be integrated in order to comply with the new system situation.

7. **Faster computation times are necessary:** As mentioned in Sect. 2.5.1, in static routing problems, a large amount of computational time is assumed to be available for generating the initial tour plan with the demanded solution quality. In contrast, the available time in dynamic problems is limited so that a quick adaptation of the tour plan according to occurred changes in the relevant information is required (see also aspect 1). Hence, adaptations have to be carried out by sophisticated solution methods which provide a good solution quality in a short amount of time.
8. **Indefinite deferment mechanisms are essential:** In dynamic problems, it may happen that the service of a particular request is theoretically indefinitely postponed. Specifically, this happens if a request possesses an unfavorable geographical location and new requests keep arriving close to the current vehicles' positions. Hence, mechanisms which prohibit an indefinite deferment of individual pending requests are required. For example, time window constraints or a nonlinear objective function which generates higher penalty costs for excessive waiting times can be utilized for this purpose.
9. **Objective function may be different:** In static routing problems, the considered objective often pursues the minimization of operative transportation costs, travel distances, or travel times. In addition, different objectives are considered in specific dynamic routing problems. Specifically, in real-world dynamic routing applications it can be observed that the higher the number of dynamic events, the more important customer-oriented objectives become (see also Sect. 2.6.5).
10. **Time constraints may be different:** In dynamic routing problems, the unexpected occurrence of new requests may lead to situations in which it becomes very difficult or impossible to service all pending requests while fulfilling given hard time window constraints. Hence, in dynamic routing problems, time constraints are often soft so that service of requests is also allowed after the end of the corresponding time windows, but at additional penalty costs. Using hard time window constraints in dynamic routing problems allows for modeling business applications in which customers are treated as unserviced if their time windows are not met (see also Sect. 2.4.12.6).
11. **Flexibility to vary vehicle fleet size is lower:** In static problems it is often possible to utilize an arbitrary number of vehicles in order to meet request time window constraints. In contrast, in dynamic routing problems, a small available fleet size often results in a lower flexibility for efficiently integrating requests which dynamically arrive during the execution of the transportation process. This may lead to the violation of time window constraints and lower quality of service.
12. **Queueing considerations may become important:** In dynamic routing problems with a high number of dynamically arriving requests (also called heavy traffic scenarios), situations may occur in which the arrival rate of new requests exceeds a certain amount beyond which vehicles cannot service pending

requests without significant delays. Psaraftis states that in this case, classic solution approaches for routing problems result in a poor system performance so that the utilization of methods from the area of queueing theory is recommendable. Bertsimas and van Ryzin evaluate the efficiency of different request dispatching strategies in light and heavy traffic scenarios by using queueing theory insights in their approaches. Their results will be discussed in the literature review in Sect. 4.5.1.

Besides Psaraftis' twelve aspects, Bianchi (2000) proposes another approach to distinguish between static and dynamic routing problems. In this approach, a routing problem (or, in general, an optimization problem) is classified as dynamic if it is uncertain how parts of the relevant information are revealed or updated during the execution of the transportation process. Hence, impacts of occurring changes in the problem data are not always known with certainty. As mentioned above, if the tour plan currently in execution is required to be adapted, this has to be carried out while time in the system elapses so that the solution process is concurrently executed with occurring changes in the relevant information. Since it is not sufficient to only determine an initial solution once, only a strategy or policy, i.e., a solution method, which is used to solve the problem during the execution of the transportation process can be specified in advance. If this is not the case, a routing problem is deemed not dynamic according to this classification. Therefore, routing problems in which one of the following two characteristics is present are not dynamic according to the author:

- The problem data changes over time, but these changes are known with certainty in the relevant information.
- The relevant information is constant, but some parts of it are unknown at the beginning of the execution of the transportation process and are only revealed during its execution, for example when a vehicle reaches a customer location. Moreover, for all of these uncertain parts, prescribed probabilistic information is given as stochastic knowledge in advance. Hence, an a priori solution which is only changed in minor ways during the execution of the transportation process can be determined in advance (see Sect. 2.5.3.2.1).

Another approach for characterizing static and dynamic routing problems is proposed in the work of Ghiani et al. (2003). The authors state that a routing problem is static if all types of its relevant information are independent of the current time during the execution of the transportation process. If this is not the case, then the routing problem is dynamic. Moreover, if all changes in the relevant information which occur during the execution of the transportation process are known in advance with certainty, the routing problem is called *deterministic*, otherwise *stochastic*. Consequently, four types of routing problems arise in this characterization approach:

- In static deterministic routing problems, no changes in the relevant information occur during the execution of the transportation process so that the problem data is constant. Hence, only the generation of the initial tour plan before the execution of the transportation process is considered in problems of this type. Problems of this type are also denoted as static routing problems which are presented in Sect. 2.5.3.1.1.

- In static stochastic routing problems, some parts of the relevant information are only known with uncertainty before the execution of the transportation process. This uncertain relevant information often belongs to customer attributes, e.g., demand of customers or whether customers are required to be serviced at all. Characteristic for these types of routing problems is that the uncertain relevant information is given by random variables that follow prescribed probability distributions which makes the problem stochastic. Moreover, the prescribed probability distributions are often available as stochastic knowledge before the execution of the transportation process. Since the values of these random variables do not change during the execution of the transportation process, the problem is characterized as static. The actual values of the uncertain relevant information are revealed during the execution of the transportation process. For example, when a vehicle reaches a customer location, its actual demand is revealed. Hence, the problem data changes in this type of problem, but these changes can be considered in the initial tour plan generation due to the available stochastic knowledge. Routing problems of this type are known in the literature as variants of the Stochastic Vehicle Routing Problem which is described in Sect. 2.5.3.2.1.
- In dynamic deterministic routing problems, the relevant information remains constant, but the problem data changes over time during the execution of the transportation process. Characteristic of such routing problems is that the point in time when these changes in the problem data happen as well as their effects are known in advance with certainty. For example, routing problems in which travel times change during the planning horizon according to a function which is known in advance belong to this problem type. Routing problems of this type are described in Sect. 2.5.3.1.2.
- In dynamic stochastic routing problems, the relevant information changes at unknown time points during the execution of the transportation process when dynamic events occur. Moreover, no knowledge about these dynamic events is available. Such problems are described in Sect. 2.5.3.2.2.

Pillac et al. (2010) propose an approach similar to the previous one. Specifically, static routing problems are classified as described in Ghiani et al. (2003). In dynamic routing problems, relevant information changes during the execution of the transportation process and these changes always inhibit uncertainty. Specifically, dynamic deterministic routing problems are characterized by the fact that no information about future changes in relevant information is available. Hence, this group of routing problems is equal to the dynamic stochastic problems in the aforementioned classification. In contrast, this classification regards dynamic stochastic routing problems as real-time routing problems in which some stochastic knowledge about future events is available. Problems of this type are presented in Sect. 2.5.3.2.2.

2.5.2.3 Dynamism of the Problem, Model, and Application in Routing Problems

In Powell et al. (1995), the authors do not only deal with the type of changes of the problem data, which they denote as the *problem*, during the transportation process, but they also categorize the approach utilized to solve the given problem. Specifically, besides a *model* which is used to represent the problem in the solution approach, the *application* defines the way in which the approach solves the model during the transportation process. Hence, in a given routing application, the problem is predetermined while the model and the application are approach-dependent.

In the proposed categorization, the authors distinguish each of these three items according to the type of *dynamism* with which each of them exists in a given routing problem and the applied solution approach:

- **Dynamism of the problem:** The problem is dynamic if one or more of its parameters change over time during the execution of the transportation process according to functions of time which may be known or unknown before its execution. Two types of dynamic problems are distinguished by the authors: Problems with *dynamic data* and problems with *time-dependent data*. In the first type, the relevant information and hence the problem changes unexpectedly, i.e., the mentioned functions are unknown. For instance, this includes dynamically arriving requests, unexpected changes in traffic conditions, or unexpected changes of vehicles or driver statuses. In dynamic problems with time-dependent data, changes in the problem occur deterministically according to functions which are known in the relevant information before the execution of the transportation process.
- **Dynamism of the model:** The model represents the problem in a specific solution approach. It is dynamic if it explicitly considers changes in the problem during the execution of the transportation process whereas a static model does not consider any occurring changes in the problem. Furthermore, models are subdivided into deterministic and stochastic models so that four types of models exist. Specifically, a *deterministic static model* can be used to model static routing problems in which the relevant information does not change and also the problem is static. A *stochastic static model* represents a routing problem in which uncertain parts in the relevant information are considered by generating a robust tour plan with regard to expected changes in the problem that occur according to available probability distributions. By using a *deterministic dynamic model*, changes which occur in the problem over time are considered but the model does not consider the dynamic structure of the problem. Specifically, when solving the model, the possible occurrence of further changes in the problem is not considered so that only the currently deterministically known parts of the relevant information are considered, even in situations where stochastic knowledge is available. In contrast, in *stochastic dynamic models*, decisions are made on the basis of relevant information known with certainty and do additionally consider available stochastic knowledge about expected future events.
- **Dynamism of the application:** The application defines how the solution approach is applied in order to solve the model. The application is *dynamic* if the

related model is repeatedly solved thereby considering changes in the problem during the execution of the transportation process. In contrast, the application of an approach is *static* if it does not adapt the solution during the transportation process even when the problem has changed.

In order to illustrate examples of different solution approaches according to the utilized model and the application, the authors discuss a routing problem in which a vehicle has to be guided from an origin to a destination location in a road network. The objective is to find the route with the minimum expected travel time. It is assumed that the traffic conditions change over time with uncertainty. Hence, the considered problem is dynamic. Moreover, it is assumed that stochastic knowledge is available in advance by prescribed probability distributions describing the expected changes in the travel time. The authors present different approaches for using a specific model and a specific application for this given dynamic problem:

- **Deterministic static model and static application:** In such an approach, the route is calculated once by using one particular static representation of the problem in the model, for example by using average expected travel times for the roads in the road network. Since the application is static, occurring changes in the problem during the execution of the transportation process are not considered so that no adaptation of the solution is performed. In dynamic routing problems, approaches of this type are likely to result in infeasible tour plans during the execution of the transportation process (see Sect. 2.5.3.2). Therefore, the utilization of this type of approach is only advisable for static routing problems (see Sect. 2.5.3.1.1).
- **Deterministic dynamic model and dynamic application:** In approaches of this type, the model is repeatedly solved during the execution of the transportation process due to the dynamic application. The model considers changes that have already occurred in the problem, but the available stochastic knowledge about future expected changes is not considered. In this book, such approaches are denoted as *reactive real-time control approaches* and *deterministic real-time control approaches* (see Sect. 2.5.3.2.2).
- **Stochastic dynamic model and static application:** These approaches consider changes that have already occurred in the problem as well as future expected changes. However, since the application is static, only one initial solution is computed which is not adapted when changes in the problem occur during the execution of the transportation process. Hence, the aim is to generate an initial tour plan which is robust, i.e., which is likely to represent a good solution during the execution of the transportation process even when the problem changes according to the available prescribed probability distributions. In dynamic routing problems, such a priori optimization-based approaches are denoted as Stochastic Vehicle Routing Problems (see Sect. 2.5.3.2.1).
- **Stochastic dynamic model and dynamic application:** Comparable to the previously described approach type, the model considers changes in the problem that have occurred as well as expected future changes. Moreover, the model is repeatedly solved during the execution of the routing problem so that occurred

and expected changes in the traffic conditions are considered. In this book, approaches of this type are denoted as *pro-active real-time control approaches* (see Sect. 2.5.3.2.2).

2.5.3 A Unified Approach for Classifying Routing Problems with Regard to Characteristics and Processing of Relevant Information

In the previous section, different approaches in the literature for characterizing relevant information and different types of routing problems were described. Based on the results of these approaches, in particular those by Powell et al. (1995) and Psaraftis (1995), we present a unified approach for classifying routing problems according to RDOPG-specific attributes in this section. This approach gives a structured overview about the type of relevant information, i.e., constant or changing, and considers the availability and processing of occurring changes in the relevant information in detail. Moreover, in case of changing relevant information, it describes the type of stochastic knowledge available, if any.

The proposed classification is illustrated in Fig. 2.6. The resulting items of this classification represent individual classes of routing problems. According to the main distinction concerning whether the relevant information is constant or comprises uncertain changes during the transportation process, the type of problem data denoted as the problem is identified. If changing relevant information exists, the availability of resulting changes in the problem is presented. Moreover, the proposed classification considers in which way the problem is processed in order to coordinate the transportation process. Finally, depending on the quality of information, i.e., the type of stochastic knowledge which is available about expected future events, appropriate variants of VRPs known in the literature are described. In what follows, characteristics for each of these classes and general solution approaches are presented.

2.5.3.1 Routing Problems with Constant Relevant Information

In routing problems with *constant relevant information*, all relevant information is exactly known before the execution of the transportation process and does not change during its execution (see Fig. 2.7). In such routing problems, the *evolution of information* is *static* and the constant relevant information is *globally available*. Moreover, a *centralized processing* of this relevant information is used. In doing so, a *centralized coordination* of the transportation process is applied. Since only the initial tour plan generation process is considered in such routing problems, solution approaches make often use of a *static deterministic model* and a *static application*. Since the relevant information is constant and no future changes occur, the *quality of information about future events* is *deterministically known*.

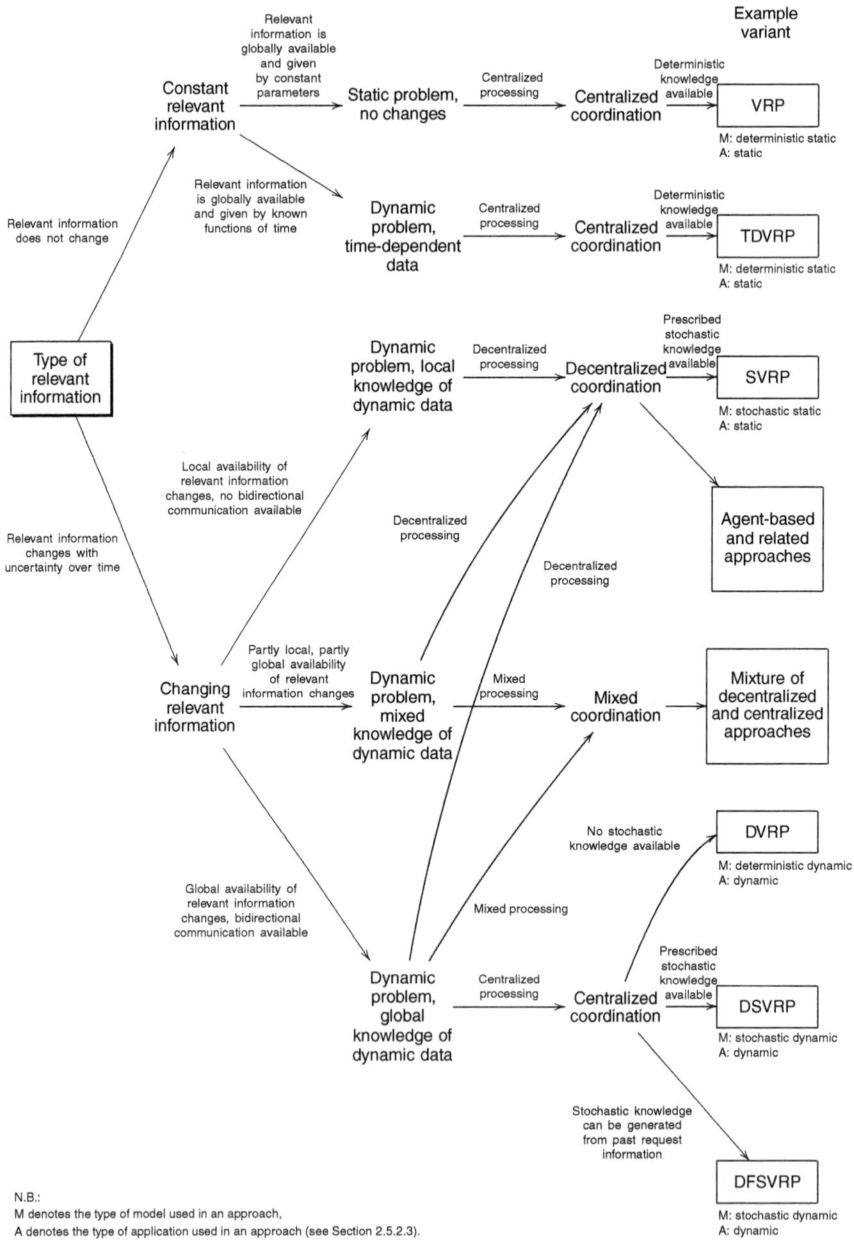

Fig. 2.6 A new classification of routing problems according to properties and processing of relevant information

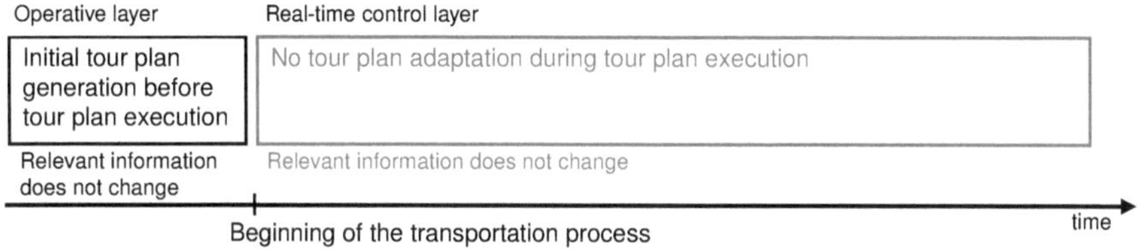

Fig. 2.7 The transportation process of routing problems with constant relevant information

As illustrated in the classification in Fig. 2.6, two types of routing problems with constant relevant information exist. In Powell et al. (1995), they are distinguished according to the dynamism of the problem, i.e., whether the problem is static or dynamic.

2.5.3.1.1 Static Problem: Static VRPs

In routing problems in which the constant relevant information is given by *constant parameters*, the problem data also remains constant during the planning horizon so that a *static problem*, in which *no changes* occur, is present. One example of this routing problem class is the CVRP which was described in Sect. 2.3.2.

2.5.3.1.2 Dynamic Problem: Time-Dependent VRPs

In routing problems of this class, the existing problem data changes during the execution of the transportation process. However, since these changes occur according to known functions of time which are available in the relevant information before the execution of the transportation process and since the relevant information does not change, all problem data changes are deterministically known. Therefore, a *dynamic problem with time-dependent data* is present. An example routing problem is the Vehicle Routing Problem with Time-Dependent Travel Times (TDVRP, see Sect. 2.4.9).

2.5.3.2 Routing Problems with Changing Relevant Information

In addition to the previously discussed routing problems with constant relevant information, there are routing problems in which the relevant information changes during the execution of the transportation process. What is important is that these changes cannot be predicted with certainty. Hence, a *dynamic problem with dynamic data* in which changes in the problem data cannot be predicted with certainty is present. According to Psaraftis, the *evolution of information* is *dynamic* and the *quality of information about future events* is not *deterministic* and depends on the type of available stochastic knowledge.

As mentioned in Sect. 2.5, uncertain changes in the relevant information are brought about by one of the following two circumstances:

1. **Occurrence of dynamic events:** Since the time point of occurrence and the impact of dynamic events cannot be predicted with certainty, dynamic events may lead to an outdated tour plan with regard to the system situation generated by the dynamic event. For example, a dynamic event by which a new request arrives in the system requires an integration of this request into the tour plan. Moreover, dynamic events may render the tour plan infeasible, e.g., if a vehicle breaks down, its assigned requests need to be reassigned to other vehicles. An overview about different types of dynamic events is presented in Sect. 2.6.2.
2. **Presence of incomplete relevant information:** In addition to routing problems in which the relevant information is changed by dynamic events, there are routing problems in which parts of the relevant information are unknown during the initial tour plan generation process. In general, these unknown parts are random variables whose actual values are revealed during the execution of the transportation process. They follow prescribed probability distributions which are usually available in the initial tour plan generation process.

What both circumstances have in common is that the problem data changes with uncertainty during the execution of the transportation process so that specific *adaptations* of the tour plan may become necessary. Therefore, the dimension of the *availability of changes in the relevant information* is important since the resulting availability of changes in the problem mainly determines the types of tour plan coordinations applicable during the execution of the transportation process. In our classification, there are three different values for this dimension:

- **Local availability of relevant information changes:** Local availability of relevant information changes is present if only one particular vehicle is informed about changes that have occurred and there is no possibility of informing a central dispatching center due to the lack of bidirectional communication equipment (see Sect. 2.6.3) or since such a central dispatching center does not exist in the considered routing problem. Due to the *local knowledge of dynamic data*, occurring changes cannot be centrally processed so that a *decentralized processing* of changes and therefore a **decentralized coordination** of the tour plan is inevitable. Hence, the vehicle driver who is informed about the changes is in charge of processing appropriate adaptations of the vehicle's tour. In VRPs, these types of routing problems are represented by variants of the Stochastic Vehicle Routing Problem presented in Sect. 2.5.3.2.1.
- **Global availability of relevant information changes:** In such scenarios, a centralized dispatching center is available and bidirectional communication devices in the vehicles allow communication between the dispatching center and the vehicles (see Sect. 2.6.3). In this case, a *global knowledge of dynamic data* exists in the dispatching center either by centrally collecting changes (e.g., customers directly call the dispatching center) or by vehicles transmitting locally revealed changes to the central dispatching center. This allows for different types of co-

ordination of the transportation process which, in what follows, are explained by the example of adapting the tour plan in case of newly arriving requests:

- **Decentralized coordination:** In this case, some or all of the vehicles are informed about the new requests and autonomously decide which new requests will be serviced by each vehicle; this results in a *decentralized processing* of changes in the relevant information. In our classification, such routing problems are denoted as *agent-based/related approaches*. This type of approaches is not discussed any further in this book. For more information, we refer for example to the work of Fischer et al. (1996), Kohout and Erol (1999), Weiss (2000), Perugini et al. (2003), Cubillos et al. (2007), Bloos et al. (2009), Hülsmann et al. (2009), and Schönberger (2010).
- **Centralized coordination:** By making use of a *centralized processing* of changes in the relevant information, the dispatching center centrally adapts the tour plan and transmits resulting changes to the vehicles involved. In the considered example, the new requests are centrally integrated into the tour plan and the vehicles to which these new requests are assigned are informed after the tour plan adaptation is complete. Routing problems which belong to this class are variants of the Dynamic Vehicle Routing Problem. They are described in detail in Sect. 2.5.3.2.2.
- **Mixed coordination:** In systems applying a mixed tour plan coordination, some types of changes in the relevant information are *centrally processed*, e.g., newly arriving requests, while other changes are *locally processed*, e.g., rerouting of a vehicle in case of traffic congestions. Such approaches can be realized by utilizing a combination of the methodologies for decentralized and centralized tour plan coordination.

• **Partly local, partly global availability of relevant information changes:** This scenario is present if specific types of changes in the relevant information are globally available and other types are only locally available. In such scenarios, tour plan adaptations can be realized either by using a **decentralized coordination** or a **mixed coordination**.

In what follows, we present variants of Stochastic Vehicle Routing Problems which utilize a decentralized coordination and variants of dynamic VRPs using a centralized coordination of the transportation process. Furthermore, the type of stochastic knowledge which is available in the individual routing problems is described.

2.5.3.2.1 Decentralized Coordination: Stochastic Vehicle Routing Problems

In Stochastic Vehicle Routing Problems (SVRP), the relevant information available during the initial tour plan generation process is incomplete so that some parts are unknown (as mentioned in aspect 2 in Sect. 2.5.3.2). In SVRPs, such unknown relevant information can comprise uncertain servicing of customers, uncertain customer demands, stochastic travel times, or stochastic service times (cf. Cordeau et al. 2007a). As mentioned above, an important aspect is that these unknown parts are

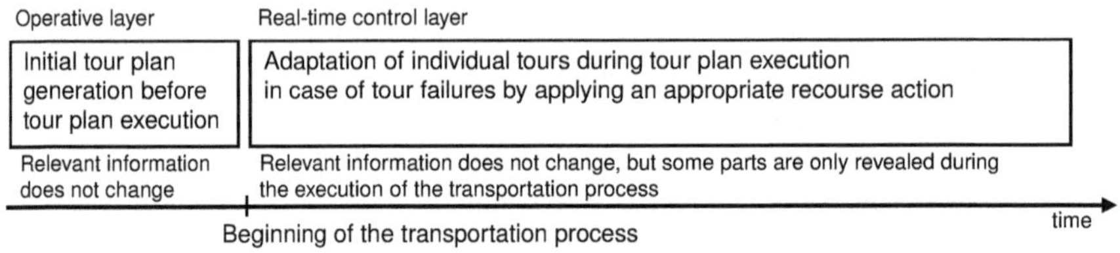

Fig. 2.8 The transportation process in Stochastic Vehicle Routing Problems

represented by random variables. Their values are unknown in the initial tour plan generation but do not change during the execution of the transportation process. They are revealed to individual vehicles during the transportation process which is discussed in detail below. Moreover, these random variables follow prescribed probability distributions which are known and can be utilized as stochastic knowledge during the initial tour plan generation. With regard to the classification of Psaraftis (1995), the *quality of information about future events follows prescribed probability distributions (probabilistic information)*. Hence, the unknown parts of the relevant information are given by distributions known to be correct with certainty (cf. Richter 2005, p. 65).

The transportation process in SVRPs is illustrated in Fig. 2.8. A decentralized coordination is applied during the execution of the transportation process since no technical equipment for a bidirectional communication is available. The initial tour plan is centrally generated irrespective of this lack of technical equipment since all vehicles are assumed to be located at the depot before the execution of the transportation process. During its execution, the individual vehicles can be required to adapt their tour when uncertain relevant information is revealed. Note that such an adaptation is usually only performed if changes in the relevant information make a tour infeasible; this is denoted as a *tour failure*. For example, this happens when the demand of a customer exceeds the currently available capacity of the assigned vehicle.

In case of a tour failure, adaptations denoted as *recourse actions* or *corrective actions* are required in order to re-establish a feasible tour (cf. Gendreau et al. 1996). Due to the decentralized coordination, recourse actions are limited to local adaptations of the vehicle's tour. Moreover, since often no computational equipment is assumed to be available in vehicles, local adaptations are often carried out by utilizing predefined *policy-based operations* consisting of specific rules which can be directly applied by the drivers. Due to these restrictions, recourse actions often result in suboptimal tour plan adaptations which generate high additional costs and therefore lead to a significantly reduced solution quality. Since the minimization of operational costs is usually pursued in SVRPs, the reduction of costs generated by recourse actions during the execution of the transportation process is desired. This is achieved by including the available stochastic knowledge in the initial tour plan generation process which is carried out by sophisticated solution methods. Due to the fact that only policy-based tour plan adaptations are performed during the execution of the transportation process, SVRPs are also denoted as *a priori optimization problems* (cf. Powell et al. 1995).

In order to integrate the available stochastic knowledge into the initial tour plan generation process, SVRPs are often formulated as *stochastic programs*. According to Dror (1989), stochastic programming techniques can be distinguished into *wait and see* as well as *here and now* approaches. While in the first type of approaches a solution is calculated *after* the disclosure of some uncertain relevant information, approaches of the second type compute a solution *before* the uncertain relevant information is revealed. Since the a priori solution of an SVRP is generated before unknown information is revealed during the execution of the transportation process, SVRPs belong to the second type of approaches. For SVRPs, two main solution techniques exist in the literature (cf. Gendreau et al. 1996):

- **Chance Constrained Program (CCP):** In this type of solution technique, the aim is to generate the initial tour plan so that the *chance* of a tour failure during the execution of the transportation process is below a defined threshold. A drawback is that CCPs neglect the costs of a recourse action that are generated when a tour failure nonetheless occurs during the execution of the transportation process.
- **Stochastic Program with Recourse (SPR):** This type of solution technique pursues the generation of an a priori tour plan P^{a} with minimum total expected costs $c_1(P^{\mathrm{a}}) + c_2(P^{\mathrm{a}})$. In this formulation, $c_1(P^{\mathrm{a}})$ describes the costs which occur in the tour plan according to the relevant information known with certainty. Moreover, utilizing the available stochastic knowledge, $c_2(P^{\mathrm{a}})$ represents the estimated costs of recourse actions which are expected to be required during the execution of the transportation process. In general, SPRs require more complex solution methods than CCPs but the considered objective function has a higher practical relevance since it explicitly considers expected additional costs.

As mentioned above, different recourse actions can be performed in case of a tour failure. The best choice for a recourse action depends on the way in which the uncertain relevant information is revealed during the execution of the transportation process. For example, according to Gendreau et al. (1996), a customer's demand can become known when service of the previous customer has been completed or when the vehicle has arrived at the corresponding customer request location. Moreover, applicable recourse actions also depend on the type of uncertainty in the relevant information which is present in the considered SVRP. In the literature, the following types of uncertainty are conducted in SVRPs (cf. Gendreau et al. 1996, Toth and Vigo 2002, p. 332, and Cordeau et al. 2007b, p. 45 et seq.):

- **Vehicle Routing Problem with Stochastic Demands (VRPSD):** In VRPSDs, the demands of customer requests are unknown. Therefore, infeasible tour plans can occur due to capacity constraints of vehicles. This SVRP variant was first studied by Tillman (1969) and Golden and Stewart Jr. (1978) and is by far the most studied variant of the SVRP. More studies on the VRPSD can be found among others in the work of Dror (1989), Secomandi (2000), Laporte et al. (2002), Novoa et al. (2006), Christiansen and Lysgaard (2007), Mendoza et al. (2009), and Secomandi and Margot (2009).
- **Vehicle Routing Problem with Stochastic Customers (VRPSC):** Each customer request $i \in R$ has an occurrence probability p_i and therefore requires no

service with a probability of $1 - p_i$. If more customers than originally expected demand service, infeasible tour plans can arise due to capacity constraints of vehicles or time window violations. This variant has been first studied by Jézéquel (1985) and Jaillet (1987). Note that the set of potential customer requests is known in advance. This is in contrast to Dynamic Vehicle Routing Problems considered in Sect. 2.5.3.2.2 in which new customer requests can theoretically occur anywhere in the considered service region as well as at unknown time points.
- **Vehicle Routing Problem with Stochastic Travel Times (VRPSTT):** In this SVRP variant, travel times between the relevant locations are given by random variables which can lead to unfavorable solutions. For example, in Lambert et al. (1993) a VRPSTT is considered in which a late service of requests results in penalty costs.
- **Vehicle Routing Problem with Stochastic Service Times (VRPSST):** In this variant of the SVRP, the time required for servicing customer requests is uncertain. If a vehicle is required to spend more time at a customer request location than planned, postponed service of consecutive requests can occur which can lead to violations of time window constraints and penalty costs. The VRPSST is considered among others in Laporte et al. (1992), Hadjiconstantinou and Roberts (2002), and Verweij et al. (2003).

Moreover, combinations of the described types of uncertainty in the relevant information are possible. A frequently studied variant is the Vehicle Routing Problem with Stochastic Customers and Demands (VRPSCD) which is also known as the Probabilistic Vehicle Routing Problem (PVRP, cf. Larsen 2000, p. 28). It was first studied by Jézéquel (1985) and Jaillet (1987). Another SVRP variant which considers two types of uncertainty denoted as the Stochastic Vehicle Routing Problem with Stochastic Travel Times and Stochastic Service Times is analyzed in the work of Laporte et al. (1992) and Kenyon and Morton (2003).

As mentioned above, the VRPSD is the most studied SVRP variant in the literature. Hence, various policy-based recourse strategies have been developed for the VRPSD. In what follows, some of them are described. One of the first recourse strategies proposed for the VRPSD lets a vehicle return to the depot whenever the request assigned next in a vehicle's tour cannot be serviced since the currently remaining vehicle capacity is too low. Laporte and Louveaux (1990) and Dror et al. (1993) describe other recourse strategies in which preventive breaks are performed at strategic positions in the tour. For example, when a vehicle is near the depot, it delivers the currently loaded goods there. Another policy-based recourse strategy is presented by Novoa et al. (2006). In this strategy, more flexibility is integrated by allowing vehicles to postpone the next assigned request and to service other requests first before returning to the depot if it is likely that those other requests can be serviced with the currently remaining capacity. Further recourse strategies for the VRPSD are discussed in Dror (1989), Gendreau et al. (1996), and Cordeau et al. (2007b).

Besides policy-based recourse actions, more complex recourse strategies are based on a reoptimization, i.e., adaptation, of the remaining tour of each vehicle when a tour failure occurs. As an advantage, this can improve the system performance. However, such recourse strategies complicate the estimation of costs for expected recourse actions during the initial tour plan generation (cf. Gendreau et al. 1996). In the work of Dror (1989), the author introduces a formulation based on a Markov Decision Process (MDP, cf. Puterman 2005) for solving a single-vehicle VRPSD. In this approach, the solution is reoptimized whenever the vehicle reaches the next request where its demand is revealed. At each customer request, the decision is made whether to service the request now or to service it later. In the work of Dror (2005), the author states that reoptimization-based approaches represent the most promising approaches for solving VRPSDs. According to Secomandi and Margot (2009), recourse actions should be divided into reactive actions which are only taken when a tour failure occurs and pro-active actions in which decisions are taken by anticipating the likelihood of a tour failure. Furthermore, they state that the opportunities given by technological advances are especially beneficial to reoptimization-based approaches.

2.5.3.2.2 Centralized Coordination: Dynamic Vehicle Routing Problems

The application of a centralized coordination of the transportation process during its execution enables the utilization of *real-time control approaches*. Real-time control approaches are able to monitor the execution of the transportation process and perform tour plan adaptations simultaneous to its execution (cf. Bock 2010). Figure 2.9 illustrates the coordination of the transportation process in dynamic vehicle routing problems. As can be seen, the tour plan currently in execution is required to be adapted during its execution. Such adaptation activities belong to the real-time control layer of the transportation process (see Sect. 1.2). The utilization of real-time control approaches allows the handling of all types of dynamic events which are described in Sect. 2.6.2.

An important factor is the decision *when* an adaptation of the tour plan shall be performed. Specifically, it is necessary to define how the complete dynamic problem is divided into subproblems which are solved during the planning horizon (cf. Bianchi 2000). For example, the tour plan can be adapted whenever a new dynamic event arrives in the system. This is denoted as single event optimization (SEO, cf. Bianchi 2000). Another possibility is to adapt the transportation process in specific time intervals (cf. Bock 2010). Further characteristics of real-time control approaches will be discussed in Sects. 4.7.4 and 5.1. More information on the implemented real-time control approach will be given in Sect. 5.2.

In the literature, different variants of dynamic vehicle routing problems centrally coordinated by real-time control approaches can be found. They can be distinguished according to the type of available stochastic knowledge, i.e., the quality of information about future events (see Sect. 2.5.2.1):

1. **Stochastic knowledge is unavailable:** If no stochastic knowledge is available, real-time control approaches can only react after new dynamic events have arrived in the system during the execution of the transportation process. In the

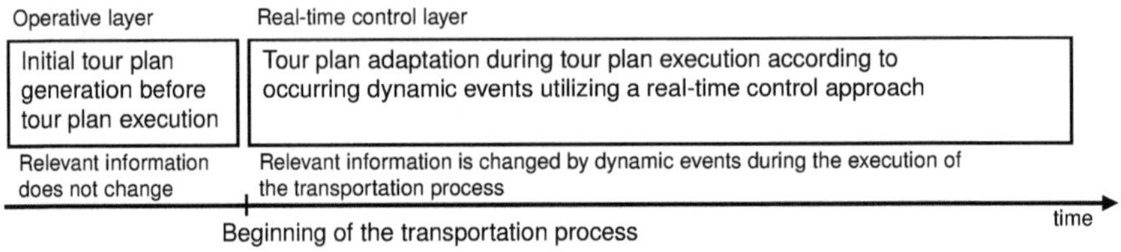

Fig. 2.9 The transportation process in dynamic vehicle routing problems with a centralized coordination

literature, such routing problems are classified as variants of the Dynamic Vehicle Routing Problem (DVRP, cf. Psaraftis 1988).

2. **Stochastic knowledge is probabilistic:** In this case, stochastic knowledge is available and *follows prescribed probability distributions* which usually describe all uncertain elements of the relevant information. Since occurring dynamic events change the relevant information according to the known probability distributions, the utilization of this type of stochastic knowledge is likely to provide significant improvements on the solution quality. In the literature, these approaches are often denoted as variants of the Dynamic Stochastic Vehicle Routing Problem (DSVRP, cf. Hvattum et al. 2006 and Ichoua et al. 2006).
3. **Stochastic knowledge is forecasted:** In many practical applications, stochastic knowledge is not directly available. However, if historical data is available, forecasted stochastic knowledge can be generated by analyzing this data. For example, by applying sophisticated methods on request arrivals of elapsed days including arrival time and location, stochastic knowledge can be generated out of this historical data. Such historical data is denoted as *past request information* in what follows. Since the stochastic knowledge is forecasted, in this book, we denote such problem scenarios as variants of a *Dynamic Forecasted Stochastic Vehicle Routing Problem* (DFSVRP) to which also the considered RDOPG applications belong.

   Note that in routing problems which forecast stochastic knowledge it is assumed that common characteristics observed in the past provide good predictions of future events. Hence, there exist situations in which future dynamic events may possess significantly different attributes as predicted. Since the stochastic knowledge is *known with uncertainty* only, in the worst case, the utilization of forecasted stochastic knowledge can lead to an inferior system performance compared to applying a deterministic real-time control approach which only considers information known with certainty. Moreover, it is advantageous if the request data of the considered routing application possesses specific characteristics. We denote them as *structural quality* of a request data set in what follows. Hence, in practice, it is advantageous to analyze existing real-world request data according to their suitability with regard to generating stochastic knowledge which can be advantageously applied. We will discuss this further in the computational experiments in Chap. 8.

With regard to Powell et al. (1995), in the first case, a *deterministic dynamic model* is utilized. Specifically, it is dynamic because it is updated during the execution of the transportation process but since it only considers changes which have already occurred, it is deterministic. In the latter two cases, the utilized model is a *stochastic dynamic model* since additional stochastic knowledge about expected future events is utilized.

As mentioned in Sect. 2.5.3.2.1, tour plan adaptations required by unexpected changes in the relevant information may lead to significant changes in the currently executed tour plan. Although the discussed real-time control approaches are able to perform a centralized adaptation of the tour plan by elaborated solution methods, dynamic events can nonetheless cause high additional costs. With regard to RDOPG applications in which new customer requests arrive during the execution of the transportation process, the arrival of many unexpected customer requests may require the dispatching of additional vehicles. If no additional vehicles are available (see aspect 11 of Psaraftis' approach in Sect. 2.5.2.2), vehicle tours are likely to require significant adaptations. This negatively influences the execution of the transportation process but also results in higher travel distances and longer request response times which often leads to a lower solution quality. Therefore, the aim is to coordinate the transportation process by real-time control approaches in a way that provides an increased level of *tour plan flexibility*. In doing so, tour plans provide more possibilities for adapting the transportation process according to changes in the relevant information that have occurred without causing significant distortions. This increased flexibility can also be regarded as a technique for increasing the robustness of tour plans towards uncertain dynamic events.

Depending on the availability of stochastic knowledge and the pursued objective, two types of approaches are useful in providing an increased level of tour plan flexibility:

- **Flexibility-increasing approaches without using stochastic knowledge:** Tour plan flexibility can be increased by least-commitment strategies such as *waiting strategies*. By postponing decisions to a later point in time, it is likely that better decisions with regard to travel distance can be made since more relevant information is available then. In what follows, such real-time control approaches are denoted as *deterministic real-time control approaches*.
- **Flexibility-increasing approaches using stochastic knowledge:** Stochastic knowledge offers additional possibilities for realizing the aforementioned waiting strategies. Moreover, waiting strategies can be extended to *relocation strategies* by actively repositioning idle vehicles to subregions in which future requests are likely to arrive soon. Furthermore, *request assignment strategies* can be used for efficiently coordinating utilized vehicles. Specifically, by estimating future vehicle positions resulting from specific request assignments, expected impacts, e.g., request response times, can be calculated for expected future requests. This information can be directly utilized for evaluating the solution quality of possible adaptations during the transportation process. Such real-time control approaches are denoted as *pro-active real-time control approaches* in this book.

In theory, stochastic knowledge is available for all types of dynamic events that have been presented in Sect. 2.6.2. However, in real-world applications, some dynamic events can be predicted better than others. For example, since in many real-world applications recurring arrival patterns of requests can be observed, the arrival of new requests has been very often considered in approaches in the literature. Besides, dynamism arising from changes in travel times is nowadays also well predictable due to floating car data (FCD, cf. Ehmke et al. 2009, 2010). Furthermore, traffic jams are automatically detectable by traffic sensors which are commonly installed at highways and major roads nowadays (cf. Ghiani et al. 2003 and Attanasio et al. 2007). In contrast to both these sources of dynamism, predicting the location and time of vehicle breakdowns is rather impossible. Hence, in the literature, most research deals with dynamic events arising from new request arrivals or changing travel times.

Real-time control approaches which do not provide an increased tour plan flexibility only react towards newly arriving dynamic events. Hence, we denote such approaches as *reactive real-time control approaches*. In the literature review, we will mainly focus on approaches which provide an increased flexibility in the transportation process. Specifically, details on approaches utilizing waiting strategies will be discussed in Sect. 4.3.1. Moreover, relocation strategies will be presented in Sect. 4.3.2 and request assignment strategies will be described in Sect. 4.3.3.

2.6 Dynamic Routing Problems Which Utilize a Centralized Coordination

In the previous section, different variants of routing problems and their characteristics were presented. In accordance with the considered RDOPG applications, this section deals with specific dynamic routing problems in detail, in particular those with a centralized coordination of the transportation process utilizing real-time control approaches (see Sect. 2.5.3.2.2). First, *typical objectives in dynamic routing problems* are described. Afterwards, *dynamic events in dynamic routing problems* by which the relevant information is changed are presented. In addition, *technologies utilized in real-time control approaches* are introduced. In order to classify dynamic routing problems according to their difficulty level with regard to finding good solutions, a measure denoted in the literature as the *degree of dynamism* is described. Based on this measure, practical dynamic routing problems are introduced using a *three-echelon classification of dynamic routing problems*. Finally, two measures for *evaluating the performance of solution approaches for dynamic routing problems* are presented.

2.6.1 Typical Objectives in Dynamic Routing Problems

In general, three different objectives (or combinations of these objectives) are pursued in dynamic routing problems (cf. Larsen et al. 2007):

- **Minimization of distribution costs:** The minimization of costs which are generated during the execution of the transportation process (see Sects. 2.4.12.1 and 2.4.12.2) is the most widely used objective in static routing problems. Furthermore, this objective plays an important role in specific dynamic routing problems, in particular those in which the transportation process is less time-critical and only a small part of the relevant information changes during the execution of the transportation process.
- **Maximization of service quality:** If a high quality of customer service is pursued, dynamically arriving customer requests are required to be quickly serviced. As described in Chap. 1, high request response times lead to customer inconvenience and may impose significantly high costs on the service company, for example due to penalty costs incurred by SLA contracts. This type of objective is considered in the RDOPG applications which are investigated in this book.
- **Maximization of system throughput:** This objective is applied in dynamic routing problems in which the service of as many customers as possible within a given planning horizon is pursued. In such applications, it may happen that the service of an isolated customer request is postponed for a long time in favor of servicing a larger amount of other requests which are closely located to each other. Hence, as mentioned by Psaraftis in aspect 8 of his approach in Sect. 2.5.2.2, it is reasonable to integrate methods which ensure that no customer is postponed indefinitely.

Note that the presented objectives cannot be fulfilled to their full extent at the same time. For example, tour plans which are generated according to the objective of maximizing quality of customer service usually produce additional travel activities during the day which makes them suboptimal from a distribution costs perspective. This has also been addressed by Richter (2005, p. 200 et seq. and p. 224). In Sect. 2.6.5, we present various real-world routing problems and their pursued objective.

2.6.2 Dynamic Events in Dynamic Routing Problems

In dynamic routing problems, dynamic events change the relevant information during the execution of the transportation process. In general, the following four types of dynamic events are considered in the literature:

- **New requests:** Each newly arriving request has a location in the service area. Depending on the considered routing problem, a new request comprises additional attributes such as demands, time windows, lateness penalties, and service times. With the arrival of a new request, an adaptation of the tour plan in order to integrate this request is required. In some applications, it may be allowed to reject new requests, often at additional penalty costs. One approach in which this applies is the approach of Yang et al. (2004). Such a rejection may be beneficial if, at request arrival, it can already be determined that a request cannot be feasibly serviced (e.g., due to violations of hard time window constraints) or if the costs of servicing the request are higher than the costs of its rejection. However, in applications which aim at maximizing quality of service, for example measured by the number of serviced customers or the minimization of request response times, all arriving requests usually need to be serviced. This is considered among others in Bent and van Hentenryck (2004b) and Kleywegt et al. (2009). According to Pillac et al. (2010), the arrival of new requests is the most studied type of dynamic events in the literature.

- **Changes in request attributes:** In some dynamic routing problems, attributes of known requests change or, similar to SVRPs, are only revealed when the vehicle reaches the customer location. For example, if the demand of a request turns out to be different from what was expected, the capacity of the assigned vehicle may not be sufficient. In this case, the request needs to be reassigned to another vehicle. This type of dynamism is conducted in Wu et al. (2004) and Hvattum et al. (2007).
- **Traffic congestions and road blockages:** This type of dynamic events causes either increased travel times or a complete blockage of affected roads. The length of such impacts may be known or unknown. Since the service of requests can be delayed by such dynamic events, re-routing of vehicles to non-congested roads or the reassignment of requests to other vehicles may be advantageous. This type of dynamic event is considered in Fleischmann et al. (2004b), Ferrucci (2006, p. 44), Attanasio et al. (2007), Barceló et al. (2007), Haghani and Yang (2007), Cortés et al. (2008), and Bock (2010).
- **Vehicle disturbances:** This results in a decreased efficiency or a complete failure of a vehicle due to a breakdown or a driver drop-out. As a consequence, pending requests may be required to be reassigned to other vehicles. Moreover, if the affected vehicle is transporting requests when a complete failure occurs, its loaded requests need to be picked up by other vehicles at the affected vehicle's current location. This type of dynamic events is studied in Ferrucci (2006, pp. 44–45), Li et al. (2009a, 2009b), Bock (2010), and Mu et al. (2010).

2.6.3 Technologies Utilized in Real-Time Control Approaches

Approximately 20 years ago, technical limitations often prohibited the application of real-time control approaches in transportation processes. In addition to new developments in computational technology, advances in communication and positioning technology which have been achieved in the last two decades also support the practicability of real-time control approaches.

In this section, we describe technologies which are used in on-board units of vehicles and those utilized in computer systems in dispatching centers in order to enable the coordination of the transportation process in real-time. After describing these technologies separately, we present an exemplary implementation of a system establishing a bidirectional communication between the vehicles and the dispatching center. Moreover, we present an exemplary architecture of a real-time control approach which makes use of the previously described bidirectional communication system. In doing so, it is illustrated how the presented technologies can be jointly used in order to efficiently realize real-time control approaches in real-world dynamic routing problems.

2.6.3.1 Recent Advances in On-Board Technologies

In order to reasonably adapt the tour plan during its execution, the current vehicle positions are required to be provided to the dispatching center. In particular, if positions of traveling vehicles are available in real-time at the dispatching center, a newly arriving request which is close to a vehicle can be directly serviced by this vehicle. In doing so, the required detour and the request response time can be reduced, compared to returning to this request later. For this purpose, on-board systems are utilized.

On-board systems allow the transmission of different types of vehicle data, such as its current position, to the dispatching center in real-time during the execution of the transportation process. The automatic determination of the current vehicles' positions and velocities became possible with the introduction of the Global Positioning System (GPS) in 1995. It consists of 24 satellites which are arranged in 6 orbital planes with 4 satellites each (cf. Kaplan and Hegarty 2006, p. 3). These satellites broadcast the current time as well as their current position which can be received and further used by GPS-capable devices. By utilizing the transmitted data of individual GPS satellites, GPS devices can determine their current position with an accuracy of a few meters using the GPS operation mode called Standard Positioning Service (SPS) which provides sufficient precision for routing applications. SPS is free of charge for commercial and non-commercial use and is utilized by millions of GPS devices worldwide (cf. Kaplan and Hegarty 2006, pp. 3–4). For more information on GPS and other satellite-assisted positioning systems such as GALILEO, see Kaplan and Hegarty (2006).

Besides GPS-capable on-board vehicle units, a data transmission technology is required for establishing an electronic communication between the vehicles and the dispatching center. In the last ten years, significant technological advances have emerged in this area. According to Larsen (2000, p. 12 et seq.), in the year 2000, dedicated radio based communication systems were favorable compared to mobile phone communication systems which had low setup costs but high operational costs. Over the years, alternative tariff models have made electronic communication via smartphones much more economical. In the year 2011, mobile phone contracts providing Internet access at a flat rate are available for less than 30 euros per month in Germany (cf. Telefónica Germany GmbH & Co. OHG 2011). Moreover, data transfer rates in mobile phone networks has evolved from GSM

(Global System for Mobile Communications, before Groupe Spécial Mobile) to GRPS/EDGE (General Packet Radio Service and Enhanced Data Rates for GSM Evolution) to UMTS (Universal Mobile Telecommunications System). UMTS provides data transfer rates with HSDPA (High Speed Downlink Packet Access) and HSUPA (High Speed Uplink Packet Access). These are comparable to fixed line Internet connections (cf. Halonen 2003, p. xxvi). The use of mobile Internet-based data transmission technologies in combination with low priced but powerful GPS-enabled smartphones and other mobile devices allows for the continuous automated transmission of current vehicles' positions to the central dispatching center in real-time at reasonable costs.

Moreover, vehicle on-board units can be used to provide information about current activities to the dispatching center during the execution of the transportation process. For example, when a service worker is at a customer request, information about the progress of the service activity can be efficiently transmitted. If unexpected difficulties occur during a service activity, the dispatching center can identify delays in scheduled tour plan activities early and can perform appropriate adaptations in time. This is especially advantageous in applications pursuing customer service quality aspects in which service times required at customer locations are uncertain. Furthermore, on-board units can provide a request management system which displays changes in the vehicle's tour to the driver and allows the driver to communicate with the dispatching center, e.g., confirm changes in the tour. Moreover, request management systems can be combined with integrated navigation systems which automatically update driving route instructions in case of a tour plan adaptation or when traffic conditions change. Note that on-board units can exist in different variants, for example installed units, GPS-capable smartphones, or tablet computers with a mobile Internet connection.

2.6.3.2 Recent Advances in Dispatching Center Technologies

Besides the previously mentioned on-board technologies which allow for the tracking and monitoring of vehicle activities in real-time, there has been an advancement in the technologies of dispatching centers in recent years. Specifically, the coordination of vehicles in real-time has become possible due to the development of sophisticated real-time control approaches which make use of elaborated solution methods providing a high solution quality within a short amount of computational time (see Sect. 4.7.4 and Chap. 5). Hence, tour plan adaptations can be carried out quickly after dynamic events have occurred in the system. This allows real-time control approaches to coordinate the transportation process under real-world restrictions.

In order to efficiently handle occurring dynamic events in a real-time control approach, an electronic transmission of unexpected changes in the relevant information (see Sect. 2.5) is advisable. For dynamically arriving customer requests, this can be realized by electronically recording new customer requests in a call center (see also Fig. 1.1). Besides day- and time-dependent travel time profiles, different information sources like traffic flow sensors automatically provide updated

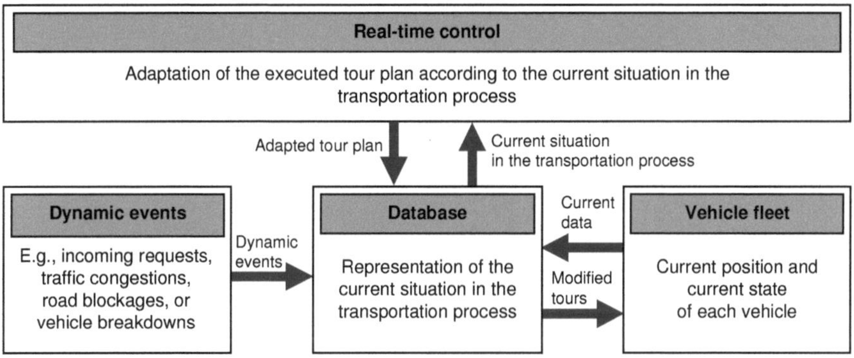

Fig. 2.10 Example architecture of a real-time control approach (cf. Bock 2010)

travel times on roads and highways. In order to quickly calculate travel distances and travel times between considered relevant locations and vehicles, the utilization of efficient shortest path approaches in combination with a Geographic Information System (GIS) which provides road network data is advantageous. Apart from commercially available road networks, OpenStreetMap is a comprehensive community-based road network data project to which users can make contributions (cf. Haklay and Weber 2008, OpenStreetMap 2011). For more information about existing approaches in the literature which make use of GIS systems, see also Sect. 4.7.5 in the literature review.

2.6.3.3 Interplay of Utilized Technologies and Information Flow in Real-Time Control Approaches

The joint use of the described technologies allows for the efficient application of real-time control approaches. In Fig. 2.10, an exemplary architecture for a real-time control approach is presented (cf. Bock 2010). In this architecture, the current system situation is stored in a central database which is supplied with new information by two systems. The first system transmits changes in the relevant information arising from dynamic events (see Sect. 2.6.2) while the second one establishes the bidirectional communication between the dispatching center and the vehicles. It provides current information about the vehicles to the database, e.g., their positions and the current activities of the service workers. Moreover, by using this system, vehicles are informed when changes in their tour have been realized by the real-time control. The real-time control adapts the current tour plan by transforming the current system situation from the database into the considered problem data. After the tour plan adaptation has been completed, the generated tour plan is stored in the database and changes in the tours are transmitted to the vehicles.

In order to realize the bidirectional communication system, Larsen (2000, p. 14) describes a communication architecture based on an Internet data connection. As illustrated in Fig. 2.11, in this architecture, vehicles are equipped with the previously

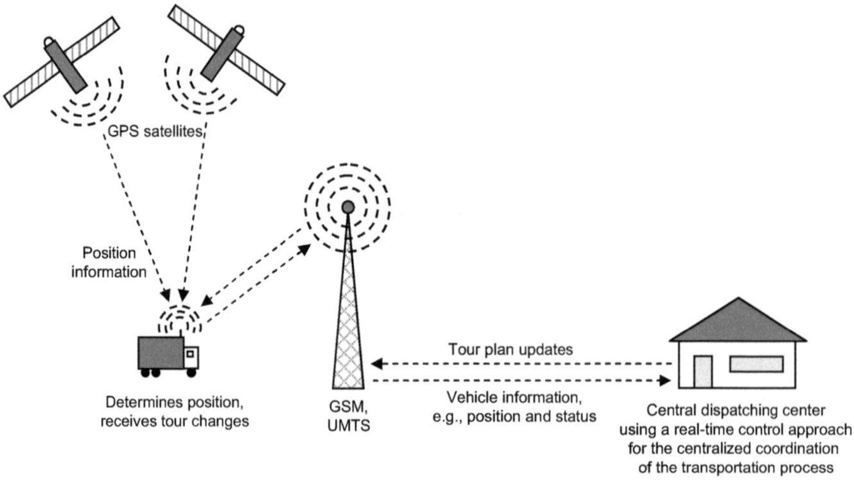

Fig. 2.11 Example architecture for a bidirectional communication system between the vehicles and the central dispatching center (cf. Larsen 2000, p. 14)

described GPS-enabled on-board units. Using this data connection, position and status are sent from the vehicles to the central dispatching center and changes in the tour plan are transmitted to the vehicles.

To sum up, it can be stated that using a real-time control approach by utilizing the described systems can significantly support dispatchers by providing decision support for tour plan adaptations. Specifically, by assessing various tour plan adaptation possibilities in a short amount of time and by enabling vehicle en-route diversion activities, the solution quality can be significantly increased (see also Sect. 8.5). Detailed information about real-time control approaches will be given in the literature review in Sect. 4.7.4 and in Chap. 5. Moreover, further research work on the utilization of communication technologies in dynamic routing problems will be described in the literature review in Sect. 4.7.5.

2.6.4 The Degree of Dynamism

Each dynamic routing problem comprises a specific difficulty in coordinating the corresponding transportation process with regard to the considered objective. We denote this as the *difficulty level* in what follows. This difficulty level depends on many factors. For example, the type of dynamic events considered in a dynamic routing problem and the considered objective are crucial attributes. Moreover, various instances of the same dynamic routing problem can also have a significantly different difficulty level. Since a higher difficulty level makes it more complex to attain a highly qualitative coordination of the transportation process, this imposes higher requirements on the solution method utilized for adapting the tour plan. Note

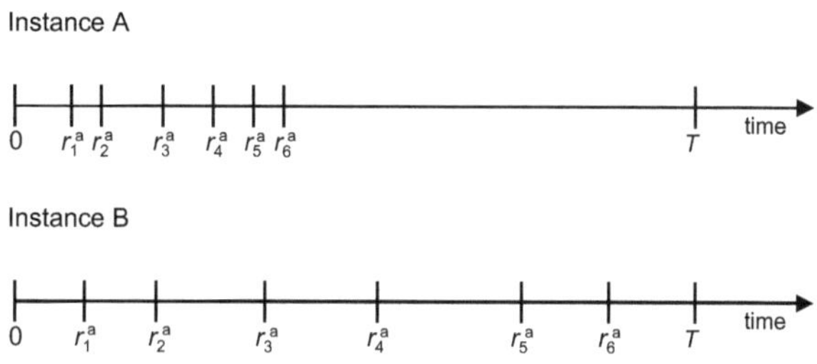

Fig. 2.12 Different arrival times of dynamic requests in two instances of a dynamic routing problem with the same degree of dynamism (cf. Larsen 2000, p. 57)

that awareness of this difficulty level is beneficial. For example, this information can be used for assessing the performance of a solution method when applied to different problem instances comprising individual difficulty levels.

In the literature, an approach for determining such differences in difficulty levels arising in dynamic routing problems is the *degree of dynamism (dod)*. This measure considers dynamic routing problems in which dynamism arises by newly arriving requests which is, as mentioned in Sect. 2.6.2, the most studied type of dynamic events in the literature. The degree of dynamism exists in three versions. Its basic version for dynamic routing problems without time windows was introduced by Lund et al. (1996). For defining the *dod*, it is assumed that the execution of the transportation process starts at time point 0 and new requests can arrive up to time point T. Furthermore, n_{stat} denotes the number of requests which are known during the initial tour plan generation process while n_{dyn} defines the number of dynamically arriving requests. With $n_{\text{all}} = n_{\text{stat}} + n_{\text{dyn}}$, the *dod* is defined as:

$$dod = \frac{n_{\text{dyn}}}{n_{\text{all}}}$$

The *dod* ranges from 0 to 1 where instances with a higher difficulty level have a higher *dod*. Specifically, if all requests are known in the initial tour plan generation process, it holds that $dod = 0$ indicating a static routing problem. In contrast, if all requests just arrive during the execution of the transportation process, it holds that $dod = 1$ so that a completely dynamic routing problem is present. This points to the fact that an instance of a dynamic routing problem in which only a small fraction of all requests dynamically arrives can be considered to be less challenging and easier to solve than an instance of the same dynamic routing problem in which all customer requests dynamically arrive. In what follows, we use the term *problems which are easier to solve* which refers to problems that make it easier to attain a solution with a good solution quality in reasonable time. Clearly, the opposite is true for problems which are more difficult to solve.

Despite its simplicity, other characteristics which also have a significant impact on the difficulty level are not considered in the *dod*. For example, it does not consider the time points at which dynamic requests arrive in the system. As illustrated in Fig. 2.12, two instances with significantly different request arrival properties can inhibit the same *dod*. Specifically, if the minimization of travel distance is the pursued objective, instance A is easier to solve than instance B since all dynamic requests arrive early in instance A (cf. Larsen 2000, p. 57). Hence, after the arrival of the last request, the remaining routing problem is a static problem. Since at this point in time it is likely that many other requests are still pending in instance A, a tour plan with a high solution quality can be generated according to this large amount of still changeable decisions. In contrast, in instance B, additional changes in the relevant information are also revealed late. Hence, at this point in time it is likely that the majority of requests have already been serviced.

In order to consider the mentioned differences in request arrival times, Larsen (2000, p. 58) extends the degree of dynamism to the *effective degree of dynamism* denoted as *edod*. In the *edod*, each dynamic request $i \in \{1, \ldots, n_{\text{dyn}}\}$ arrives at time point r_i^{a} in the system. The *edod* is then defined as follows:

$$edod = \frac{\sum_{i=1}^{n_{\text{dyn}}} \left(\frac{r_i^{\text{a}}}{T}\right)}{n_{\text{all}}}$$

The effective degree of dynamism represents the ratio of the arrival times of dynamic requests to the latest possible time point T at which they could arrive. In accordance with the previously mentioned objective of minimizing travel distances, if dynamic requests arrive early, the *edod* has a low value while later request arrivals result in higher *edod* values. Similar to the *dod*, the *edod* is a measure between 0 and 1 in which larger values indicate problem instances which are more difficult to solve.

If time windows are considered in dynamic routing problems, it is reasonable to consider them in the calculation of the degree of dynamism since tight hard time windows limit dispatching possibilities making it difficult to attain good solutions. In case of soft time windows, this leads to less possibilities for servicing requests without penalty costs compared to problem instances with relaxed request time window constraints. If the objective of distance minimization is considered, requests can be fulfilled in more distance-optimized ways. Therefore, such problem instances can be considered to be easier to solve than those with tight time windows. In order to consider such time window characteristics of dynamic routing problems, Larsen (2000, p. 59) proposes a third version of the degree of dynamism denoted as the *effective degree of dynamism with time windows* ($edod_{\text{tw}}$). It is defined as

$$edod_{\text{tw}} = \frac{1}{n_{\text{all}}} \sum_{i=1}^{n_{\text{dyn}}} \left(\frac{T - (r_i^{\text{l}} - r_i^{\text{a}})}{T} \right) = \frac{1}{n_{\text{all}}} \sum_{i=1}^{n_{\text{dyn}}} \left(1 - \frac{r_i^{\text{re}}}{T} \right)$$

where $r_i^{\text{re}} = r_i^{\text{l}} - r_i^{\text{a}}$ denotes the available *reaction time* of request i and represents the time period from the arrival time of request i to the end of its time window. Hence, the existence of larger reaction times provides more possibilities for dispatching and servicing pending requests which leads to a low $edod_{\text{tw}}$ and indicates relaxed system situations. In Fig. 2.13, the reaction times of two instances of dynamic routing problems, each comprising two dynamic requests, are illustrated. As can be seen in this figure, the reaction times are larger in instance A than in instance B making instance A easier to solve.

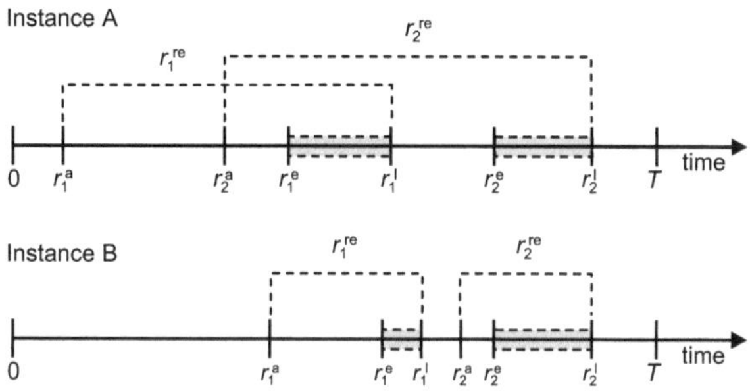

Fig. 2.13 The effective degree of dynamism with time windows (cf. Larsen 2000, p. 60)

Note that in dynamic routing problems in which the minimization of customer inconvenience is pursued, it holds that $r_i^l = r_i^e = r_i^a \ \forall i \in \{1, \ldots, n_{\text{dyn}}\}$ since customer inconvenience represented by penalty costs directly starts upon request arrival in the system. In such applications, the reaction time r_i^{re} is 0 for all dynamically arriving requests. Since almost all customer requests are dynamic in RDOPG applications considered in this book, these applications have an $edod_{\text{tw}}$ close to 1 making them difficult to solve with a high solution quality.

Further Extensions to the Degree of Dynamism The described *dod* and its variants provide efficient measures for determining the difficulty level of routing problems in which the minimization of travel distance or comparable objective functions is pursued. With regard to RDOPG applications in which the minimization of customer inconvenience is pursued, further extensions are possible out of which two are described in what follows. For example, considering Fig. 2.12, a different difficulty level for the two instances illustrated arises. Under this objective, instance A can be considered to be more difficult to solve than instance B since in instance A, the dispatcher has to deal with a large amount of newly arriving requests in a short amount of time. This can be complex, especially if customer inconvenience depends on request response times, good and fast tour plan adaptations are required. Hence, with regard to RDOPG applications, it is advantageous to additionally evaluate other request arrival characteristics. Specifically, the distribution of request inter-arrival times is another important aspect. As illustrated in Fig. 2.14, two instances with the same *edod* can have a significantly different request arrival distribution. If the minimization of customer inconvenience is pursued, instance A is more difficult to solve than instance B since, as mentioned above, the dispatcher is required to make good decisions for a large amount of requests at the same time. Moreover, in dynamic VRPs, the fleet size is often restricted (see aspect 11 of Psaraftis' approach presented in Sect. 2.5.2.2). Hence, a large number of requests shortly arriving one after the other causes longer request response times and hence higher customer inconvenience values compared to evenly distributed request arrivals.

Another possible extension which is relevant for RDOPG applications is mentioned by Pillac et al. (2010). The authors state that the geographical distribution of newly arriving requests and resulting travel times between requests should be considered since they significantly determine the complexity of dynamic routing problems in which the minimization of the request response times is pursued. Related to

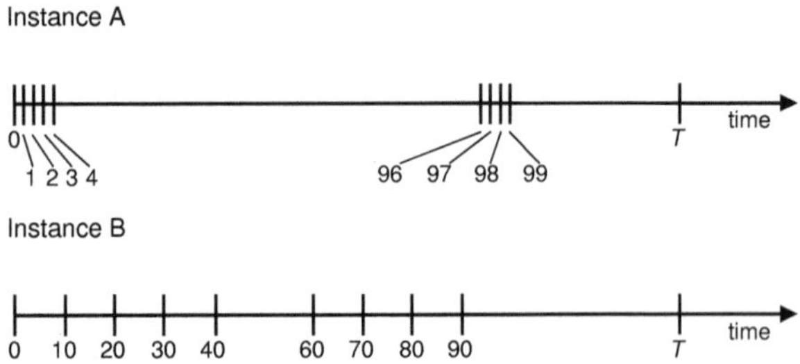

Fig. 2.14 Two instances with the same effective degree of dynamism of 0.4 and significantly different inter-arrival times

this, we will propose a new measure for characterizing differences in dynamic routing problems with regard to the applicability of stochastic knowledge in pro-active real-time control approaches in Sect. 8.9.

2.6.5 Three-Echelon Classification of Dynamic Routing Problems

In Larsen (2000, p. 60 et seq.) and Larsen et al. (2007) a two-dimensional classification for distinguishing real-world dynamic routing problems is proposed according to their degree of dynamism and the pursued objective. In this classification, dynamic routing problems are, according to the percentage of dynamically arriving requests, subdivided into *weakly*, *moderately*, and *strongly* dynamic systems. Hence, these echelons represent routing problems with a low, medium, and high degree of dynamism. Moreover, the pursued objective ranges from the minimization of distribution costs to the maximization of service quality. As mentioned in Sect. 2.6.1, these two objectives usually contradict each other so that both objectives cannot be completely fulfilled at the same time. Nevertheless, there exist dynamic routing problems in which both objectives are considered with a similar priority.

The proposed classification is illustrated in Fig. 2.15. As can be seen, there is a strong correlation between the degree of dynamism and the pursued objective in real-world dynamic routing problems, i.e., the higher the degree of dynamism, the more important service quality aspects become in the considered routing problem. In what follows, we describe characteristics of each echelon and present corresponding real-world applications (cf. Larsen 2000 and Ghiani et al. 2003).

2.6.5.1 Echelon I: Weakly Dynamic Systems

Weakly dynamic systems are characterized by only a small number of customers dynamically arriving during the execution of the transportation process. According to the authors, the majority, i.e., at least 80 % of all customer requests are already known during the initial tour plan generation process (cf. Larsen et al. 2007). Hence, routing problems in this echelon have a *dod* < 0.2. In what follows, we present typical real-world routing problems which belong to this echelon:

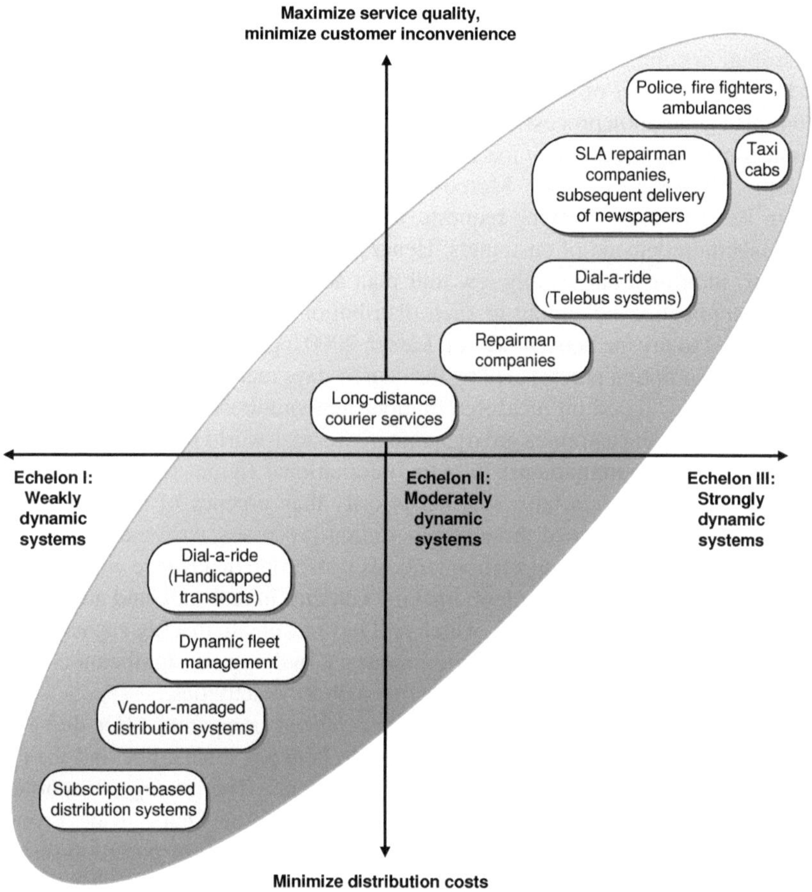

Fig. 2.15 Three-echelon classification of dynamic routing problems (cf. Larsen 2000, p. 63, Larsen et al. 2007)

- **Subscription-based distribution systems:** In general, customers subscribe to this type of service so that customer requests and their demands are often known in advance. The authors mention that less than 5 % of all customers request an additional service or make changes in their pre-ordered demands. In such systems, the tour plan is usually only adapted in longer time intervals, e.g., on a monthly basis, in order to save coordination effort and increase efficiency by utilizing driver learning curves. In order to service the few spontaneous customer requests, some slack time is included in the tour plan. One real-world example of subscription-based distribution systems is the delivery of groceries to private households on a regular basis.
- **Vendor-managed distribution systems:** In such distribution systems, companies manage the inventory of their customers and automatically replenish their inventory when necessary. In order to determine which customers are required to

be serviced during the next planning horizon which is usually the next day, the distribution company estimates customers' inventory levels using specific methods. Hence, the set of pending requests can be well estimated during the initial tour plan generation process. However, since the actual consumption rates of customers may vary from what has been predicted, a small percentage of customers additionally require service. Moreover, new customers may demand urgent service. Besides newly arriving requests, another source of dynamism is given by the uncertain demand of customers. Hence, capacity limitations of vehicles may require additional but usually few tour plan adaptations during the day. A real-world application belonging to such distribution systems is the distribution of heating oil to private households (cf. Larsen 2000, pp. 16–17). In this application, the oil consumption is estimated by the degree-days method which estimates consumption rates based on measured outdoor temperatures. The author states that about 20 % of all customers are dynamic in this real-world problem.
- **Dynamic fleet management:** In many international freight forwarding companies, cost aspects often have a higher priority than urgency of requests. Due to the long-haul character of these transportation processes, customers are usually required to negotiate transportation requests some time in advance with the transportation company. Nevertheless, tracking vehicles in real-time and allowing the re-routing of traveling vehicles, which still have available loading capacities, can help to integrate dynamically arriving requests. In doing so, significant cost savings can be achieved on long-distance transportation activities.
- **Dial-a-ride for handicapped people:** Specialized companies offer dial-a-ride services for transporting handicapped people between their home and their desired destination, usually for medical appointments. The arising transportation processes can be modeled as dynamic variants of the Dial-A-Ride Problem (DARP, see Sect. 2.3.3.3). Since customers usually book transportation services in advance, only few customers request a service on the same day. Hence, it is possible to generate cost-efficient tour plans which also provide a high customer service quality by meeting the customers' desired appointment times and minimizing their ride times. For more information on related routing problems, see, e.g., Cordeau and Laporte (2003, 2007) and Cordeau et al. (2007a).

2.6.5.2 Echelon II: Moderately Dynamic Systems

This echelon consists of real-world applications in which the percentage of dynamically arriving customer requests during the execution of the transportation process is noticeable. Nevertheless, there is a considerable proportion of requests still known before the execution of the transportation process (cf. Larsen et al. 2007). The following real-world applications belong to the class of moderately dynamic systems:

- **Long-distance courier services and (express) parcel services:** This routing application occurs in companies which offer services for transporting mail and parcels. Besides regular customers, there are customers who demand transportation services on short notice so that the vehicle tours are required to be replanned each day. Furthermore, some companies accept late-in requests. In this case, customers may demand goods to be picked up at their location after the vehicles have already started to collect goods so that their tours require real-time adaptations. In order to realize long-distance transportation services, different logistical structures can be used. For example, shipments are collected at the customer locations and are transported to a hub. From this hub, the shipments are transported to individual destination hubs from which they are transported to their final destinations. For more information on such logistical structures, see, e.g., Bock (2004, p. 48 et seq.) and Crainic and Kim (2007).

Besides standard transportation services, several companies offer next-day express services which guarantee the delivery of shipments to national and international destinations by a specified time point the next day, e.g., 9:30 am, 10:30 am, or 12:00 am. Moreover, very urgent freights can also be delivered on the same day (cf. DHL Express Germany GmbH 2010). In order to efficiently coordinate such time-critical transportation processes, the utilization of sophisticated real-time control approaches is advantageous.

- **Repairman companies with scheduled customers:** Specialized companies provide repair services and maintenance for their customers, e.g., for maintaining automated teller machines (cf. Larsen 2000, p. 15). A characteristic of this type of repairman problems is that most requests are known in advance since they arise from maintenance contracts in which regular service is specified in known time intervals. In addition to customer requests known in advance, some urgent customer requests which require an immediate service can occur during the planning horizon. Other repairman companies mainly dealing with urgent requests which are not known in advance are described in the next echelon.

2.6.5.3 Echelon III: Strongly Dynamic Systems

This echelon comprises real-world routing problems in which only few or no requests are known before the execution of the transportation process. In these applications, the *dod* amounts to at least 80 % (cf. Larsen et al. 2002). Furthermore, arriving customer requests often require an urgent service. Due to primarily customer-oriented objectives which often consider customer inconvenience depending on the request response time, one possibility for improving the solution quality is the utilization of stochastic knowledge in order to utilize available vehicles more efficiently. More information about this will be given in the literature review in Sect. 4.3. The following real-world routing problems belong to the echelon of strongly dynamic systems:

- **Telebus dial-a-ride systems:** Larsen (2000, p. 64) presents a variant of a DARP which can be found in telebus systems. In such systems, buses only operate on demand, i.e., if customers request for transportation. For example, telebus systems are used on bus lines, either all day or at times where there is usually a very low demand. According to Borndörfer et al. (1997), telebuses can also be rented for handicapped people who cannot use public transportation systems. Similar to taxi cab services, most customer requests dynamically arrive during the day. However, since telebuses are usually required to be ordered some time in advance, the urgency is lower than in taxi applications described later in this section.

- **SLA repairman companies:** Besides the previously mentioned repairman companies in which most of the requests are known in advance, there exist repairman companies which mainly deal with urgent problems arising in case of severe failures. As described in Chap. 1, example applications comprise professional repairman companies which quickly resolve severe problems faced by their customers, e.g., manufacturing companies which are part of a supply chain. In such applications, repairman companies often need to comply with the regulations of SLA contracts negotiated with their customers. Failing to fulfill the SLA results in high penalty costs so that a quick reaction to dynamically arriving customer requests is of high importance. Bertsimas and van Ryzin (1991) introduce the Dynamic Traveling Repairman Problem (DTRP) and evaluate solution strategies which are described in detail in Sect. 4.5.1.
- **Subsequent delivery of newspapers:** As introduced in Chap. 1, the subsequent delivery of newspapers occurs when a subscriber has not received his newspaper in the morning as expected. In this case, a quick subsequent delivery of the missing newspaper is required in order to minimize customer inconvenience and maintain customer loyalty.
- **Taxi cab services:** Taxi drivers are often connected to a central call center where new customer requests are recorded and dispatched. By using appropriate technical equipment (see Sect. 2.6.3), individual taxi cabs can be located in real-time and pending customer requests are assigned to them (cf. Horn 2002a, 2002b). Since people frequently demand instant transportation, quality of service is also measured in this application by the resulting response time. These processes are modeled as a Dynamic Dial-A-Ride Problem (cf. Cordeau and Laporte 2007) since a customer request consists of two locations instead of one. Moreover, the service time of a request, i.e., the time needed for transporting a customer comprises some uncertainty since travel times can only be estimated.
- **Emergency services:** For emergency services such as police, fire fighters, and ambulances, typically no requests are known in advance. The solution quality of an emergency system is often measured by the achieved request response time. According to Larsen et al. (2007), typical ambulance emergency systems agree on a certain level of service quality so that, for instance, 90 % of the requests should be serviced within 5 minutes and the remaining requests within a maximum of 8 minutes. Due to the high significance of emergency services, there is a lot of ongoing research on this topic which has also been the case for the last decades. For example, Larson and Odoni (1981) propose the hypercube queuing model which has been widely used, especially in the coordination of police patrols. In Brotcorne et al. (2003), the authors study approaches for ambulance dispatching developed during the past three decades. More information on relocation strategies which have also been applied to this type of applications will be given in the literature review in Sect. 4.3.2.

As can be derived from this description, RDOPG applications, i.e., SLA repairmen companies and the subsequent delivery of newspapers (see Sect. 1.1) belong to the echelon of strongly dynamic systems.

2.6.6 *Evaluating the Performance of Solution Approaches for Dynamic Routing Problems*

In what follows, we describe two measures known in the literature for evaluating the performance of solution approaches for dynamic routing problems.

2.6.6.1 Competitive Analysis

Sleator and Tarjan (1985) introduced a general framework denoted as *competitive analysis*. In dynamic routing problems, this framework can be used to define the *competitive ratio* of a solution approach A denoted as CR_A (cf. Larsen et al. 2007). Given an instance I of the considered dynamic routing problem, $z_A(I)$ defines the objective function value of the solution generated by A applied to this instance. Furthermore, $z^*(I^O)$ represents the objective function value of the solution generated by an optimal solution approach when applied to the corresponding *offline instance* I^O, i.e., the corresponding static instance of I in which the optimal solution approach has access to all relevant information before the execution of the transportation process. Furthermore, the set M comprises all theoretically possible instances of the considered dynamic problem. The competitive ratio of solution approach A is defined as:

$$CR_A = \sup_{I \in M} \frac{z_A(I)}{z^*(I^O)}$$

By definition, CR_A represents the worst-case ratio of solution approach A compared to an optimal offline solution approach. Therefore, the competitive ratio is a measure for evaluating the worst-case performance of A on all instances of a given dynamic routing problem. One issue with applying the competitive ratio is that up to now many dynamic routing problems can only be empirically evaluated by simulation-based methods. Hence, for such problems, it is not possible to evaluate all theoretically possible instances $I \in M$. Nevertheless, there exist specific dynamic routing problems to which the competitive ratio has been successfully applied (cf. Bertsimas and van Ryzin 1991, 1993a, 1993b, de Paepe 2002, and Angelelli et al. 2007). In these problems, specific assumptions, e.g., about request arrivals and their distribution, are made. In case of dynamic events, plan adaptations are carried out by strategy-oriented approaches which allow the competitive ratio to be applied since the effects of the applied adaptation methods on the solution quality can be analytically studied. Strategy-oriented approaches will be described in the literature review in Sect. 4.5.1.

2.6.6.2 The Value of Information

In order to evaluate the performance of solution approaches on dynamic routing problems which cannot be evaluated by the competitive ratio previously introduced, Mitrovic-Minic et al. (2004) propose a measure denoted as the *value of information*. It evaluates the worth of having complete knowledge of all relevant information before the execution of the transportation process. Given a solution approach H (often a heuristic approach) and an instance I of the considered dynamic routing problem, $z_H(I)$ represents the objective function value of the solution generated by H for problem instance I. Furthermore, $z_H(I^{\mathrm{O}})$ represents the objective function value of the solution which has been generated by H on the corresponding offline instance of I. The value of information of solution approach H on instance I is defined as:

$$V_H(I) = \frac{z_H(I^{\mathrm{O}}) - z_H(I)}{z_H(I^{\mathrm{O}})}$$

Following this definition, the value of information represents the maximum possible gain in solution quality when H is applied on I if all information is known in advance. Hence, a larger value of $V_H(I)$ indicates instances in which it is especially beneficial when more relevant information is known before the execution of the transportation process, for example by utilizing stochastic knowledge about future expected events. Based on the value of information, we will describe in Sect. 8.2.6 a measure which is used in our computational results for evaluating differences in the performance of the real-time control approaches developed in this book.

Although the value of information gives an upper bound on improvements in solution quality which are attainable if more information, e.g., suitable stochastic knowledge, is known in advance, it does not consider specific characteristics which are present in the application or in test instances. Moreover, note that the estimation of this measure is also solution approach-dependent. In this book, we will propose new measures which take into account specific characteristics present in the considered routing problem. Hence, these measures allow for estimating the improvements which are additionally attainable just by analyzing the request data without requiring empirical assessment of the attained solution quality in the online and offline case and therefore allows for an approach-independent estimation. Moreover, these measures allow for identifying specific characteristics of request data sets which contribute to an efficient utilization of stochastic knowledge thereby assessing possible improvements and supporting estimations more precisely. These measures will be described in Sects. 8.8.2 and 8.9.

2.7 A General Classification Scheme for Routing Problems

In order to capture all attributes of routing problems, several authors have developed a general classification scheme (cf., e.g., Bodin and Golden 1981, Stumpf 1998, Richter 2005, and Eksioglu et al. 2009). In this section, a variant of this scheme

Table 2.2 General classification scheme for routing problems including RDOPG-specific attributes (in accordance with, e.g., Bodin and Golden 1981, Stumpf 1998, Richter 2005, and Eksioglu et al. 2009)

Attribute	Possible values
Requests	
Type	Pickup only
	Delivery only
	Pickup and delivery on the same tour (with backhauls, see Sect. 2.4.2)
	Combined requests (Pickup and delivery)
Location	At nodes
	At arcs/edges
	Mixed
Geographical distribution	Distributed with a pattern
	Randomly distributed
Degree of dynamism	Low
	Moderate
	High
Percentage of regular customers	None
	Few
	Medium
	High
	All
Time restrictions	No time windows
	One-sided time window
	Two-sided time window
	Fixed time points for service
	Hard time windows
	Soft time windows
	Multiple time windows
Divisibility	Non-divisible requests
	Arbitrarily divisible requests
	Restricted divisible requests (e.g., minimum amount per part)
Transshipment	No transshipment
	Transshipment possible
Request size	Full-truckload (FT)
	Less-than-truckload (LTL, e.g., piece goods)
Types of transportation objects	Goods
	Services
	Persons

Table 2.2 (Continued)

Attribute	Possible values
Size of service area	Urban
	Rural
	Regional
	National
	International
Sequence of requests	Arbitrary
	Predecessor-/Successor-restrictions
Selection of requests	Service of all requests
	Possibility to select requests to be serviced, i.e., requests can be rejected
Frequency of request occurrence	Once
	Semi-periodic
	Periodic
Fleet	
Size	One vehicle
	Multiple vehicles, restricted fleet size
	Multiple vehicles, unlimited fleet size
Consistency	Homogeneous (one type of vehicles)
	Heterogeneous (different types of vehicles)
Capacity	No capacity constraints
	One-dimensional capacity constraints
	Multi-dimensional capacity constraints
Location	One depot
	Multiple depots
	Freely distributed
Number of tours per vehicle and planning horizon	One tour
	Multiple tours
Time restrictions	No restrictions
	Maximum duration of tour
Loading restrictions	Arbitrary loading
	Restricted set of goods can be loaded
Number of drivers per vehicle	(Number)
Driver to vehicle assignment	Fixed
	Free
Vehicle reliability	Reliable
	Vehicle slowdowns possible
	Vehicle breakdowns possible

Table 2.2 (Continued)

Attribute	Possible values
Network	
Type of network	Road network (real distances)
	Coordinate system (Euclidean distances)
Data	Symmetric (undirected graph)
	Asymmetric (directed graph)
	Mixed
Travel times	Constant
	Time-dependent known
	Dynamic with uncertainty
Planning horizon	
Length	One period
	Multiple periods
	Open-ended
Type	Once
	Continuous
	Repeating
Tour plan	
Type	Daily changing tour plan
	Fixed tour plan
	Fixed tour plan with slack time for flexibility
Type of tours	Open tours (without return to depot)
	Closed tours (with return to depot)
Restrictions	Maximum tour duration
	Maximum tour length
	Worker-dependent restrictions (e.g., maximum working time, driving time restrictions)
Relevant information	
Type of relevant information	Constant
	Changing
Type of problem	Static
	Dynamic, changes deterministically known
	Dynamic, changes comprise uncertainty
Availability of changes	Global
	Local
	Mixed
Types of considered dynamic events	New requests
	Changes in request attributes

Table 2.2 (Continued)

Attribute	Possible values
	Traffic congestions and road blockages
	Vehicle disturbances
Type of data	Real-world data
	Designed data
	Both real-world and designed data
Stochastic knowledge	
Type of stochastic knowledge	None
	Prescribed probability distributions are available
	Forecasted by analyzing historical data, e.g., past request information
Increasing tour plan flexibility	
Methods used for increasing tour plan flexibility	None
	Waiting strategies
	Relocation strategies
	Request assignment strategies
Objective	
Considered objective	Minimization of travel length
	Minimization of travel time
	Minimization of operational costs
	Minimization of the number of utilized vehicles
	Minimization of the number of unserviced customers
	Minimization of response time
	Minimization of customer inconvenience
Applied approach	
Processing of changes in the relevant information	Centralized
	Decentralized
	Mixed
Type of applied methods	Exact
	Heuristic
	Simulation
	Real-time control methods
Type of model	Deterministic static
	Stochastic static
	Deterministic dynamic
	Stochastic dynamic
Type of application	Static
	Dynamic

extended by attributes which are relevant for the considered RDOPG applications is presented. The resulting classification scheme is illustrated in Table 2.2.

The RDOPG applications considered in this book will be characterized according to this general classification scheme in Sect. 3.2.3.

2.8 Summary

In this chapter, we discussed different characteristics and variants of routing problems. First, an introduction to classic static routing problems was given. Since RDOPG applications can be modeled as a DVRP variant in which new requests dynamically arrive, variants of the VRP and the DVRP made up the main part of this chapter. Specifically, first static VRPs which differ from each other in terms of problem attributes, such as considered type of requests, number of depots, available vehicles, and pursued objective function, were described. After this description, dynamic routing problems in which the relevant information changes during the execution of the transportation process were discussed and classifications known in the literature which deal with various differences between static and dynamic routing problems were described. With regard to attributes specific for RDOPG applications, a unified approach for classifying routing problems based on insights from known classifications in the literature was introduced. In this unified approach, the type of relevant information and corresponding changes that occur in the problem data during the execution of the transportation process as well as the availability and processing of these changes were considered. Moreover, depending on the type of available stochastic knowledge, different variants of VRPs known in the literature were described. After this description, in accordance with the considered RDOPG applications, dynamic routing problems in which a centralized coordination of the transportation process is performed were considered in detail. This included typical objectives pursued in dynamic routing problems and a description of dynamic events considered in dynamic routing problems. Moreover, a measure denoted as the degree of dynamism which has been proposed by Lund et al. (1996) and Larsen (2000) was described. This measure is used for determining the difficulty level of solving individual problem instances of dynamic routing problems. Based on this measure, a three-echelon classification proposed by Larsen (2000) was introduced which includes the characterization of various real-world dynamic routing applications. In order to evaluate the performance of dynamic routing approaches, two measures known in the literature were presented, in particular the competitive ratio and the value of information. Finally, a general classification scheme for characterizing routing problems was presented and extended with regard to additional characteristics which are relevant for RDOPG applications.

Chapter 3
The Considered RDOPG Applications

In Chap. 1, different real-world problems which belong to RDOPG applications considered in this book were introduced. In this chapter, we introduce the *problem description of RDOPG applications* in detail. Following this description, *characteristics of RDOPG applications* are discussed.

3.1 Problem Description of RDOPG Applications

In the considered RDOPG applications, a daily distribution process has to be controlled in real-time. In this process, a set of $\mathcal{R}$ customer requests $R = \{1, \ldots, \mathcal{R}\}$ that occur dynamically during the day needs to be serviced by a set of $\mathcal{K}$ homogeneous vehicles $K = \{1, \ldots, \mathcal{K}\}$ which are originally located at one depot to which they return after their tour has been completed. During the execution of the transportation process, the relevant information is changed by dynamic events by which new requests dynamically arrive. As mentioned in Chap. 1, the vehicle tours consist of delivery activities, i.e., vehicles deliver goods that are loaded onto the vehicles at the depot before the execution of the transportation process. All of the arriving requests have to be accepted and serviced by the available vehicles. In other words, they cannot be rejected or subcontracted. Moreover, vehicle capacity constraints can be regarded as negligible since a sufficient amount of the goods to be delivered is loaded onto the vehicles at the depot before the transportation process starts. The considered planning horizon has a length of up to one working day and the considered service region consists of a medium to large-sized urban area.

As mentioned in Chap. 1, the goods to be delivered are denoted as *perishable goods* because they have to be delivered as quickly as possible once they are required by customers. Therefore, the considered objective pursues the minimization of the *total customer inconvenience*. This total customer inconvenience consists of the sum of all individual customer request inconvenience values which depend on the request response times, i.e., the time period between request arrival and start of service (see also Sect. 2.4.12.7). In the marketing literature, two main approaches to

F. Ferrucci, *Pro-active Dynamic Vehicle Routing*, Contributions to Management Science, DOI 10.1007/978-3-642-33472-6_3, © Springer-Verlag Berlin Heidelberg 2013

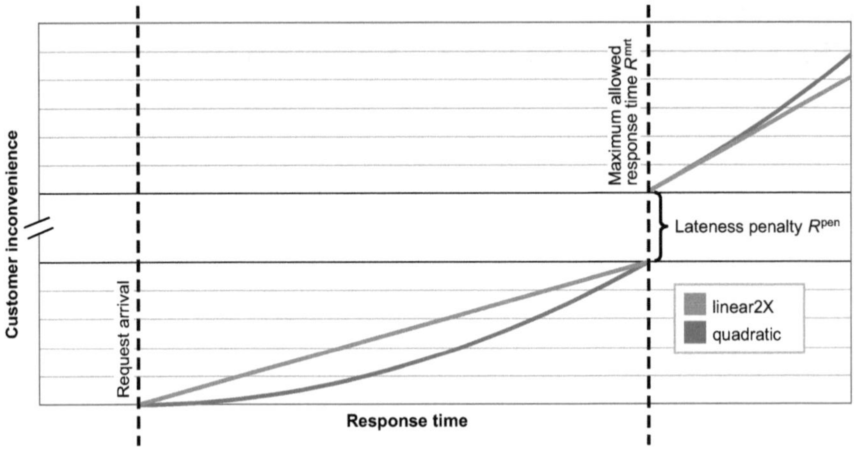

Fig. 3.1 Illustration of the linear2X and quadratic customer inconvenience

model customer inconvenience dependent on elapsed response time can be found. Davis and Maggard (1990) propose that customer inconvenience increases linearly with resulting response time. In contrast, Kristensen et al. (1992) derive a quadratic dependency. Since both approaches are reasonable for the considered RDOPG applications, we evaluate both alternatives in this book. Further information on other approaches that utilize customer inconvenience aspects in the objective function can be found in the literature review in Sect. 4.6. Additionally, a maximum request response time denoted by R^{mrt} has to be considered for each request, otherwise, high penalty costs denoted by R^{pen} occur. In this case, the loss of the corresponding customer is likely, which should be prevented under all circumstances. Moreover, this integrates an effective method for avoiding an indefinite deferment of individual customer requests (see aspect 8 of Psaraftis' approach presented in Sect. 2.5.2.2) which is desirable if the quality of customer service aspects is to be considered. The evolution of customer inconvenience used in RDOPG applications is illustrated in Fig. 3.1.

As mentioned in Sect. 1.3, we investigate the subsequent delivery of newspapers as the exemplary real-world RDOPG application in this book. Nevertheless, note that other related RDOPG applications can be modeled by setting the problem parameters accordingly. One important characteristic of the subsequent delivery of newspapers is that the service time which is required for servicing a request can be considered to be short and equal for all customer requests. In this book, we assume a service time of $R^{st} = 60$ seconds for all requests. Other related real-world problems for RDOPG applications can be found in Sect. 1.1. In what follows, further elements which are part of the considered RDOPG problems are introduced.

3.1.1 Existence of a Real Urban Road Network

In order to enable an accurate mapping of vehicle movements on roads and to calculate the resulting travel times between customer requests and current vehicle positions accordingly, a real-world urban road network is used. In this road network, different road classes which allow for modeling individual travel speeds are present. Moreover, one-way roads which restrict possible travel paths exist. Note that using a road network enables a much higher degree of realism compared to assuming that vehicle movements are performed on Euclidean distances between relevant locations. The utilized road network will be introduced in detail in Sect. 5.3.1.

3.1.2 Presence of Vehicle On-Board Units and a Central Dispatching Center

In order to enable tracking of vehicle positions and to inform vehicles about performed tour adaptations in real-time, on-board communication units are assumed to be available in the deployed vehicles. This allows a bidirectional communication between the vehicles and the dispatching center which performs a centralized coordination of tour plan adaptations during the execution of the transportation process (see Sect. 2.6).

3.1.3 Availability and Utilization of Past Request Information

In the considered RDOPG applications, past request information is available. This past request information consists of arrival time and location of all requests that have arrived in the system on elapsed working days. Note that such information is available in many real-world applications. In our proposed approach, this past request information is used to generate stochastic knowledge in order to anticipate future requests which are expected to occur with a high probability during the next planning horizon (see also Sect. 2.5.3.2.2).

With regard to the considered objective of minimizing the total customer inconvenience, an improved solution quality can be achieved by quickly responding to dynamically arriving customer requests. This quick response can be attained by distributing the available vehicles over the considered service region so that they are efficiently prepared to service expected future request arrivals. Specifically, by guiding vehicles into *request-likely areas* in which new requests are likely to appear and keeping vehicles there for a specified time period, newly arriving requests can be serviced faster. Note that a request-likely area is represented by a subregion of the service region during a certain time period only since request arrival rates may change during the planning horizon. The generated stochastic knowledge represents such request-likely areas in the proposed pro-active real-time control approach.

Note that in our considered problem setting, the stochastic knowledge is generated by solely analyzing and examining the available past request information. In particular, information about expected future request arrivals is only derived from this data. Since no prescribed probability distributions are available for generating stochastic knowledge, the underlying distributions of the available past request information are unknown to the approach which forecasts the stochastic knowledge. This is representative of many practical applications. The proposed forecasting approaches for generating and integrating stochastic knowledge into the pro-active real-time control approach will be presented in Chap. 6.

3.1.4 Allowing Vehicle En-Route Diversion

In the considered problem setting, allowing en-route diversion of vehicles enables a reassignment of the first request in a vehicle's tour to which the vehicle is currently traveling. This increases dispatching flexibilities and enables a more direct reaction to newly arriving requests (cf. Ichoua et al. 2000). Specifically, by allowing vehicle en-route diversion, a newly arriving request which is close to the current position of a vehicle can be serviced quickly since the corresponding vehicle only needs to perform a small detour on its way to its currently assigned first request. Ichoua et al. (2000) and Richter (2005, p. 219) have shown that this can be advantageously applied if the considered objective function comprises quality of customer service aspects, as is the case in the studied RDOPG applications. Note that due to the use of a real road network, vehicles can only change their direction at the next road junction. However, since the utilized urban road network is very dense, individual road parts are quite short so that the next road junction can often be quickly reached by a vehicle.

3.1.5 Integration of Vehicle Scheduling Decisions

As mentioned in Sect. 3.1.3, by making use of stochastic knowledge, vehicles can be kept in request-likely areas for specific time periods. Since the length of these time periods is determined by the proposed pro-active real-time control approach according to the current request situation, vehicle scheduling decisions become a part of the considered problem. Therefore, the problem considered in RDOPG applications belongs to the class of Vehicle Routing and Scheduling Problems (VRSP, see Sect. 2.4.6) if the available stochastic knowledge, i.e., the pro-active real-time control approach, is utilized.

3.2 Characteristics of RDOPG Applications

After the problem considered in RDOPG applications has been described, in this section we examine further characteristics of RDOPG applications. First, *differences from other dynamic routing problems* are discussed. After that, the *computational complexity* of RDOPG applications is derived. The proposed classification approach for routing problems described in Sect. 2.5.3 is used for *classifying RDOPG applications with regard to characteristics and processing of relevant information*. Moreover, a *general classification of RDOPG applications* is presented according to the classification scheme for routing problems which was introduced in Sect. 2.7.

3.2.1 Differences from Other Dynamic Routing Problems

As mentioned above, RDOPG applications consider the distribution of perishable goods without transshipment to dynamically arriving customer requests using a known fleet size located at one depot. According to the different variants and characteristics of routing problems which have been presented in Chap. 2, RDOPG applications can be modeled on the basis of an Asymmetric Dynamic Forecasted Stochastic Vehicle Routing Problem with Soft Time Windows (ADFSVRPSTW) which is a specific variant of the Dynamic Vehicle Routing Problem with Soft Time Window constraints (DVRPSTW, cf. Gendreau et al. 1999a). The considered problem is asymmetric since travel times between relevant locations can depend on the direction of travel due to one-way streets in the road network. Furthermore, the problem is dynamic since new requests arrive during the execution of the transportation process. In the problem, stochastic knowledge is available which is solely generated from available past request information resulting in forecasted stochastic knowledge. Moreover, the considered distribution of goods from one depot to customer locations can be modeled by an uncapacitated VRP (see Sect. 2.3.2). In order to model the considered objective function, soft time windows with variable and fixed penalty costs are used. Specifically, variable penalty costs directly starting upon request arrival represent customer inconvenience while fixed high penalty costs occur when the maximum response time R^{mrt} is not fulfilled.

Differences from other dynamic routing problems arise from the structure of customer requests. An analysis of the real-world request data of the subsequent delivery of newspapers (see Sect. 8.2.4) has shown that only a negligible part of the customers is known before the execution of the transportation process. The largest part of all customer requests dynamically arrive during the execution of the transportation process. Specifically, over 90 % of all customer requests are dynamic. Hence, in comparison to other DVRPs with a low or moderate degree of dynamism, the considered RDOPG applications are determined by a high degree of dynamism (see Sect. 2.6.4). Moreover, the utilized objective function distinguishes RDOPG applications from many other real-time routing problems with soft time windows. As mentioned in Chap. 1, in many other approaches a significant time period between the request arrival and the start time of penalty costs incurred by violations of soft time window restrictions exists. In the $edod_{\mathrm{tw}}$ presented in Sect. 2.6.4, this time period is denoted as reaction time. Since customer inconvenience directly starts with the arrival of the request in the system, dynamic requests have a reaction time of 0 and therefore no *grace period*. Owing to the absence of such a grace period, appropriate computational experiments have revealed a value of information (see Sect. 2.6.6.2) of about 10 which is high compared to other dynamic routing problems. For example, Gendreau et al. (1999a) state that a value of information between 1.025 and 1.041 was attained, depending on the utilized request data scenario. Note that we were able to achieve similar results for the value of information by allowing a grace period of 30 minutes between request arrival and the start of customer inconvenience in our problem setting. Due to the high value of information, it can be concluded that using additional available information about expected future events is likely to result in a significantly increased solution quality. Hence, in the computational experiments in Chap. 8, the impact of utilizing stochastic knowledge in the considered RDOPG applications will be evaluated in detail.

3.2.2 Computational Complexity

In the previous section it was described that RDOPG applications can be represented as a variant of the DFSVRP. As mentioned in Sect. 2.5.3.2.2, the DFSVRP is a DVRP which is extended by stochastic knowledge. According to Larsen (2000, p. 5), "the DVRP is a richer problem compared to the conventional static VRP. If the problem class of VRP is denoted $\mathcal{P}(\mathcal{VRP})$ and the problem class of DVRP is denoted $\mathcal{P}(\mathcal{DVRP})$, then $\mathcal{P}(\mathcal{VRP}) \subset \mathcal{P}(\mathcal{DVRP})$". Since the VRP belongs to the class of strongly NP-hard combinatorial problems (see Sect. 2.3.4), the considered RDOPG applications also belong to the class of strongly NP-hard combinatorial problems.

3.2.3 Classifying RDOPG Applications with Regard to Characteristics and Processing of Relevant Information

In this section, we characterize the considered RDOPG applications with regard to our proposed classification for categorizing routing problems presented in Sect. 2.5.3. According to this classification, RDOPG applications have the following attributes:

- **Type of relevant information and type of problem:** During the execution of the transportation process, dynamic events occur by which new customer requests arrive in the system. Since these dynamic events and their impacts are not deterministically known in advance, **changing relevant information** is present. These changes in the relevant information lead to modifications of the problem which cannot be predicted in advance during the execution of the transportation process. Hence, a **dynamic problem** is present.

- **Availability of changes in the relevant information and type of stochastic knowledge:** As mentioned in Sect. 3.1.2, available on-board units in the vehicles allow for a bidirectional communication with the central dispatching center. At this dispatching center, new requests are centrally collected so that a **global knowledge of dynamic data** is present. Moreover, as mentioned in Sect. 3.1.3, past request information is available so that **stochastic knowledge can be generated from past request information**.

As mentioned above, the RDOPG applications are modeled by a variant of a DFSVRP. Hence, the applied real-time control approaches which will be presented in Chaps. 5 and 6 has the following attributes:

- **Type of transportation process coordination:** Adaptations to the tour plan are carried out by the dispatching center. Hence, a **centralized coordination** of the transportation process is applied.
- **Type of model:** The occurring changes in the relevant information during the execution of the transportation process update the model so that a **deterministic dynamic model** is present in the real-time control approach presented in Chap. 5. Moreover, in the pro-active real-time control approach, future expected events derived from the available stochastic knowledge are also considered in the model. Hence, a **stochastic dynamic model** is present.
- **Type of application:** The problem represented by the stochastic dynamic model is controlled in real-time during the transportation process. This results in a **dynamic application**.

In Fig. 3.2, the considered RDOPG applications are categorized according to the classification for routing problems proposed in Sect. 2.5.3.

3.2.4 General Classification of RDOPG Applications

In Sect. 2.7, a general classification scheme for routing problems was introduced. In Table 3.1, RDOPG applications and the applied pro-active real-time control approach are classified according to this classification scheme.

3.3 Summary

In this chapter, the problem of coordinating transportation processes of RDOPG applications in real-time was described in detail. One of the main characteristics which distinguishes RDOPG applications from several other dynamic routing applications is the objective function of minimizing customer inconvenience. Customer

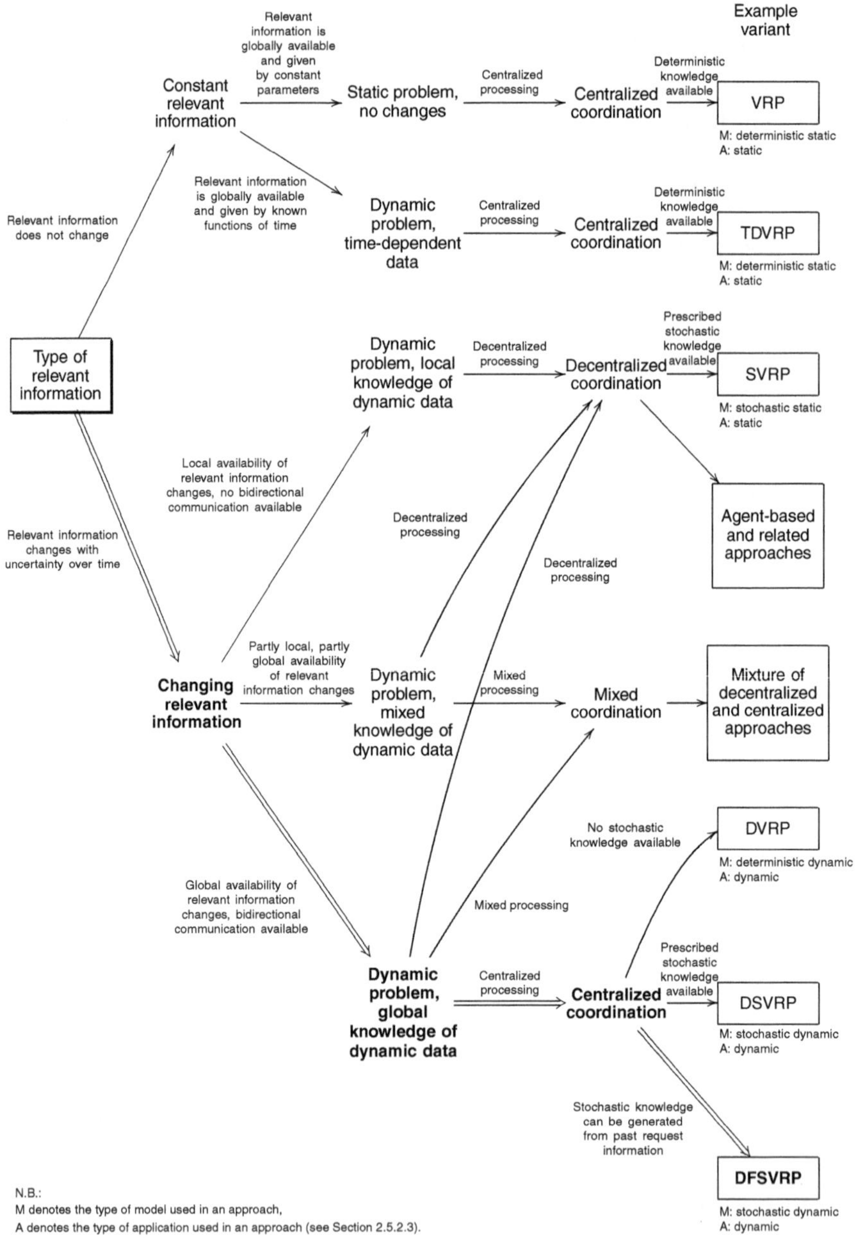

Fig. 3.2 RDOPG applications categorized according to the proposed classification of routing problems

Table 3.1 The general classification scheme for routing problems applied to RDOPG applications

Attribute	Possible values
Requests	
Type	Delivery only
Location	At nodes
Geographical distribution	Distributed with a pattern
Degree of dynamism	High (over 90 %)
Percentage of regular customers	None
Time restrictions	Soft time windows
Divisibility	Non-divisible requests
Transshipment	No transshipment
Request size	Less-than-truckload (LTL, e.g., piece goods)
Types of transportation objects	Goods (material and immaterial, i.e., services)
Size of service area	Urban
Sequence of requests	Arbitrarily
Selection of requests	Service of all requests
Frequency of request occurrence	Once
Fleet	
Size	Multiple vehicles, restricted fleet size
Consistency	Homogeneous (one type of vehicles)
Capacity	No capacity constraints
Location	One depot
Number of tours per vehicle and planning horizon	One tour
Time restrictions	No restrictions
Loading restrictions	Arbitrary loading
Number of drivers per vehicle	1
Driver to vehicle assignment	Fixed
Vehicle reliability	Reliable
Network	
Type of network	Road network (real distances)
Data	Asymmetric (directed graph)
Travel times	Constant
Planning horizon	
Length	One period (i.e., one day)
Type	Repeating (from day to day)

Table 3.1 (Continued)

Attribute	Possible values
Tour plan	
Type	Daily changing tour plan
Type of tours	Closed tours (with return to depot)
Restrictions	None
Relevant information	
Type of relevant information	Changing
Type of problem	Dynamic, changes comprise uncertainty
Availability of changes	Global
Types of considered dynamic events	New requests
Type of data	Both real-world and designed request data
Stochastic knowledge	
Type of stochastic knowledge	Forecasted; past request information is available
Increasing tour plan flexibility	
Methods used for increasing tour plan flexibility	Waiting strategies Relocation strategies (pro-active real-time control approach) Request assignment strategies (pro-active real-time control approach)
Objective	
Considered objective	Minimization of customer inconvenience
Applied approach	
Processing of changes in the relevant information	Centralized
Type of applied solution method	Exact solution methods used in the proposed approach for generating stochastic knowledge out of the past request information Pro-active real-time control approach utilizing simulation and heuristic solution methods for adapting the tour plan during the transportation process
Type of model	Stochastic dynamic, using available stochastic knowledge
Type of application	Dynamic (Real-time control approach utilizing simulation)

inconvenience directly starts upon arrival of a customer request in the system and therefore can be distinguished from penalty costs which only occur if a request is serviced late. According to results from the marketing literature and in order to evaluate different effects, customer inconvenience is modeled as a linear and a quadratic function. Moreover, a maximum allowed response time is defined in order to avoid an indefinite deferment of individual customer requests. Another important aspect is that vehicles move on a real road network and vehicle en-route diversion is allowed during the execution of the transportation process. The availability of past request information allows for the application of appropriate forecasting methods for generating stochastic knowledge out of it. This is a prerequisite for applying the proposed pro-active real-time control approach to RDOPG applications. After the problem description, various attributes of the considered RDOPG applications, including their problem complexity, were described. Finally, RDOPG applications were characterized according to the proposed classification approach for routing problems (see Sect. 2.5.3) as well as to the general classification scheme which was presented in Sect. 2.7.

Chapter 4
Review of the Literature Related to the Considered RDOPG Applications

Routing problems are a vital research area, not least due to their practical relevance. Recent advances in computational power and available technological equipment have opened up new areas for research and applications in practice. As illustrated in Sect. 3.2.1, RDOPG applications can be modeled as a variant of a Dynamic Vehicle Routing Problem (DVRP). Hence, this literature review focuses on DVRPs that are controlled by real-time methods in which changes in the relevant information are centrally available and centrally processed (see Sect. 2.5.3). For static VRPs, different extensive surveys exist in the literature. We refer the interested reader to the recent and detailed reviews of exact and approximated approaches for static VRPs by Laporte (1992a), Toth and Vigo (2002), Cordeau et al. (2005), Baldacci et al. (2007), and El-Sherbeny (2010). Moreover, for a recent and thorough survey of Dynamic Pickup and Delivery Problems (DPDPs) as well as of Dynamic Dial-A-Ride Problems (DDARPs) including recent results from competitive analysis, we refer to Berbeglia et al. (2010). Further details on static and dynamic DARPs can be found among others in Cordeau and Laporte (2003, 2007) and Cordeau et al. (2007a). Besides DVRP based approaches, approaches of other types of routing problems will be presented in this review if they contain specific additional relevant aspects.

In this review, the terms real-time, dynamic, and online will be synonymously used. The presented literature review is based on the comprehensive surveys by Psaraftis (1988, 1995), Powell et al. (1995), Séguin et al. (1997), Gendreau and Potvin (1998), Bianchi (2000), Larsen (2000), Cordeau and Laporte (2003, 2007), Ghiani et al. (2003), Lackner (2004), Richter (2005), Flatberg et al. (2005), Ichoua et al. (2007), Cordeau et al. (2007a), Larsen et al. (2007, 2008), Eksioglu et al. (2009), and Pillac et al. (2010).

As mentioned in Sect. 2.5.3.2, solution approaches for dynamic routing problems can be divided into reactive, deterministic, and pro-active real-time control approaches. This literature review primarily deals with deterministic and pro-active real-time control approaches. These types of real-time control approaches are determined by providing additional flexibility in the transportation process. With regard to the considered RDOPG applications, this is very important since the majority of requests only arrive during the execution of the transportation process so that the

F. Ferrucci, *Pro-active Dynamic Vehicle Routing*, Contributions to Management Science,
DOI 10.1007/978-3-642-33472-6_4,

tour plan frequently needs to be adapted according to the changes in the system situation. Besides deterministic approaches which do not use stochastic knowledge, in this literature review, various pro-active approaches are presented according to the type and the utilization of stochastic knowledge.

This chapter is organized as follows. Some of the *first papers on dynamic routing problems in the literature* are introduced. This is followed by a description of *selected reactive real-time control approaches for dynamic routing problems*. Furthermore, *strategies for increasing flexibility in dynamic routing problems* which can be found in the literature are described. The main part of this literature review describes solution approaches which provide *flexibility in dynamic routing without stochastic knowledge* followed by those that allow for providing *flexibility in dynamic routing using stochastic knowledge*. Since the objective function in the considered RDOPG applications differs from many other approaches of dynamic routing, special attention is also given to approaches in the literature which utilize *related objective functions*. After that, approaches in the literature that cover *other relevant factors for RDOPG applications*, such as vehicle en-route diversion, real-time control concepts, or communication technologies, are presented. The chapter closes with a *conclusion of the literature review* in which aspects that have not yet been dealt with in the research on pro-active approaches are also described.

4.1 First Papers on Dynamic Routing Problems in the Literature

The majority of approaches which deal with dynamic routing problems have been developed during the last two decades while pro-active approaches have only attained increased attention in recent years. Nevertheless, the first solution approaches were already published more than 30 years ago. One of the first papers that deals with a dynamic routing problem was presented by Wilson and Colvin (1977). The authors conduct research on a single-vehicle DARP in which customer requests dynamically arrive during the execution of the transportation process. The requests represent customers that request a trip from a given origin to a given destination. The authors propose insertion heuristics for solving the considered problem.

Psaraftis (1980) also deals with a DARP utilizing a single capacitated vehicle. Two cases are considered in this work. First, research on the static case is conducted, in which all customers are known in advance. The author considers the minimization of a weighted combination of the time to service all customers and the total customer dissatisfaction experienced while waiting for service. The dissatisfaction of a customer is modeled by a linear function depending on the waiting and ride times. The second study considers the related dynamic routing problem in which customers dynamically appear and demand an instant service when they arrive in the system. In this case, the objective solely consists of minimizing customer dissatisfaction. Note that this type of objective function is similar to the one considered in the RDOPG applications in this book since no grace period between customer request arrival and start of customer dissatisfaction exists (see also Sect. 3.1). The author develops a dynamic programming approach with a special priority rule called maximum position shift (MPS) which prohibits an indefinite postponement of customer requests. Benchmark instances with up to 10 customers are examined and optimally solved. The author notes that, due to the MPS rule, the size of the search space is restricted in comparison to the size of the search space that occurs when the MPS rule is not used. Therefore, the tested benchmark instances could be optimally solved. However, since the search space exponentially grows, the author encourages the comparison of the performance of exact approaches to heuristic approaches. Furthermore, the author suggests the use of the presented exact method as a subroutine in multiple-vehicle variants of the considered problem.

Another early contribution to the dynamic routing literature is the work of Bell et al. (1983). In the described real-world problem, the delivery of industrial gases from a depot to a set of customers is considered. The objective function pursues the maximization of the sum of delivered gas quantities. The tour plan is generated for the next two to five days but the focus is on the part of the tour plan that is executed on the current day. The problem is partly static and partly dynamic since all requests of the current day are known in advance while additional requests arriving during the day are only required to be serviced on the days that follow. However, dynamism also arises on the current day since the quantity of the industrial gas to be delivered to the individual customers can only be estimated with uncertainty. The exact amount of gas that a customer needs can only be determined while filling the tank at the customer's location so that adaptations of the tour plan may be required if estimated delivery quantities are not sufficient. The investigated problem utilizes a real road network for calculating mileage and toll costs involving a heterogeneous fleet and time windows of customer requests. To solve the problem, the authors propose a two-step solution method. In the first step, a set of possible one-day tours is determined. In the second step, tours are selected by decomposing the MIP problem using a Lagrangian relaxation into a knapsack problem for each vehicle tour and embedding the relaxed problem into a Branch-and-Bound (B&B) procedure. According to the authors, the integration of the proposed approach in a real-world delivery process has achieved savings of 6 % to 10 % in operational costs while increasing the productivity of the deployed vehicles.

4.2 Selected Reactive Real-Time Control Approaches for Dynamic Routing Problems

In the last decade, a lot of research on DVRPs and related routing problems has been published and reactive real-time control approaches have been developed for these problems. In place of this magnitude of various research work, we present some research proposing reactive real-time control approaches in this section. Research work proposing pro-active real-time control approaches will be presented in the literature review in the sections that follow.

Pankratz (2002) considers a DPDP for the long-distance freight forwarding in which multiple orders can be simultaneously loaded onto the vehicles. Outsourcing of requests to subcontracted companies (at additional costs) is allowed and vehicles can be immediately reused after their tour has ended. Dynamism is considered by newly arriving requests during the day. In order to solve the routing problem, a heuristic solution approach based on a Genetic Algorithm (GA) is proposed. For future research, the author suggests considering additional sources of dynamism such as vehicle breakdowns and traffic jams.

In Lackner (2004) four metaheuristics are developed and applied to the capacitated Dynamic Vehicle Routing Problem with Time Windows (DVRPTW). These four metaheuristics consist of two evolutionary strategies, an Ant Colony Optimization based algorithm, and a Simulated Annealing metaheuristic. In the considered problem, dynamism arises from newly incoming requests. The author analyzes the behavior of the implemented metaheuristics under different degrees of dynamism (see Sect. 2.6.4) ranging from 10 % to 90 % as well as under different scenarios that differ in the arrival rate of customer requests, in particular one to nine new requests per minute. A single event optimization strategy is used, i.e., an adaptation of the tour plan by the specified metaheuristic is directly performed whenever a new customer request arrives. Vehicle en-route diversion is not allowed so that a vehicle can receive new orders after servicing the currently first assigned request. Computational results based on the Solomon instances are assessed under the considered objective function which pursues the minimization of the number of rejected customers, traveled distance, as well as lateness occurring when servicing customer requests. The computational results show that the implemented evolutionary strategies perform best among the tested metaheuristics. As an outlook on future research, the author proposes the combination and parallelization of different metaheuristics.

Richter (2005) examines the adaptation of classic construction heuristics to the Dynamic Traveling Salesman Problem (DTSP) as well as to the DVRP in their capacitated versions. The author argues that his main reasons for adapting construction heuristics to the dynamic case are short required computational times, which are advantageous in real-time environments, as well as an easy understanding of how these heuristics operate. In his work, the author also assesses the effects on the solution quality if vehicle en-route diversion is allowed. In an extensive computational study, scenarios with 50 to 200 customer requests as well as with a degree of dynamism between 20 % and 80 % are evaluated using two different objective functions, specifically the minimization of travel distance and the minimization of customer waiting times, i.e., response times. Richter concludes that the utilization of vehicle en-route diversion allows a significant improvement of the solution quality if the minimization of customer waiting time is pursued.

Bock (2010) considers a multi-modal transportation problem. In this complex transportation problem, a large variety of dynamic events are considered. These types of dynamism include the arrival of new requests, breakdowns and slowdowns of vehicles, and traffic jams. The objective is to minimize the overall operative transportation costs. In order to solve the problem, a real-time control concept which regards the concurrency of tour plan adaptation and tour plan execution is proposed.

In this real-time control approach, a specifically designed improvement heuristic is applied that makes use of different neighborhood operators which are selected depending on the quality of solutions found in previous iterations. Computational results show that the application of the proposed real-time control approach, which utilizes the complete execution time of the transportation process for adapting the tour plan, yields the best results among the evaluated real-time control approaches.

After introducing the first papers on dynamic routing problems in the literature as well as selected research work dealing with reactive real-time control approaches, the literature review continues with describing basic strategies which are used in dynamic routing approaches in order to increase the flexibility in the transportation process. Afterwards, publications in the literature that make use of these strategies are presented.

4.3 Strategies for Increasing Flexibility in Dynamic Routing Problems

In this section, we illustrate basic strategies that are used in order to increase the flexibility which is available during the execution of the transportation process. Specifically, with regard to the considered RDOPG applications, flexibility according to possibilities of integrating dynamically arriving requests is considered. As mentioned in Sect. 2.5.3.2, an increased flexibility is advantageous when dealing with dynamic events that occur during the transportation process. These strategies can be distinguished into *waiting strategies*, *relocation strategies*, and *request assignment strategies*. The successive description of approaches that can be found in the literature provides an explanation about which of these strategies are used in the individual approaches.

4.3.1 Waiting Strategies

In static VRPs, the common procedure is that a vehicle directly moves to its next assigned request after it has finished servicing its current request. If time windows are used in an approach and a vehicle arrives at a request location prior to the start of the assigned time window, it is often allowed to wait at this request location until the time window opens. In contrast, in dynamic routing problems in which the objective of travel distance minimization is pursued, waiting strategies can be advantageously used to keep vehicles at their last serviced request location for some additional time after service of the request has been finished. Therefore, waiting strategies can be especially beneficial in applications in which time windows of requests open in the far future in relation to the arrival times of requests in the system. In such cases, a time period denoted as *slack time* occurs during which a vehicle would need to wait anyway, for example at the last serviced request location or at the next scheduled request location.

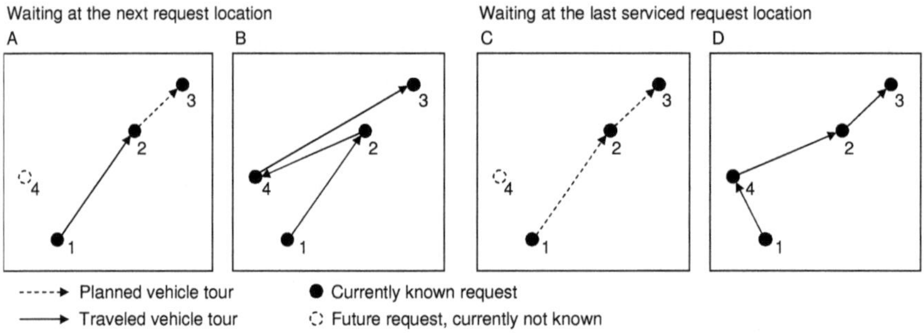

Fig. 4.1 Execution of a transportation process using different waiting strategies

In Fig. 4.1, the effects of waiting at the last request and at the next request location are illustrated. Specifically, when a vehicle directly travels to the next request location defined by request 2 and waits there (Picture A) and a new request (denoted by request 4) arrives, the vehicle has to perform a long additional travel distance (Picture B). In contrast, by waiting at the last serviced request location defined by request 1 (Picture C), the decision concerning which request should be serviced next can be postponed to a later time point. If a new request at a location close to the vehicle's last serviced request location arrives as the vehicle is still waiting there, this new request only requires a small amount of additional travel distance to be serviced (Picture D). Hence, new requests which arrive near the current vehicle waiting position can be efficiently serviced with regard to the considered objective of minimizing the total travel distance. This is especially beneficial in applications in which no vehicle en-route diversion is allowed since, in this case, no re-routing of the vehicle is allowed when request 4 arrives even if the vehicle has just left request 1 in order to travel to request 2. Note that if waiting strategies are used, the decision space of the routing problem is extended since it has to be decided *how long* a vehicle should wait at a specific location. Hence, the VRP is extended to a Vehicle Routing and Scheduling Problem (VRSP, see also Sect. 2.4.6).

As illustrated above, waiting strategies can be used even when no stochastic knowledge is available. If stochastic knowledge is available, waiting strategies can additionally improve the flexibility by deciding where and how long vehicles should wait according to the stochastic knowledge about expected request arrivals in individual subregions in the near future.

Note that although it is likely that applying waiting strategies will have positive effects on the solution quality in many cases, there are also situations in which waiting strategies lead to a worse solution quality. For instance, consider a situation where a large number of new requests with tight hard time windows occurs close to a customer request which is to be serviced next in a vehicle's tour. If only a limited number of vehicles which are already highly utilized is available and vehicles are ordered to wait at their last request location, this may lead to a larger number of unserviced customer requests since they cannot be serviced within their time windows. Furthermore, waiting strategies can also lead to a worse system performance in specific applications. For example, in the considered RDOPG applications in which

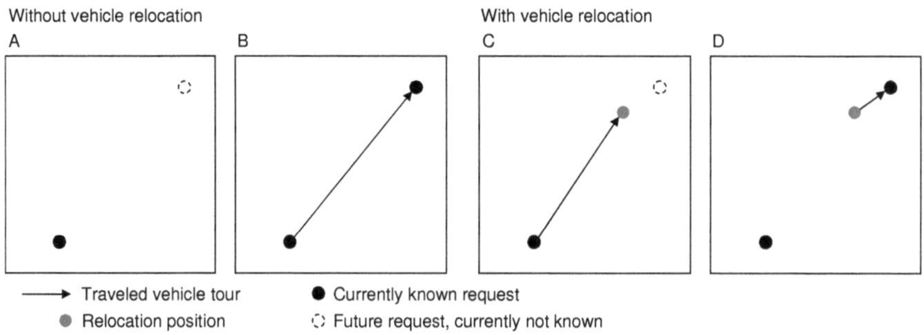

Fig. 4.2 Execution of a transportation process with and without using a relocation strategy

dynamically arriving requests can be directly serviced upon arrival and the objective is to minimize customer inconvenience, it should be carefully decided whether a vehicle should wait when it has pending requests. Specifically, this decision should be taken according to the probability that new requests will arrive soon and in close vicinity of the considered vehicle. However, if this probability is not taken into account so that the vehicle waits in any case, the customer inconvenience of pending requests will further increase. Therefore, approaches which directly service pending requests may achieve a better solution quality in this case.

4.3.2 Relocation Strategies

In dynamic routing problems in which quality of service aspects are pursued, relocation strategies enable potential solution improvements by actively repositioning vehicles to strategic locations where future requests are expected to arrive. Relocation strategies were first assessed by Powell (1988). By using relocation strategies, new arriving requests in close vicinity of the strategic waiting positions can be serviced faster. Relocation activities can be beneficial in scenarios where a quick response to new arriving requests is mandatory and significantly more important than the total travel distance. For example, this is true for ambulance relocation problems (cf., e.g., Gendreau et al. 2001, Haghani and Yang 2007, and Schmid 2012) and, as mentioned above, in dynamic routing problems in which the solution quality is mainly determined by customer-related quality aspects. Specifically, such aspects may comprise inconvenience aspects so that a fast reaction to newly arriving requests is pursued (as considered in Larsen 2000 and the RDOPG applications in this book) or aspects that consider the maximization of the feasibly serviced customers within hard time windows (cf., e.g., Bent and van Hentenryck 2007). The utilized strategic relocation positions may be constant or dynamically determined during the day and can be the same or individual for the vehicles.

In Fig. 4.2, the difference between letting an idle vehicle wait at its last serviced request location and relocating it to a strategic location is illustrated in an application in which the minimization of customer inconvenience which depends on request response time is pursued. If the vehicle waits at the location of its last serviced request (Picture A) and a new request arrives, the vehicle may need to travel for a much longer time before it can reach and start servicing the new customer request if the new request is far away from the last serviced request location (Picture B). This may result in a long response time and therefore a high customer inconvenience. In contrast, if relocation strategies are used, an idle vehicle is relocated to a strategic waiting position in whose vicinity new requests are expected to arrive in the near future (Picture C). If a new request arrives close to this waiting position, the vehicle requires less travel time so that it can service the new request much faster (Picture D) which results in significantly reduced customer inconvenience. Clearly, the use of relocation strategies can only be utilized if stochastic knowledge about expected future request arrivals is available. Approaches that make use of relocation strategies are presented in Sect. 4.5.

The previously introduced waiting strategies are strictly connected to relocation strategies. Specifically, when vehicles are relocated to certain positions, it needs to be decided how long each vehicle should wait at this relocated position under different circumstances, e.g., while the vehicle is idle or if newly arrived requests have been assigned to the vehicle. This decision also depends on the urgency of the new request and the travel time required from the current vehicle's position to the location of the new request.

4.3.3 Request Assignment Strategies

The above described waiting and relocation strategies are mainly applied to vehicles which are currently not ordered to directly service any requests, i.e., vehicles which can be seen as being currently *idle*. In contrast, request assignment strategies are utilized to increase flexibility in the transportation process for *busy* vehicles by directly influencing the assignment and sequencing of requests on vehicles, i.e., by actively impacting the tour plan generation and adaptation process.

If no stochastic knowledge is available, request assignment strategies can increase the flexibility in tour plans by restricting the detailed planning process to short-term decisions since it is assumed that decisions in the long-term will be re-adapted due to other dynamically arriving requests (cf., e.g., Mitrovic-Minic et al. 2004). Furthermore, simultaneously storing parts of different valid tour plans (cf. Gendreau et al. 1999a) or simultaneously storing various complete tour plans (cf. Bent and van Hentenryck 2004b) makes it possible to make quick decision about good request assignments. For example, if hard time windows are present, this enables fast determination about whether or not new pending requests can be feasibly integrated into the currently executed tour plan.

If stochastic knowledge is available, request assignments to vehicles can be performed based on estimated objective function value outcomes. By considering expected request arrivals in the tour plan generation process, more flexible tour plans can be created. Specifically, it is likely to be advantageous if the assignment and sequencing of requests to vehicles is performed under consideration of the resulting vehicles' positions and the estimated response times concerning expected future requests. For example, if available stochastic knowledge reveals that request locations are uniformly distributed, requests can be assigned to vehicles in a way that uniformly scatters the vehicles over the considered service area. In this case, it is likely that expected future requests are serviced within a shorter response time. Moreover, vehicle tours can be generated in such a way that additional time is reserved in specific subregions at certain time intervals. This additional time can be used for integrating expected future requests in the vehicles' tours without requiring already planned requests to be rescheduled. Similar to the above strategies, this strategy is beneficial if the objective pursued considers quality of service aspects such as in RDOPG applications. As a result, by using stochastic knowledge directly in the tour plan generation process, future positions of vehicles can be advantageously influenced. Hence, effects that are similar to the ones achieved if idle vehicles are relocated can also be achieved for vehicles which are currently busy.

In the following sections, recent publications in the literature which consider methods for increasing tour plan flexibility in order to allow a better integration of expected future requests are presented. In some applications, no stochastic knowledge can be utilized either because stochastic knowledge with sufficient forecasting quality is not available or because technical or other restrictions prohibit the use of available stochastic knowledge. Based on this distinction, research work on approaches which do not use stochastic knowledge is presented. Afterwards, approaches which use stochastic knowledge are described.

Unless stated otherwise, all approaches consider applications that service requests by visiting them in a restricted Euclidean plane with Euclidean distances and travel paths between request locations. Furthermore, the single source of dynamism is, by default, represented by newly arriving requests during the execution of the transportation process. The considered planning horizon usually has a length of one working day and vehicle en-route diversion is not allowed. Moreover, the considered problems generally have one depot from which the vehicles start their tour and to which they return after completion of their assigned tour.

4.4 Flexibility in Dynamic Routing Without Stochastic Knowledge

In this section, we present approaches in the literature which provide an increased flexibility in dynamic routing problems without using stochastic knowledge. According to our classification in Sect. 2.5.3.2.2, we denote them as deterministic approaches.

In the work of Kilby et al. (1998), the authors consider a DVRP where dynamism arises from newly arriving requests. The considered routing problem is heuristically solved and a specific waiting strategy is applied, that is, if a vehicle has currently no further requests assigned, the vehicle waits at its last serviced request location for further requests which are assigned to its tour. The objective consists of minimizing the total travel distance regardless of the number of used vehicles. Computational results are evaluated on modified test instances of Taillard, Christofides, and Fisher. Besides varying degrees of dynamism, the authors evaluate different lengths of the commit horizon, i.e., the time period for which unchangeable decisions are made during the transportation process. The authors test their approaches on the DVRP case in which goods are picked up at customer locations as well as on the case in which goods are delivered from the depot to customer locations (see Sect. 2.3.2). The authors state that in the static case, solving both variants results in the same solution quality. In contrast, in the considered dynamic problem, the delivery variant is more costly than the collection variant. This can be explained by the fact that a request which needs to be collected can be serviced by any vehicle with free capacity while the delivery of a request can only be performed by the vehicle which is transporting the specific good. Moreover, this cost gap increases the higher the degree of dynamism. With regard to the evaluated commit horizon, the authors state that the use of longer commit horizons results in a worse solution quality. For future research, the authors propose, among other research fields, investigating the integration of past request information into the transportation process.

The work of Gendreau et al. (1999a) considers a Dynamic Vehicle Routing Problem with Soft Time Windows (DVRPSTW) with a limited fleet size. Requests may be serviced late, i.e., after the end of the assigned time window, but at additional penalty costs. The objective aims at minimizing the weighted sum of travel distance and linear lateness costs incurred by late serviced customers. The dynamic problem is solved as a series of static problem instances consisting of the currently pending requests. Each static problem instance is solved by a Parallel Tabu Search metaheuristic that has been proposed by Badeau et al. (1997). In order to increase flexibility in the transportation process, the authors utilize the concept of an adaptive memory in which tours of different tour plans that have been previously found during the search process are stored. The advantage of this variety of stored tours is that when new requests arrive, the variety can be used to achieve a quick construction of a new tour plan of high quality which integrates the new requests. Specifically, a new tour plan is constructed by randomly selecting tours from the adaptive memory while the probability that a tour is selected depends on its solution quality. Since this typically leads to a tour plan which is constructed out of tours from different solutions, this approach combines strengths of Tabu Search and GA metaheuristics. Computational results on modified Solomon instances show that the proposed Parallel Tabu Search metaheuristic produces superior results compared to other evaluated solution methods. The authors state that it would be interesting to integrate stochastic knowledge about future events such as the time-space distribution of requests into the solution approach in order to improve decision making during the execution of the transportation process.

The adaptive memory concept described in the previous approach is generalized in the Multiple Plan Approach (MPA) for a DVRPTW proposed by Bent and van Hentenryck (2004b). Since this problem considers hard time windows, requests need to be serviced within given time windows. Otherwise, requests cannot be serviced any more and are counted as unserviced customer requests. The objective is to maximize the number of serviced requests (see also Sect. 2.4.12.6). The MPA generates and maintains several tour plans which are consistent with the current system situation while deleting incompatible ones. Individual tour plans are generated by a solution method for VRPs using Simulated Annealing and Large Neighborhood Search which has been proposed by Bent and van Hentenryck (2004a). The authors suggest the continuous application of the solution method in the dynamic problem setting. Specifically, since decisions are only required at specific events such as at the completion of the service of a request or at the arrival of a new request, the intermediate time can be used for improving tour plans stored in the MPA. This is advantageous since it allows a quick reaction based on high-quality tour plans if fast decisions are required. Besides the stored tour plans in the MPA, one additional tour plan denoted as the *distinguished plan* which is executed in the system is maintained and incrementally generated during the transportation process. Specifically, whenever a vehicle has completed the service of its currently assigned request, the distinguished plan is extended by selecting the next request to be serviced by the respective vehicle. This request selection is made according to the stored tour plans in the MPA by using a dedicated *ranking function*. In their work, the authors test two ranking functions. The first ranking function called *distance* selects the next request for a vehicle by the least travel distance criteria which is used in the corresponding approach variant called MPA^{d}. The second ranking function is called *consensus* and selects the next request for a vehicle based on similarity aspects. Specifically, for vehicle v, it is evaluated how often request r is assigned to vehicle v as the next request in the tour plans stored in the MPA. After evaluating all requests, the request which is assigned most is selected as the next request for vehicle v. The MPA approach variant using the consensus ranking function is called MPA^{c}. As can be seen, both MPA variants iteratively determine the tour for each vehicle. Specifically, on the basis of the corresponding ranking function the next request to be serviced by the vehicle is selected whenever a vehicle has completed its current activity. Computational results show that MPA^{c} significantly outperforms MPA^{d} with regard to the number of requests that can be feasibly serviced within their time windows. Furthermore, the authors extend the MPA to Multiple Scenario Approaches which use stochastic knowledge. These approaches are described in Sect. 4.5.5.

Fleischmann et al. (2004b) study an urban single-load Dynamic Pickup and Delivery Problem with Soft Time Windows (DPDPSTW) in which dynamism arises from newly incoming requests and changes in travel times. Since no requests can be rejected, all requests have to be serviced. Requests can be serviced also after the end of their time window but at high penalty costs. The objective pursues primarily the minimization of request delays and, secondarily, operational transportation costs. Since the transportation process is carried out on a real road network, shortest path computations are required when travel time information changes during the execution of the transportation process. Updates in the travel time information are provided by a traffic management center and shortest paths are calculated using a modified version of Dijkstra's algorithm. If appropriate, vehicles are allowed to wait

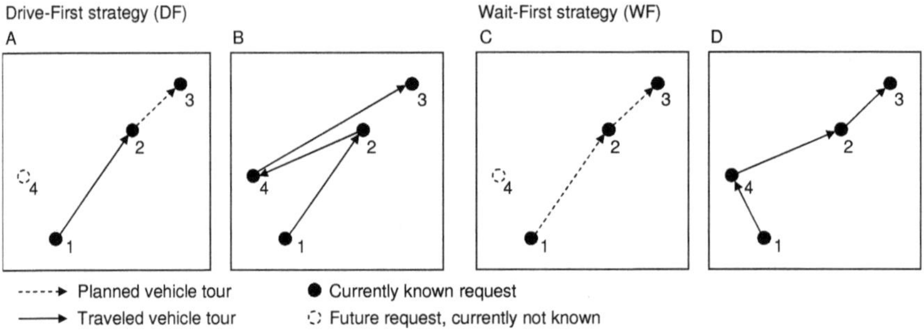

Fig. 4.3 Execution of a transportation process using Drive-First and Wait-First strategy

at their last serviced request location and service of requests may be postponed. Hence, a waiting strategy is used in this approach. For assigning the next request to a vehicle when it has finished its current activity, the authors test three different strategies: assignment rules, an assignment algorithm, and an insertion heuristic in various settings. The assignment rules determine the next activity whenever a vehicle has finished the completion of its current activity. Here, only the corresponding vehicle is considered and the decision with the least cost is chosen. In case of the assignment algorithm, an assignment problem is solved, i.e., for each vehicle the next request is determined. Since no vehicle en-route diversion is allowed, this assignment determines the next request for each vehicle after the respective vehicle has completed its current activity. The third tested strategy based on an insertion heuristic solves the dynamic problem by maintaining a complete tour plan so that all pending requests are always assigned to vehicles' tours. The authors state that test results on instances with 500 requests reveal that the assignment algorithm performs best among the tested strategies. Specifically, higher degrees of dynamism lead to a bigger improvement compared to the other evaluated strategies. The authors direct future research possibilities to the evaluation of real-time methods on the multiple-load DPDP.

Mitrovic-Minic and Laporte (2004) evaluate four different waiting strategies for a DPDP with hard time windows and multiple vehicle starting locations. Moreover, the available fleet size is not restricted. The pursued objective considers the minimization of total travel distances. The first tested waiting strategy is Drive-First (DF) which is equal to a policy in which vehicles directly travel to the next request after service of the current request is completed. Since no vehicle en-route diversion is allowed, this strategy directly commits vehicles to the next assigned requests. Therefore, requests which arrive after departure and before service of the next assigned request begins may lead to an additional detour when further requests arrive (see Pictures A and B in Fig. 4.3).

The second strategy, Wait-First (WF), forces vehicles to wait at their current location up to the point in time at which a departure at the last serviced request location leads to an arrival at the next scheduled request location directly at the end of its time window, i.e., at the latest allowed time point. This strategy provides the biggest flexibility in terms of integrating new arriving requests and leads to the best results in terms of total travel distance (see Pictures C and D in Fig. 4.3). However, due to the resulting tight schedule the authors state that it happens often that new arriving requests require the use of additional vehicles. Because of this, the authors provide two further strategies called Dynamic Waiting (DW) and Advanced Dynamic Waiting (ADW). For these strategies, the authors introduce a dynamic partitioning of the pending requests into service zones. Here, each service zone contains a number of consecutive request locations that are close to each other. What the waiting strategies DW and ADW have in common is that for traveling between requests within the same service zone, the DF strategy is applied. At the end of a service zone, DW applies the WF strategy whereas ADW uses only a certain part of the maximum available waiting time. The authors define this part as the ratio of the service zone time span to the sum of time spans of all service zones that are assigned to the tour of a vehicle. The authors test their strategies with instances of between 100 and 1,000 requests. Computational results show the superiority of the waiting strategy ADW with regard to total distance and number of required vehicles, especially if an elaborated Tabu Search solution method is applied for adapting the tour plan.

In their subsequent work, Mitrovic-Minic et al. (2004) also consider a DPDP with hard time windows. An unrestricted fleet size is assumed since, in the considered real-world problem, a very large fleet size is available and fixed costs for vehicle deployment are paid by the drivers. The objective consists of minimizing the total travel distance. The authors propose a concept called double horizon that generalizes short-term rolling horizon approaches. This generalization is applied in order to prevent unwanted long-term effects on the tour plan execution process if only good short-term decisions are made. The double horizon concept is implemented by using a different objective function for the short-term horizon and the long-term horizon. Specifically, for the short-term horizon, the objective used is identical with the one that is pursued in the problem while the objective of the long-term horizon is to maintain tours in such conditions which facilitate the integration of future arriving requests. This is achieved by selecting tour plans which exhibit large slack times in the distant future in their tours. Both horizons are optimized using a simplified version of the Tabu Search metaheuristic proposed by Gendreau et al. (1998) (later published in Gendreau et al. 2006). The authors apply some of the waiting strategies described in Mitrovic-Minic and Laporte (2004). Computational results are carried out on 100 and 500 customer request instances generated using real-world data from two courier service companies in Canada. The results show the potential that can be achieved by applying the double horizon approach compared to classic single rolling horizon approaches. Furthermore, the authors introduce the definition of the *value of information* that assesses the system performance of a given solution approach in the dynamic case compared to the static case (see also Sect. 2.6.6.2).

Following the concept of waiting strategies, de Oliveira et al. (2008) propose an approach for a capacitated DVRPTW considering the objective of minimizing travel distance. The authors evaluate four different metaheuristics: a Simulated Annealing algorithm, a Greedy Randomized Adaptive Search Procedure heuristic, a Memetic Algorithm, and a hill climber heuristic. In order to increase flexibility, the authors integrate a vehicular waiting time heuristic (VWTH). This heuristic is used to determine appropriate waiting times for vehicles at serviced request locations. As in the previously described approaches, the purpose of the VWTH is to enable more possibilities for vehicles to integrate future arriving requests by letting them wait for an appropriate amount of time. Consistent with the results of Mitrovic-Minic and Laporte (2004), the computational results show that using the VWTH can significantly improve the solution quality.

Another concept denoted as request buffering was introduced to dynamic routing problems by Pureza and Laporte (2008). The authors study an uncapacitated DPDP with hard time windows. In order to prioritize the quality of service, a hierarchical objective function is applied. Specifically, the main objective is to minimize the number of unserviced requests followed by the number of used vehicles and finally the total travel distance. Dynamism arises from newly arriving requests as well as from random travel times. Furthermore, time-dependent travel times are present. Since uncertain travel times may lead to longer travel activities, the authors adapt the WF strategy proposed by Mitrovic-Minic and Laporte (2004) to a new strategy called WE. This new strategy schedules vehicle departures in a way that a vehicle arrives at each customer request location at the start of its time window but not earlier. Since travel activities are carried out on the Euclidean plane, traveling from location i to location j is always assumed to be performed by using the direct Euclidean connection between both locations defined by the arc (i, j). Furthermore, since time-dependent travel times are modeled by modifying the travel time of an entire arc, the triangle inequality is not valid in this problem. Therefore, the authors extend the WE strategy by integrating travel time information based on fastest paths between locations. Specifically, this concept tests whether the destination can be reached faster by traveling via an intermediate location instead of using the direct arc between the current location and the destination. In their approach, the authors construct fastest paths with at most one intermediate location. This time can be used as additional slack time that increases flexibility for integrating further requests. Additionally, another strategy called Request buffering (RB) is proposed. This strategy postpones the insertion of newly arrived but less urgent requests to a later time. Since no vehicle en-route diversion is allowed in the considered application, decisions become unchangeable once taken. Hence, this deferment of requests leads to situations in which a larger number of pending requests, on whose basis better routing and scheduling decisions can be taken, is present. This is in line with the observations by Larsen regarding the effective degree of dynamism (see Sect. 2.6.4). The authors evaluate instances with 100, 200, and 400 requests with varying degrees of dynamism. Computational results show that using a combination of fastest paths and the RB strategy yields the best results among the tested strategies.

In the work of van de Klundert and Wormer (2010), the authors consider a real-world application, which is called the After Salesman Problem (ASAP). This problem arises from after-sales service demands of customers and has similarities to the considered RDOPG applications. A real-world case study is conducted at the largest service organization in the Netherlands in which customer requests are required to be serviced within a maximum response time which is defined by SLA contracts.

Besides dynamically arriving customer requests, uncertainty arises from uncertain service time durations which are considered to be given by stochastic variables only. In this problem, a service area with Euclidean distances is considered. In the evaluations, the main pursued objective consists of maximizing responsiveness by minimizing the average waiting time of the dynamically arriving customers which is comparable to applications in which the minimization of a linear customer inconvenience function is pursued. In order to solve the problem, the authors propose three solution approaches. The first is based on a periodically performed multiple-vehicle First-come, first-served (FCFS) strategy. The second approach called matching periodically selects the next request for idle vehicles. It is modeled as a linear assignment problem which is optimally solved by the Hungarian Method in polynomial time. The third approach called set partitioning periodically assigns all requests to vehicle tours forming a complete tour plan. In order to solve this set partitioning problem, an approach similar to the one proposed by Krumke et al. (2002) is applied (see also Sect. 4.6). In the computational results, the authors show that in instances with high request arrival rates, the FCFS strategy performs worst since this strategy does not take advantage of the possibility to service requests which have arrived later but are close to a vehicle. Out of the tested approaches, the set partitioning strategy performs best. In another computational study, the authors assess the impact of the interval length after which the reoptimization, i.e., the adaptation of the tour plan, is performed. According to their results, in light traffic conditions a more frequent reoptimization leads to small improvements while in heavy traffic conditions, i.e., if high request arrival rates are present, longer intervals result in small improvements since better decisions can be taken when more pending requests are available. However, this may trigger additional costs due to a postponement of the service of waiting customers, thus inconveniencing them more. Furthermore, the authors consider that solving individual problem instances during the execution of the transportation process to optimality does not guarantee that the routing problem considered during the whole planning horizon, denoted as the end-of-day objective, is also optimally solved. Therefore, the impact of using a different objective function—for solving the individual problem instances—on the solution quality is examined. For instance, the authors show that if the aim is to maintain a defined SLA for the customers by servicing customers within a maximum allowed response time, it is beneficial to integrate information about the request response times into the objective function in order to allow the solution method for discriminating between the quality of different solutions in which all requests fulfill the SLA requirements (see also Sect. 2.4.12.6). Moreover, they suggest to use a lower time limit for the SLA in the problem instances than required in the end-of-day objective in order to maintain some buffer time for exceeding this limit in the dynamic transportation process and still maintaining the demanded end-of-day SLA. Furthermore, effects of allowing vehicle en-route diversion are assessed. Computational results show that in heavy traffic scenarios, allowing vehicle en-route diversion yields an improvement of over 20 % of the solution quality. Moreover, in another study in which the effects of using additionally available knowledge about the expected request service times are assessed, results indicate that this knowledge can yield significant improvements on the solution quality. For future research, the integration of stochastic knowledge in the considered application is suggested. Note that this research question is conducted in the RDOPG applications considered in this book.

4.5 Flexibility in Dynamic Routing Using Stochastic Knowledge

In this section, we focus on the presentation of approaches that use stochastic knowledge, hence increasing flexibility in the transportation process and providing possibilities to improve the solution quality. Since dynamism in the considered RDOPG applications arises from newly arriving requests, in what follows, we focus on approaches which use stochastic knowledge in order to expect future request arrivals. According to Ichoua et al. (2007), the use of stochastic knowledge in approaches which consider dynamically arriving requests is motivated by knowledge that a human dispatcher has. Specifically, having attained valuable knowledge from experience, dispatchers typically have knowledge about patterns regarding arrival times of new requests and their location. This allows them to make predictions about when and where new requests are likely to arrive during the day and allows them to manage the vehicles more efficiently in anticipation of these future arriving requests. Therefore, if stochastic knowledge is available and utilized in real-time control approaches, additional significant improvements are often attainable when compared to the previously illustrated methods for increasing flexibility without using stochastic knowledge. According to our classification in Sect. 2.5.3.2.2, we denote such approaches as pro-active approaches.

In the presented approaches, stochastic knowledge is either manually entered into the system based on human experience or automatically forecasted by appropriate approaches that analyze past request information. Bent and van Hentenryck (2004b) mention several articles in the literature in which authors suggest the utilization of stochastic knowledge in order to anticipate future requests. This stochastic knowledge can either be given by analyzing past request information using appropriate methods (as suggested in Kilby et al. 1998 and Gendreau et al. 1999a, 2006) or by prescribed probabilistic distributions that have been entered into the system in some other way (cf., e.g., Gendreau et al. 1996 and Yang et al. 2000). Furthermore, in the previously described thesis of Richter (2005) and in the work of Bock (2010), the integration of stochastic knowledge is also suggested.

According to Bianchi (2000), approaches for dynamic routing problems can be differentiated by the way the dynamic problem is divided into individual problem instances which are solved by a defined solution method during the execution of the transportation process. Specifically, *strategy-oriented approaches* evaluate different ways of dividing the dynamic problem while the solution method utilized for solving these individual problem instances is often assumed to be available and is not analyzed in detail. Specifically, it is often assumed that a solution approach which optimally solves the given individual problem instance is available. This type of studies often analytically derives lower bounds and other properties of specific solution strategies for dynamic routing problems. However, due to the computational complexity, not all types of dynamic routing problems can be analytically evaluated. In contrast, *heuristic-oriented approaches* utilize a defined strategy for dividing the dynamic problem into subproblems and evaluate in detail the system performance of one or more solution methods applied to solve the individual problem instances. Studies which develop such approaches often empirically provide computational results on given problem instances for complex dynamic routing problems.

This section starts by presenting the *strategy-oriented approaches*. In the following subsections, heuristic-oriented approaches which make use of stochastic knowledge are presented. We refer to this type of approaches as real-time control approaches which reoptimize individual problem instances of the transportation process during its execution. First, *approaches using exact solution methods* are illustrated. This is followed by the presentation of *stochastic modeling based approaches* (cf. Pillac et al. 2010). These approaches often use multi-stage models in which the stochastic knowledge is directly incorporated which results in models which are often very hard to solve to optimality. Due to complexity reasons, a large part of approaches deals with heuristically inspired solution methods. Accordingly, heuristic solution *approaches with manually provided stochastic knowledge*. After that, *sampling-based approaches* which reduce complexity issues by generating scenarios through sampling probability distributions are described.

4.5.1 Strategy-Oriented Approaches

As mentioned above, strategy-oriented approaches utilize available stochastic knowledge in one or more strategies in order to assess the impact of each strategy on the solution quality (cf. Bianchi 2000). Specifically, such approaches perform tour plan adaptations according to a specific strategy (or rule). Strategy-oriented evaluations for dynamic routing problems using stochastic knowledge were first assessed by Bertsimas and van Ryzin. Their studies are mainly based on the Dynamic Traveling Repairman Problem (DTRP) which is introduced in Bertsimas and van Ryzin (1991). In its simplest form, the DTRP is related to the DTSP. The DTSP was introduced by Psaraftis (1988) and is defined as follows. In a considered service area, a set of nodes is given forming a complete graph. New requests are independently generated at each node of the graph according to a Poisson process with a known intensity parameter. The requests have to be serviced by a single salesman which requires a deterministically known travel time on the arcs between the nodes. At each node, a known service time is required for servicing the request. The aim is to minimize a given objective, for example, the sum of customer request waiting times.

In the DTRP, a single repairman (also denoted as vehicle in what follows) travels in the Euclidean plane with a constant velocity and needs to service dynamically arriving customer requests. In contrast to the DTSP, requests in the DTRP can occur at an arbitrary location within the considered service area. None of the requests are known in advance so that the problem has the highest possible degree of dynamism (see Sect. 2.6.4). As the name of the problem suggests, the DTRP represents a problem in which time-critical orders arrive dynamically and are required to be serviced as quickly as possible. According to the authors, this problem belongs to the set of real-world applications in which the minimization of customer waiting times is more important than operative travel costs. Therefore, the objective is to minimize the average expected system time of requests which consists of the expected waiting time, i.e., the request response time, and the service time required at the request. Note that due to the considered objective function, the DTSP and the DTRP are related to the RDOPG applications considered in this book if a linear customer inconvenience function, depending on request response times, is utilized (see also Sect. 3.1).

In the first studied DTRP variant in Bertsimas and van Ryzin (1991), dynamic requests appear according to a Poisson process with a given intensity parameter λ. Stochastic knowledge is given since it is known that request arrivals follow the described Poisson process and the additionally given information that newly arriving requests are independently and uniformly distributed in the considered Euclidean plane. Furthermore, service times of requests are independently and identically distributed according to a known distribution. Hence, in this approach, stochastic knowledge is given by prescribed probability distributions. In their study, the authors derive system characteristics of the described problem which include lower bounds on the performance by using mathematical methods. Specifically, they present a model formulation of the problem that allows the combination of approaches from different areas such as queuing theory, geometrical probability, combinatorial optimization, and simulation. In order to apply results from queuing theory, the authors model the problem as a stochastic queue median problem (SQMP). According to Kendall's notation (cf. Kendall 1953 and Tijms 2003, p. 341 et seq.), this problem is represented by the notation $M/G/1$ since the requests that arrive according to a Poisson process have exponentially distributed inter-arrival times and one vehicle is available for servicing the pending requests. The aim of the study is to determine the best strategy for the repairman among a set of tested strategies under light and heavy traffic conditions, i.e., low and high request arrival rates. The authors test several strategies for sequencing the pending requests on the vehicle. In the First-come, first-served (FCFS) policy, all requests are serviced in their sequence of arrival. When the vehicle is idle, it is relocated to the median of the service region where it waits for further requests. Therefore, a relocation and a waiting strategy is used. The stochastic queue median policy (FCFS-SQM policy) behaves similar to FCFS but always repositions the vehicle at the median after it has finished service of the current request before it continues to service other waiting requests. The Partitioning policy (PART) partitions the service region into m^2 subregions in which arriving requests are serviced in an FCFS manner. If no requests are pending in the subregion which is currently serviced by the vehicle, the vehicle continues to service requests in the next adjacent subregion. In this strategy, the sequence in which subregions are considered is predetermined. In another strategy denoted as the TSP strategy, requests are queued and grouped into sets of a predetermined size n. When a new set of n requests has been generated, an optimal TSP solution for the requests in this set is calculated. Afterwards, the requests are serviced according to this optimal solution. If more than one set is pending at the same time, the sets are considered in FCFS order. The Space Filling Curve (SFC) policy services requests ordered by their arrival in repeated clockwise sweeps of a circle that covers the Euclidean service region. In the Nearest Neighbor (NN) policy, the request to be serviced next is determined by choosing the one closest to the vehicle's current position.

The authors evaluate the performance of the given strategies in light and heavy traffic scenarios by providing appropriate lower bounds. They show that the FCFS-SQM policy is optimal in case of light traffic, i.e., in scenarios with low request arrival rates. In scenarios with increased request arrival rates, the FCFS and FCFS-SQM policy become unstable. In this case, the authors indicate that the PART policy performs well. If a heavy traffic condition is present, the authors show that the TSP, SFC, and NN strategies are the best policies. Specifically, the SFC strategy reduces the average request time in the system by about 15 % when compared to the TSP policy. Furthermore, the NN strategy reduces the average request system time by an additional 10 % compared to the SFC strategy. On the other hand, the authors note that the TSP and SFC policies tend to be fairer by considering the request arrival sequence more than the NN strategy and they additionally offer provable performance guarantees. Furthermore, the authors identify that the heavy traffic policies achieve asymptotically the same behavior so that their performance distinguishes from each other only by a constant factor. For future research, the authors suggest extending the problem to the multiple-vehicle case as well as considering capacitated problem variants. Furthermore, other policies could also be investigated.

In Bertsimas and van Ryzin (1993a), the authors extend the previously considered problem to the multiple-vehicle case in which m vehicles are available for servicing requests in the Euclidean plane. They analyze two variants of the problem, in particular the uncapacitated case and the capacitated case in which each vehicle can only service q requests before it is required to intermediately travel to the depot. In this work, the authors present strategies for the single-depot case but they mention that the results can be extended to a multi-depot scenario if certain symmetry conditions exist. When dividing the problem into m subregions, m independent $M/G/1$ queueing systems (cf. Kendall 1953 and Tijms 2003, p. 341 et seq.) arise to each of which the FCFS-SQM policy can be applied. Hence, the authors modify the FCFS-SQM policy of the single-vehicle case to the FCFS-mSQM policy in which each of the m vehicles is positioned at one of the m median locations in the considered service area. Similar to the FCFS-SQM policy, each vehicle returns after completion of a request to its assigned median before it services the next request in FCFS arrival order. As also mentioned in Bertsimas and van Ryzin (1991), in light traffic scenarios, the authors prove that this policy is optimal. For heavy traffic scenarios, the authors propose different strategies and lower bounds. The randomized assignment (RA) policy divides the Poisson request arrival process into m Poisson subprocesses by randomly assigning requests that arrive in the original Poisson process to the m subprocesses. Each subprocess is assigned to one vehicle that services delegated requests in the whole service area. The authors show that this strategy provides a constant performance factor guarantee but that this factor increases with m. Another policy that possesses the same dependency of m is a $G/G/m$ version (cf. Kendall 1953 and Tijms 2003, p. 341 et seq.) of the TSP policy which is an extension of the previously described TSP policy for the single-vehicle case. In this policy, requests are queued in sets of size n. In an FCFS manner, the next set is then assigned to the next available vehicle which services the requests of this set in sequence of an optimal TSP tour. The authors also present two variants of the TSP policy with a performance that is independent of the number of available vehicles. In the first policy variant called the modified $G/G/m$ policy, the service area is subdivided into k subregions where each subregion forms a wedge of area A/k with A denoting the whole service area. In each subregion, arriving requests are queued into sets of size n/k. Whenever a set is complete, the next available vehicle services the requests assigned to this set, in sequence of an optimal TSP tour while pending sets are queued in an FCFS manner. The authors state that this strategy can also be used well in the single-vehicle case since it halves the request system time that the best strategy evaluated in Bertsimas and van Ryzin (1991) requires. In the second policy variant, the authors utilize the same subregions as in the previous policy but arriving requests are queued in each subregion. In specified time intervals, all requests which are queued in the subsequent region are considered in an optimal TSP tour which is added to the queue of pending tours which are again processed in an FCFS manner. Since in this policy a new request (consisting of a tour of requests) arrives in the queue after a predefined amount of time, this forms a $D/G/m$ system (cf. Kendall 1953) which can be analyzed using the same lower bounds developed for the previously described $G/G/m$ systems. In another policy called the Pμ policy, the service area is partitioned into m subregions and one vehicle is assigned to each of these subregions. Requests which arrive in each subregion are serviced by the corresponding vehicle according to a chosen single-server policy. Hence, the authors conclude that for the uncapacitated case, all heavy traffic policies with a constant factor which were proposed for the single-vehicle case can be extended to the multiple-vehicle case. Furthermore, for the capacitated case, strategies from the uncapacitated case are adapted and new strategies are also proposed. Additionally, the authors discuss possibilities to adapt the presented policies to a scenario where the objective consists not only of minimizing waiting and service times, but additionally comprises the minimization of travel costs. As an outlook on future research, the authors suggest evaluating dynamic routing problems in road networks instead of using the Euclidean plane.

In Bertsimas and van Ryzin (1993b), the authors extend their findings to more generalized situations in which new requests arrive according to a general renewal process and their locations are determined by an arbitrary distribution. Furthermore, the authors state that static vehicle routing approaches can yield near-optimal policies for dynamic routing problems if they are properly adapted. Therefore, the authors state that most of the heuristics and exact methods which have been developed for static problems are not irrelevant for developing effective policies in dynamic-stochastic routing problems.

Papastavrou (1996) introduces a new strategy for the single-vehicle case of the DTRP called the generation policy. This policy positions the vehicle at the median of the service region and services requests as soon as they are available. After servicing, the vehicle returns to the median. If no further requests are pending, the vehicle waits and the strategy starts all over again. Otherwise, the currently pending requests are serviced in sequence of the corresponding optimal TSP tour. After servicing all of the requests of the generated optimal tour, the vehicle returns to the median and the generation policy is again applied. In contrast to the previously mentioned strategies, Papastavrou states that the generation policy performs well in light as well as in heavy traffic conditions. The author emphasizes that a policy that works well under all traffic conditions is advantageous for practical use since a change of the applied policy can be costly and difficult to realize. This is especially important in scenarios where the request arrival rate often changes. The author mentions that the generation strategy can be easily adapted to the multiple-vehicle case by positioning m vehicles at the m medians of the service region if the request locations are uniformly distributed or at the weighted m medians in routing problems in which a general request distribution exists.

Swihart and Papastavrou (1999) extend the research on the DTRP to a single-vehicle DPDP while considering the same objective as in the DTRP. In this research, two cases are analyzed: The unit-capacity case where the vehicle directly needs to travel from the pickup to the delivery location of the request before new requests can be transported and the multiple-capacity case in which the vehicle can transport more than one request at the same time. For the unit-capacity case, the authors derive new bounds for the system performance in the light and heavy traffic case and evaluate different policies that are comparable to the ones tested in the DTRP. Like in the DTRP, in the light traffic case, the authors prove that the SQM policy is optimal. In the heavy traffic case, different policies are tested. Specifically, in the sectoring policy, the service area is partitioned into m^2 sectors. When the vehicle has finished servicing the current request, the request that has waited longest in the current vehicle's sector is serviced next. If no request is currently pending in this sector, another sector is randomly chosen. Furthermore, the stacker crane policy is a variation of the TSP policy in which each node of a pickup and delivery pair is treated as a single location in an asymmetric TSP. This is possible since once a request has been picked up, the vehicle is required to directly travel to its delivery location. Furthermore, a nearest neighbor policy is also evaluated. In order to assess the performance of the strategies, the authors apply mathematical as well as simulation-based methods since not all strategies are mathematically tractable due to complexity reasons. Their evaluations show that while the sectoring policy has a bad performance, the stacker crane policy as well as the nearest neighbor policy perform better while the nearest neighbor policy performs best out of the tested policies. In the multiple-capacity case, the light traffic case is not analyzed any further since it is very unlikely that in this case more than one request is pending at a time point. For the heavy traffic case, two further strategies are invented, that is the Dual TSP policy and the multiple-capacity nearest neighbor policy. Computational results show that the multiple-capacity nearest neighbor strategy performs better than the Dual TSP policy. According to the authors, two further tested variants of the Dual TSP policy, i.e., the Partially Sectored Dual TSP policy and the Sectored Dual

TSP policy, yield the best results when only one sector is used so that both variants become exactly the same as the original Dual TSP policy. For future research, the authors suggest improving the proposed bounds and considering multiple-vehicle problem variants.

In his PhD thesis, Larsen (2000) extends the definition of the degree of dynamism (see Sect. 2.6.4) which had originally been introduced by Lund et al. (1996). Furthermore, the author gives detailed information about the generation of customer requests and request-specific attributes such as appropriate distributions for generating arrival times, service times, locations, and demands. Moreover, he proposes heuristic solution methods for three different dynamic routing problems. One of the evaluations is performed on the DVRPTW under different degrees of dynamism without using stochastic knowledge while the other two investigated dynamic routing problems use stochastic knowledge and are described in what follows. In the first problem, the author extends the single-vehicle case of the DTRP to the *The Partially Dynamic Traveling Repairman Problem* (PDTRP). In this problem, a given portion of all requests to be serviced is known before the execution of the transportation process. Furthermore, the considered objective function is different from the research on the DTRP described so far since it pursues the minimization of travel costs. During the execution of the transportation process, the dynamic requests arrive according to a Poisson process whose properties are known and can be used as stochastic knowledge in order to define appropriate strategies. Service times of all customers are assumed to be distributed according to a log-normal distribution. Moreover, vehicle en-route diversion is not allowed. The author evaluates the strategies FCFS, FCFS-SQM, NN, and PART on instances with varying degrees of dynamism. The results of the conducted experiments show that the NN policy performs best among the tested strategies. Utilizing the given objective, the FCFS-SQM policy performs poorly since it generates additional travel activities to the median of the service area after each completion of service of a request. Furthermore, using the NN strategy, the author states that travel costs linearly increase with the degree of dynamism. The results of this study as well as the extended definition of the degree of dynamism have also been published in Larsen et al. (2002). For future work, the author proposes utilizing stochastic knowledge that provides information about arrival rates of customer requests in specific parts of the service area as well as information about changes in these rates during the day. For this purpose, the author outlines an algorithm for integrating dummy customers based on historical request data. In the second considered application which utilizes stochastic knowledge, Larsen introduces the A Priori Dynamic Traveling Salesman Problem with Time Windows (ADTSPTW) in which soft time windows are present. Similar to the previously considered PDTRP, service times of requests are distributed according to a log-normal distribution and vehicle en-route diversion is prohibited. The objective function consists of minimizing a weighted sum of travel costs and lateness costs which occur if a request is serviced after the end of its time window. Lateness costs depend linearly on the length of the lateness. Dynamically arriving requests are generated by dividing the considered service area into subregions. In each subregion i, requests arrive according to a Poisson process with an individual intensity parameter λ_i. These intensity parameters remain constant over the considered planning horizon and are assumed to be available as stochastic knowledge. In this study, the effects of applying different strategies are examined if the available stochastic knowledge is utilized. Specifically, this knowledge is used to choose an appropriate waiting position for the vehicle if it is currently idle or if the time window of its next assigned customer request only starts in the far future. Hence, the proposed approach makes use of waiting and relocation strategies. Whenever new requests arrive, a TSP is solved by a local search heuristic in which a given solution is improved by exchanging two customer requests if this leads to an improvement of the solution. The author mentions that since the applied heuristic utilizes a running time of less than two seconds, concurrency considerations of adapting and executing the tour can be neglected due to the shortness of the required time period. In the computational experiments, different strategies are assessed to select the waiting position out of a set of predefined waiting locations where one idle point is defined per subregion i. Depending on the used strategy, the waiting point for the vehicle is: (a) the nearest (NEAREST), (b) the one in the subregion i with the highest rate parameter λ_i (BUSIEST), or (c) the waiting point of the subregion in which it is expected that the most requests can be serviced within the current *idle time* of the vehicle, i.e., the time period which is available before the vehicle departs from the selected subregion, at the latest, in order to service other currently pending requests with a time window opening far in the future (HI-REQ). Hence, the HI-REQ strategy takes into account the distance from the chosen subregion to other customers that are known but can only be serviced farther in the future. After the potential waiting location has been selected according to the used strategy, the probability that at least one more request will occur in the selected subregion within the current idle time is determined. This probability is used to decide, according to a defined threshold, whether the vehicle is relocated to this waiting point or not. Clearly, the higher this threshold, the fewer the relocation activities that are performed. As a reference, the author defines a strategy denoted as NO REPOS in which the vehicle always waits at the last serviced customer request location so that no relocations are performed. In the computational experiments, two variants of the objective function were considered. In the first variant, the minimization of travel costs is prioritized while in the second variant, the minimization of lateness costs is the topmost aim. According to first provided results, small improvements can be attained for reducing the lateness while a higher threshold reduces the total travel distance due to fewer relocation activities. The author states that more research on these policies is likely to result in significant additional solution quality improvements. In Larsen et al. (2004), the authors continue to study the previously described ADTSPTW. The authors provide extensive computational results and identify recommendations about which policy should be used depending on the objective and the degree of dynamism.

Irani et al. (2004) perform research on a variant of the DTRP in which a single vehicle services dynamically arriving requests which require no service time at their location. Each arriving request is required to be serviced until a given latest time point. In their study, the authors use a uniform distribution for modeling the length of the time period between request arrival and the latest allowed time point. The considered objective is to maximize the number of feasibly serviced requests up to the latest allowed time point. The authors develop two strategies called BATCH and Double-Gain. For both strategies, they present upper bounds of the competitive ratio as functions depending on the diameter of the considered metric space. Furthermore, several lower bounds for arbitrary deterministic strategies are presented. For future work, the authors propose evaluating the Double-Gain strategy in scenarios where the time period available for feasibly servicing requests is not uniformly distributed.

Branke et al. (2005) consider a DVRP in which exactly one additional customer request dynamically arrives during the execution of the transportation process. Vehicles have fixed tours for servicing the known requests. Requests have no time windows nor service times but the vehicles need to arrive at the depot, at the latest, at a given time point. The objective is to maximize the probability that the additional occurring request can be feasibly integrated into the fixed tour plan, i.e., so that all vehicles still feasibly arrive at the depot. For this purpose, the fixed tours comprise additional slack time which can be used to wait for the additional request as well as service it. Stochastic knowledge is available in form of prescribed probability distributions which describe the location and the arrival time of the additional request. Regarding its location, it is known that the arriving customer request is uniformly located in the considered service area. For the arrival time, two variants are considered. In the first variant, it is known that the request only arrives after a specific time point whereas in the second variant, this time point is equal to the start time of the transportation process so that the request can arrive at any point in time. In both cases, the arrival time of the additional request is uniformly distributed after the specified time point. In order to maximize the probability of feasibly integrating the additional customer request, the stochastic knowledge is utilized to make vehicles wait at suitable locations during their tours. Hence, in this problem, the positions of the vehicles can be influenced by using a waiting strategy. The considered problem is called the waiting drivers problem (WDP) and the authors show that both variants of the WDP are NP-hard. The authors evaluate different waiting strategies. In the single-vehicle case, it is proven that the optimal strategy is not to wait at all. If two vehicles are available, the authors show that waiting is beneficial and propose an optimal strategy for this scenario. For the multiple-vehicle case with more than two vehicles, due to the NP-hardness of the problem, six different heuristic waiting strategies as well as two variants of an evolutionary algorithm which improves the solution by examining different waiting time values are proposed. Computational results for the multiple-vehicle case show that, compared to a strategy in which vehicles do not wait at all, an appropriate waiting strategy can significantly increase the probability of feasibly servicing the additional customer request as well as reduce the additional travel distance required for servicing the dynamically arriving request. Specifically, one of the tested strategies called "variable" orders vehicles to service requests without waiting until the point in time at which the remaining travel time to the depot is equal to the slack time available in the tour. From this time point, the available waiting time is proportionally distributed to the remaining customers according to the remaining travel distances. By applying this strategy, the authors state that the probability of being unable to feasibly service the additional request is reduced by approximately 10 % while the average length of the additional travel distance in case of a successful insertion of the new request can be reduced by about 35 %. The authors also give insights into variants of the considered problem, in particular when a non-uniform distribution of customer requests is present as well as in the case where two or more customer requests dynamically arrive. As an outlook on future research, the authors suggest considering problem variants in which a non-uniform distribution of new customers and the arrival of multiple dynamic customers is present. Furthermore, they propose evaluating possibilities of modifying tours during the execution of the transportation process and improving the proposed evolutionary algorithm.

In Attanasio et al. (2007), a real-time management system for coordinating a fleet of same-day couriers in London, United Kingdom, is presented. The considered problem is a DPDP with dynamically arriving requests and uncertain travel times. The objective is to maximize the quality of customer service as well as the average number of serviced customers per courier. Customers are divided into three groups that are distinguished from each other according to the maximum allowed response time to newly arriving requests. Vehicle en-route diversion is allowed to service dynamically arriving requests. Stochastic knowledge about request arrivals is derived from historical data by using a time series decomposition approach. Specifically, the service area is divided into zones and for each pair of zones, the expected request arrival rate is calculated. Travel times are learned from historical data while they are also updated in real-time using online traffic information during the day. The deployed couriers can accept or reject requests that are offered to them by the implemented request management system. Requests which are not directly accepted by couriers are integrated into the tour plan by using a Tabu Search procedure. While stochastic knowledge about expected travel times directly influences the assignment of requests to couriers, stochastic knowledge about expected request arrival rates is used for relocating idle vehicles to zones in which the expected request arrival rate is high in the near future. While providing no computational results, the authors mention that the use of the system has increased the quality of service as well as courier efficiency and that it also provides reduced marginal administration costs for additional deployed couriers.

In Kleywegt et al. (2009), the authors study a dynamic and stochastic multiple-period routing problem in which new requests arrive over time but only one dynamic request arrives per period. Based on SLAs, the dispatcher can decide whether a newly arriving request should be serviced in the current period or better still in the next period. The considered objective is to minimize the average transportation cost per period over a given planning horizon. It is assumed that probability distributions of newly arriving requests are available so that stochastic knowledge can directly be utilized. The authors examine various dispatch strategies. These strategies include a two-stage myopic policy, a sampling-based policy, a multi-stage sample average approximation policy, and an optimal policy. Three different types of service areas are evaluated in the computational experiments, namely an interval, a circle, and a disk. The relative performance of the strategies is assessed and also compared to an optimal solution in the corresponding offline case where all information about future request arrivals is known in advance with certainty. Two different request location distributions are tested. In the first setting, request locations are independent and identically distributed according to the uniform distribution on the interval, and in the second setting, customer locations are independent and identically distributed according to a specific discrete distribution. Computational results provided for several instances under the different scenarios show that the multi-stage sample average approximation policy often performs best among the tested approaches. The authors propose to extend the approaches to the multiple request case per period. First results for this case can be found in the PhD thesis of Uyar (2008).

4.5.2 Approaches Using Exact Solution Methods

In the work of Yang et al. (2004), the authors consider a multi-vehicle truckload DPDP in which each vehicle can carry at most one request at the same time. Requests dynamically arrive over time and a small time period for each request is given during which a newly arriving request may be accepted or rejected. Requests have soft time windows at pickups and deliveries so that service after the end of a time window is allowed but at the expense of additional linear lateness costs. Revenue is gained if a request is accepted, otherwise penalty costs have to be paid. The objective is to maximize the total profit resulting from request profits and penalties as well as lateness and travel costs. The individual problem instances which arise during the execution of the transportation process are formulated as a Mixed Integer Programming (MIP) problem whereas, due to complexity reasons, the number of considered requests in each MIP problem instance is limited by a pre-selection routine. The authors evaluate five different policies for solving the dynamic problem. Among these policies are three local adaptation heuristics and two approaches which globally reoptimize the tour plan. The latter two approaches solve the MIP problem with a time limit of 20 seconds. Both approaches are different from each other in that one reoptimization approach makes myopic decisions based on deterministically available knowledge only while the other approach additionally uses given stochastic knowledge about expected future request arrivals. This stochastic knowledge is utilized by integrating expected opportunity costs into the objective function. These costs are based on the current and future vehicle positions resulting from assigned requests in relation to expected future request arrivals. Hence, vehicles are more likely to be deployed within the considered service area according to the expected request arrival rates. Computational results on test instances with 10 vehicles and 1,000 requests with varying degrees of system utilization show that the two reoptimization approaches perform better than the local adaptation heuristic approaches. Out of the two evaluated reoptimization-based approaches, the one which additionally uses stochastic knowledge performs best by yielding about 5 % better results compared to the myopic reoptimization approach. The authors mention that further research on methods which utilize stochastic knowledge more efficiently is likely to have a big potential for additional improvements.

4.5.3 Stochastic Modeling Based Approaches

In this section, approaches which model the dynamic routing problem by using specific stochastic formulations are presented. According to Powell et al. (1995), these models are called multistage stochastic models which directly utilize stochastic knowledge available by prescribed probabilistic distributions. For example, these models use Markov Decision Processes (MDP) or Approximate Dynamic Programming formulations.

Powell (1988) considers a truckload DPDP in which each vehicle can carry one request at a time. In this problem, requests have to be picked up at one location and require transportation to another location. After the request has been delivered, the next action for the empty vehicle has to be decided. Specifically, possible actions consist of assigning the vehicle to another pending request, keeping it at its current location, or relocating it to another location (e.g., a different city) in anticipation of future requests arriving there. Hence, waiting and relocation strategies are used in this approach. Stochastic knowledge is given by request arrival rates which vary in different subregions and may also change during the day. The considered practical application arises from a business case in which large transportation companies regularly subcontract requests. According to the author, this problem has a degree of dynamism of 60 %. Since newly arriving requests are often delegated to companies that currently have trucks close to the pickup location of a new request, it is beneficial for carriers to have trucks close to locations where new arriving requests are expected to occur in the near future. However, each company has to balance travel costs generated by relocating trucks with the expected profitability of arriving requests in each region. Moreover, companies may reject requests if they do not appear profitable. The author names this problem the dynamic vehicle allocation problem (DVA). In this work, the author gives an overview of different techniques on how to model the DVA in the single and multiple-vehicle variant. In addition to deterministic models including static single stage and dynamic time-space models, stochastic network formulations are presented. Furthermore, the problem is also modeled using an MDP formulation. The author mentions that the state space grows very quickly so that the problem becomes intractable for real-life problem instances. Moreover, a two-stage stochastic programming formulation of the DVA is described in which the first stage consists of known requests and the second stage represents the expected value of having additional vehicles in specific regions. Insights into the multistage deterministic formulation and the two-stage stochastic formulation are combined to form a hybrid model. Moreover, an overview of alternative models is given. This work provides one of the first references in the literature for using stochastic knowledge in order to relocate vehicles and decide about their waiting times at certain locations where future requests are anticipated.

In a later paper, Powell (1996) continues the research of a long-haul transportation process in which requests have to be assigned to vehicles in real-time. Besides reviewing a variety of alternative models for this dynamic assignment problem, the author develops a new model for approximately solving the stochastic and dynamic request to vehicle assignment problem. A hybrid model which consists of an assignment and a forecasting component that anticipates future requests is presented. In the proposed model formulation, costs of possibly required recourse actions are included in the arc costs of the transportation network. The author shows that the model is computationally fast enough to handle the largest problem instances which arise in the considered practical application while being flexible enough to integrate different practical requirements at the same time. Computational results carried out on experiments with 200 vehicles and 600 requests per week with a degree of dynamism of about 60 % demonstrate that the implemented anticipative model outperforms standard myopic models.

In the work of Ichoua (2001), a DVRP is considered in which vehicles that have finished servicing a request are not always directly sent to their next assigned request. Instead, if stochastic knowledge indicates that further requests are expected to arrive soon within the vicinity of a vehicle's last serviced request location, the vehicle is required to wait at its current position for a given amount of time. The results of this thesis are integrated into the approach of Ichoua et al. (2006) which is described below.

The work of Thomas and White III (2004) deals with a routing problem in which a single uncapacitated vehicle travels on a road network from an origin O to a destination D which needs to be reached by a defined time point at the latest. The arcs in the road network have different travel costs and dynamism arises from newly arriving customer requests which demand goods to be picked up and transported to destination D. A special aspect of this approach is that new requests only arise at a set of known locations, i.e., at a defined set of nodes in the road network. Requests do not have time windows nor service times. For each potentially arriving request, the probability of its arrival is given as a function of time. When a pending request is serviced, defined revenue is granted but requests may also be rejected. The objective is to determine the optimal path along the road network so that the expected profit is maximized, i.e., the sum of expected revenues from serviced requests minus travel costs. Hence, if revenues are small and travel costs are high, the optimal policy might consist of just traveling on the least cost path from O to D while neglecting all potentially arriving requests. The authors formulate the problem as an MDP and solve it using stochastic dynamic programming techniques. The solution is iteratively constructed by making a decision at each node that the vehicle visits about which node the vehicle should travel to next. The vehicle anticipates future requests by not only visiting possible request locations but also other nodes in the road network. Different experiments on a road network with 132 nodes and 202 arcs are carried out on different probability distribution settings according to the mean and variance of the applied request arrival probabilities in order to analyze the effect of the implemented anticipatory routing method. Computational experiments which are conducted in cases where one customer additionally arrives indicate that the optimal anticipatory routing policy performs well compared to a deterministic and purely deterministic version of the solution approach. This is especially true if the probability that this additional customer appears late has a high mean and a low variance value. The authors also provide computational results for the multiple request case which turns the considered problem into an NP-hard TSP. This fact can also be observed at the computational times required for calculating an optimal solution. Specifically, while less than one second is required in the single request case, an instance with seven potentially arriving customer requests requires over 8,000 seconds of computational time. Hence, the authors state that using the proposed dynamic programming approach might encounter problems in solving real-life sized problem instances so that future research should improve the computational efficiency of the applied solution methods. Furthermore the authors suggest extending the anticipatory strategy to the multiple-vehicle variant of the considered routing problem.

In Ichoua et al. (2006), the authors consider an uncapacitated DVRP where a set of given heterogeneous vehicles needs to service dynamically arriving requests with soft time windows. If service of a request is performed after the end of its time window, linear penalty costs occur. All vehicles have to return to the depot by a given time point at the latest. The objective is to maximize the number of serviced customers and, as a subordinate goal, minimizing the sum of travel and lateness costs. Vehicles move in the Euclidean plane and en-route diversion of vehicles is not allowed. Requests arrive according to time- and space-dependent Poisson processes. Specifically, the considered service area is divided into geographic zones and the considered planning horizon is divided into time periods. Each of these zones can have a different request arrival rate assigned in each time period. Therefore, request arrival rates vary in different zones of the considered service area. Stochastic knowledge is given by having knowledge about the intensity rates of the time-space Poisson processes in advance. In order to increase the ability of servicing more requests, vehicles may wait at the location of their last serviced request if the probability of further arriving requests nearby is high in the near future, even if requests that are located further away are currently pending. Hence, the VRP is extended to a VRSP (see Sect. 2.4.6) using a waiting strategy which utilizes the available stochastic knowledge. The authors formulate the problem using a dynamic programming approach which calculates an optimal solution for assigning requests and scheduling the vehicles. Since the problem is strongly NP-hard, the authors propose a vehicle-waiting heuristic to solve the problem. In the proposed heuristic that uses results from the PhD thesis of Ichoua (2001), for each idle vehicle k at time t it is decided whether the vehicle should wait for a defined amount of time δt_k or directly travel to the next request. If the probability that new requests will arrive within the neighborhood geographical zone where vehicle k is currently located during the time interval $[t, t + \delta t_k]$ is higher than a given threshold and if not too many other idle vehicles are already waiting in this zone, the vehicle waits for δt_k time units. Otherwise, it directly moves to its next assigned request. In order to determine which request is assigned to a vehicle next, the Parallel Tabu Search metaheuristic proposed in Gendreau et al. (1999a) is applied in a modified version. Specifically, if a vehicle is ordered to wait, a dummy customer is inserted at the beginning of the tour of the corresponding vehicle k stating that the vehicle is available for further activities only at $t + \delta t_k$. If a new request arrives while a vehicle is currently waiting in the associated zone, the temporarily inserted dummy customer is replaced by this new customer request. If no customer request has arrived until time point $t + \delta t_k$, the dummy customer is removed from the vehicle's tour and the vehicle continues to travel to its next assigned customer according to the best solution in the adaptive memory of the Parallel Tabu Search. Computational results on four different scenarios resulting from two different request arrival rates (27 and 36 requests per hour leading to an average of 138.1 and 187 requests per instance, respectively) combined with two different fleet sizes of 4 and 6 vehicles are performed. The proposed vehicle waiting heuristic-enabled approach is compared to the original Parallel Tabu Search procedure which does not use any stochastic knowledge. The computational experiments show that the utilization of the vehicle-waiting heuristic yields significant improvements, especially when utilized in scenarios with a smaller fleet size or more pending requests.

Thomas (2007) continues the study on applying waiting strategies in dynamic routing problems. The author considers a dynamic and stochastic routing problem with a single uncapacitated vehicle. The objective is to maximize the expected number of requests which can be serviced. In this problem, stochastic knowledge about expected future requests is used in order to decide when and where the vehicle should wait. The author utilizes an MDP formulation and derives analytical characteristics of tested policies. In the scenario in which one customer dynamically arrives, the optimal policy is completely analyzed. These results are used for defining a real-time heuristic and other heuristics that can be applied in scenarios in which multiple requests dynamically arrive. Computational results conducted on different data sets show that using a heuristic strategy which uses stochastic knowledge denoted as center-of-gravity longest wait heuristic, performs better when the percentage of dynamic customers is 25 % or less. In instances with a higher degree of dynamism, the authors state that distributing waiting times across the currently known requests is an effective heuristic strategy which has also been shown to perform well in situations where no stochastic knowledge is available (see Mitrovic-Minic and Laporte 2004 in Sect. 4.4). The authors suggest including the consideration of the multiple-vehicle case as well as time windows for requests in future research work.

Secomandi and Margot (2009) consider a single-vehicle capacitated Vehicle Routing Problem with Stochastic Demands (VRPSD). In this problem, customer requests which are known with certainty need to be supplied with a specific good. Uncertainty arises from the fact that the quantity demanded by each request is only known by prescribed probability distributions. Note that most approaches in the literature solve this kind of problem by using a priori optimization methods as described in Sect. 2.5.3.2.1. However, as mentioned in the same section, some authors state that the full potential of solving these problems can only by achieved by using real-time solution methods. Hence, the authors propose a new optimal decision strategy for solving the transportation process during its execution. Having finished the current request activity, the proposed strategy called partial reoptimization decides whether the vehicle should move to one of the pending customer requests directly or first return to the depot and perform a replenishment activity. Furthermore, split deliveries are possible so that a portion of the customer's demand can be delivered before and after the intermediate return to the depot. This decision takes the currently remaining vehicle capacity into account. Specifically, the vehicle should perform the replenishment activity next when its currently remaining load drops below a threshold. This threshold is based on various factors such as the expected demands of pending customer requests. The authors propose a new computationally efficient MDP formulation of the proposed strategy. The formulation is based on the concept of solving a restricted set of states so that at least one admissible policy, i.e., a solution in which the demands of all pending requests are entirely serviced, is determined. For instances with up to 15 customer requests, an optimal solution can be generated by using the proposed solution approach. In larger instances, memory limitations occur so that two different methods are introduced in order to select a subset of states of the MDP. Based on the restricted set of states, the policy is determined by backward dynamic programming. The performance of the approach is assessed on various instances. The authors show that on instances with up to 90 customers several versions of the proposed heuristically operating MDP approaches outperform existing heuristics in the literature and that the best performing heuristic has a gap of about 10–13 % compared to a lower bound which has been computed with perfect knowledge of customer demands given in advance. Furthermore, new lower bounds for the VRPSD are derived. With the ability of solving instances up to 15 customers to optimality, the authors show that for these cases, the proposed lower bounds are quite tight.

As has been mentioned in this section, solving real-life problem size instances of dynamic routing problems to optimality by using stochastic models is often computationally intractable. This is due to the generally used recourse function applied in dynamic programming which determines an optimal solution by first solving a large number of subproblems and then determining an optimal solution by backward recursion. The issue that frequently occurs is that the state space quickly becomes too large (cf., e.g., Desrochers et al. 1988 and Powell 2007, p. 47). In order to deal with this issue, the backward-working dynamic programming has been modified to a forward-working *Approximate Dynamic Programming* (ADP). In ADP approaches, the need to evaluate all possible states is eliminated by approximating the value function in a forward-manner (cf. Pillac et al. 2010). Further information on ADP approaches can be found in Powell (2007, 2009). According to Pillac et al. (2010), ADP has been successfully used in various field of transportation applications. One of these fields is fleet management. Here, Godfrey and Powell (2002) consider a stochastic version of a dynamic resource allocation problem in which a set of homogeneous resources is assigned to randomly arriving new requests which are required to be picked up at one location and need to be delivered to another location. Furthermore, some requests are known at the beginning of the planning horizon. Three characteristics related to routing problems make this problem different from a usual assignment problem: the integration of costs for preparing resources for servicing requests, penalty costs for late service of requests, and the occurrence of new vehicles (besides new requests) over time. It is assumed that stochastic knowledge which provides arrival rates of expected future events is available. The objective is to maximize the profit consisting of the expected reward for servicing requests minus transportation and penalty costs. The problem is solved using an ADP approach with nonlinear functional approximations which provide information about the expected value of resources in the future. Computational tests show that the proposed method significantly outperforms a rolling horizon approach which adapts the transportation process by using deterministically known information only. The authors state that by using stochastic knowledge, improvements of 2 % to 16 % can be attained where higher improvements can be achieved in scenarios with less resources or more customer requests, i.e., in scenarios with a higher resource utilization. Other applications to fleet management can be found among others in Powell and Topaloglu (2005).

Similar to the previously described work of Secomandi and Margot (2009), Novoa and Storer (2009) introduce a new approach for dealing with a single-vehicle VRPSD from a real-time control approach perspective. They formulate the problem as a stochastic shortest path problem that is related to an MDP formulation. Furthermore, the authors propose an ADP approach called the rollout algorithm. Different variants of one-step and two-step lookahead stages in the rollout algorithm are compared to an a priori solution on instances with 60 customers. The authors note that computing an optimal solution of VRPSDs is currently intractable for instances with more than 20 customer requests. The experiments show that among the tested methods, the two-step lookahead rollout algorithm performs best. Specifically, the routing costs are about 4.8 % less compared to the results achieved by the one-step rollout algorithm. Moreover, the authors compare the results obtained for the online instances with results of related offline instances in which perfect information is given in advance. For one of the two evaluated instance sets, the authors show that the best tested two-step rollout algorithm is only 3.38 % away from the estimated perfect information lower bound.

Schmid (2012) considers a practical application dealing with real-time coordination of ambulances in Vienna, Austria using a real road network with time-dependent travel times. When a new emergency request arrives, an ambulance has to be chosen among the idle and waiting ambulances and dispatched to serve the new request under the objective to minimize the response time of this request that has newly arrived. When an ambulance has finished servicing a request and becomes idle again, a decision has to be made about where to position this vehicle. For this purpose, a set of predefined waiting locations in the city are available. The aim is to relocate the vehicle to a specific waiting location so that the estimated response time of expected future requests is minimized. Due to legal regulations, the closest vehicle is always sent to a request so that the decision about which ambulance should be assigned to a request is not considered. In order to be able to quickly service expected future requests, it can also be advantageous to relocate waiting vehicles during the day for two reasons. First, the overall and region-dependent request arrival rates change throughout the day. Second, due to the existence of time-dependent travel times, the coverage rate defining the area that is reachable from a given idle position within a maximum amount of time also varies during the day. However, since another law prohibits the relocation of an ambulance from one waiting location to another while it is idle, as far as the original problem is considered this decision can only be taken for vehicles which just have finished servicing a request. In order to solve the considered problem, the author proposes a new ADP approach which derives good positioning decisions depending on the current system situation by analyzing historical real-world request data and given time-dependent travel time information. Hence, in the described approach, stochastic knowledge is automatically generated from the available past request and travel time information. Extensive computational tests on real-world problem instances show that, compared to the currently applied strategy in which ambulances have a fixed waiting location, the approach is able to reduce the average response time by 7 % while also reducing the maximum response time. In a further test study, the regulation which stipulates that the closest vehicle is always required to be dispatched is relaxed if a new request is not life-threatening. Using this relaxation, the average response time can be further reduced by 0.7 %. Future work is directed at allowing the direct en-route dispatching of idle ambulances which are currently traveling to a waiting location. Moreover, the evaluation of improvements attainable by relocating idle ambulances between waiting locations according to changing expected request arrivals during the day is also suggested.

Another type of state-based solution approaches are Hybrid Adaptive Predictive Control (HAPC) approaches. In HAPC approaches, individual states involve both integer and continuous variables. They work in discrete time steps by generating individual problem instances. A new time step is considered whenever a new relevant event occurs during the execution of the transportation process. Sáez et al. (2008) develop a HAPC approach for a capacitated multiple-vehicle DPDP that can be considered as a DDARP in the Euclidean plane. Dynamism arises from newly arriving customer requests that demand an instant transportation from an origin to a destination. In the considered problem, up to four customers can be transported in a vehicle at the same time. As known from DARPs, the objective includes quality of service factors. Specifically, the objective is to minimize the waiting and travel time of customers as well as operational costs mapped by idle travel times of vehicles. The authors analytically develop a problem-specific state space formulation which uses results published in Cortés et al. (2009). The dynamic transportation problem is carried out by solving the problem for the complete number of defined anticipation steps and, after the solution process is completed, implementing the next step of the best found solution. The authors mention the importance of the number of considered anticipation steps which is heavily problem-dependent. In order to solve the problem, the authors propose an explicit enumeration scheme as well as a GA which is evaluated in three variants that can be distinguished from each other according to the number of considered individuals and generations. Historical information about requests is available and utilized to automatically generate stochastic knowledge about expected origin-destination trip patterns using a fuzzy clustering method (FCM, cf., e.g., Xu 2005). Specifically, the FCM assigns past requests to one or more defined fuzzy zones. In fuzzy zoning, requests can belong to multiple zones through a so-called membership function which defines to what percentage a request belongs to each zone. Stochastic knowledge is then generated by assigning requests to the fuzzy zone to which this request belongs with a percentage which is at least a specified threshold. If a request does not occupy a zone with such a minimum required percentage, it is not considered during the generation of the stochastic knowledge. In doing so, only stochastic knowledge with a certain quality is generated. Several test instances are evaluated in which the available historical data is generated from patterns which have been designed according to observed request arrivals in a real-world request data set. For small problem instances, the authors show that applying the explicit enumeration scheme is a possible option for generating solutions of high quality. For instances of real-world size, the explicit enumeration scheme produces an impractical runtime so that the use of the proposed GA is advantageous. The GA generates the best results in the tested variant with the largest population size and the largest number of generations. Compared to a myopic strategy of the approach which does not use stochastic knowledge, the application of a two- and three-step ahead version of the selected GA variant results in significant improvements on tested instances with 120 customers. Specifically, the three-step ahead version is able to reduce waiting time by more than 22 %. In an outlook for future research, the authors propose that stochastic knowledge should be generated as a combination of historical data and updated information which becomes available during the execution of the transportation process. Moreover, the authors suggest extending the objective function by including hard and soft time windows and a more detailed consideration of operational costs.

In a consecutive paper, Cortés et al. (2008) consider two sources of dynamism, in particular future request arrivals as well as uncertain travel times in the Euclidean plane. For both types of dynamism, stochastic knowledge is assumed to be available in the presented HAPC approach in which it is utilized to generate expected future costs. Moreover, deterministically known time-dependent travel times are considered. Computational results show that a two-step ahead approach is able to achieve significant improvements compared to a myopic one-step approach that does not make use of the available stochastic knowledge. Furthermore, in Núñez et al. (2008), the authors extend the objective function by including operational costs. Since the minimization of total operational costs and the minimization of the previously proposed customer inconvenience aspects are to some extent contradicting aims, the authors use an evolutionary multi-objective optimization concept in order to calculate the Pareto-optimal front of solutions. Three different criteria for selecting a solution which is to be implemented in each decision step are proposed. Computational results show that the proposed approach can be efficiently used to either prioritize waiting time, i.e., minimization of customer inconvenience, or travel time and distance, i.e., minimization of operational costs, while also considering the other aim to some extent.

4.5.4 Approaches with Manually Provided Stochastic Knowledge

In this section, dynamic routing approaches in the literature in which available stochastic knowledge is manually provided are presented. In van Hemert and La Poutré (2004), the authors consider a DVRP in which each vehicle has a fixed capacity and needs to pick up requests which have to be transported to a central depot.

Pending requests may be rejected and the considered objective is to maximize the number of serviced requests while considering fixed time windows of the requests and a latest return time at the depot. The motivation for utilizing stochastic knowledge in the considered approach is that in various real-world applications request arrivals are often not uniformly distributed but appear in a clustered fashion. In their approach, the authors denote such clusters as fruitful regions which are characterized by a high probability that new requests will arrive there. In the approach, test instances are generated by first defining a set of fruitful regions, represented by a set of clusters, around which requests are generated in a subsequent step within a predefined range. In order to solve the dynamic transportation process, an evolutionary algorithm is used. It involves a strategy called anticipated move which relocates vehicles to fruitful regions in expectation of newly arriving requests. The decision whether to perform anticipated moves is directly integrated into the objective function of the evolutionary algorithm. This decision is based on the stochastic knowledge available from information about fruitful regions. Using anticipated moves, vehicles are ordered to travel to fruitful regions even if no requests have arrived there yet but are expected to do so with a high probability in the future. Benchmark results indicate that for certain parameter settings regarding time windows and vehicle capacities, the utilization of the proposed anticipated move strategy is very beneficial.

Branchini et al. (2009) consider a DVRPSTW in which the fleet consists of vehicles with the same capacity. While some customers are known in advance, dynamism arises from newly arriving requests whose location and demand is only known by probability distributions which are available as stochastic knowledge. Pending requests have soft time windows so that they may be serviced late at the expense of linear lateness costs. Furthermore, requests may also be rejected. Each request has an individual revenue and a variable lateness penalty. Tours of vehicles are dynamically adapted by a real-time control which pursues the objective of maximizing the expected profit regarding request revenues, lateness costs, fixed costs for each utilized vehicle, and travel costs. A tour plan adaptation by the real-time control approach takes place whenever a new request arrives and vehicle en-route diversion is allowed. Tours are constructed using a proposed constructive heuristic which scatters the utilized vehicles in the service area. Afterwards, the generated tours are improved by an adaptive granular local search procedure. The distribution of the vehicles over the service area is achieved by calculating so-called seed customers. These seed customers are created by including four request-specific criteria such as time window incompatibility and vehicle waiting times at requests. Each seed customer is assigned to a different vehicle tour. Due to the fixed vehicle usage costs, one aim is to determine the number of required vehicles before the execution of the transportation process. However, additional vehicles can be deployed at additional costs during the day if necessary. The real-time control approach makes use of relocation and waiting strategies. For this purpose, based on historical data, strategic waiting positions are manually identified and provided to the approach as stochastic knowledge. The performance of the approach is assessed using designed test instances and real-world problem instances that are based on transportation processes of Brazilian logistic companies. These instances are determined by a degree of dynamism between 60 % and 80 %. Different settings for the real-time control approach and the applied flexibility-increasing strategies are evaluated. Computational results show that the application of all three strategies, i.e., vehicle en-route diversion, waiting strategy, and relocation strategy in the real-time control approach which uses the proposed granular local search performs best among the tested solution approaches.

4.5.5 Sampling-Based Approaches

In comparison to the approaches presented above, sampling-based approaches utilize stochastic knowledge by *sampling* available probability distributions. This sampling results in one or more *scenarios* that contain specific realizations of the sampled distributions. When sampling is used to anticipate future request arrivals, this often generates possible future scenarios that contain different expected requests. After optimizing each of these instances, analyzing individual solutions generated for the individual instance can yield insights that lead to robust decisions according to expected future events.

Bent and Van Hentenryck are among the first to use sampling-based approaches in dynamic routing applications. The basic problem which is considered by the authors is a DVRPTW in which dynamism occurs by newly arriving requests. In the past few years, both authors have proposed several sampling-based approaches for dynamic routing problems which are known as Online Stochastic Combinatorial Optimization (OSCO, cf. van Hentenryck and Bent 2009 and Bent et al. 2010). The authors use a general framework for online optimization in all of their proposed strategies. In this framework, two black boxes are used for decision making: a problem-specific sampling procedure with a given distribution for generating stochastic knowledge resulting in sample scenarios and an optimization algorithm for solving these sampled problem instances (cf. Bent and van Hentenryck 2007). According to Bent et al. (2010), the idea of using sampled requests comes from the key observation that in many practical dynamic applications, the arising uncertainty does not depend on the *decisions* taken since request arrivals are independent of generated tour plans and vehicle activities. Hence, the authors state that in these applications it is not necessary to use more general models that explore sequences of decisions or scenario trees. Therefore, given distributions can be sampled independent of the decisions taken so that expected future scenarios can be provided. As a consequence, the proposed framework provides significant computational advantages compared to the previously mentioned general models. The proposed framework is depicted in a basic form in Algorithm 4.1.

The proposed framework can be used for different online optimization problems, but may require problem-specific adaptations. For example, in dynamic routing problems with hard time windows, it is necessary to check if a new arriving request can be feasibly integrated into the tour plan. The dynamic problem is considered as a discrete-time model over a given time horizon $H = [\underline{H}, \overline{H}]$. In what follows,

Algorithm 4.1 Basic online optimization (H) (adapted from Bent and van Hentenryck 2004c)

```
1:  R ← ∅
2:  w ← 0
3:  t ← H̲
4:  while t ≤ H̄ do
5:      R ← AVAILABLEREQUESTS(R, t) ∪ NEWREQUESTS(t)
6:      r ← CHOOSEREQUEST(R, t)
7:      t ← SERVICEREQUEST(r, t)
8:      w ← w + w(r)
9:      R ← R \ {r}
10: end while
```

this basic model is considered for a dynamic routing problem in which no vehicle en-route diversion is allowed. Note that this basic formulation only considers routing problems in which a single vehicle is present since the described framework can only make a new decision after the previous one has been completed. The tour plan is iteratively generated. Specifically, when the vehicle has finished servicing its currently assigned request at $t \in H$, for each pending request $r \in R$ a weight $w(r)$ that represents the gain according to the given objective function, if request r is chosen for service at time t, is assigned. Note that this weight may change over time. At the beginning of the process, t is set to the beginning of the considered time horizon denoted by $\underline{H}$. While the time horizon has not ended, new requests that are available up to the current time are added to the set of pending requests R. As mentioned above, one request from this request set is chosen to be serviced next. After this selection, the request is serviced (line 7) and the time t is advanced to the point in time when service of request r is completed. Furthermore, the realized gain $w(r)$ is added to the objective function value of the solution denoted by w and the request is removed from the set R. In Bent et al. (2010), the authors make several extensions to this basic framework, among others, for multiple decisions in order to extend the problem to the multiple-vehicle case.

Regarding the basic strategies in DVRPs for providing increased flexibility as presented in Sect. 4.3, it can be observed that the described framework considers choosing the next request as the only type of decision so that the definition of the function CHOOSEREQUEST(R,t) is the essential task for applying this framework. An implementation of this function may or may not use stochastic knowledge. The authors mention two implementations of this function that do not use stochastic knowledge. Specifically, CHOOSEREQUEST-G is a greedy heuristic that simply selects the request that possesses the highest objective function gain out of the pending requests. In another implementation called CHOOSEREQUEST-LO, it is assumed that a function called OPTIMALSOLUTION exists which can compute an optimal solution for the set of currently pending requests. The first request of this solution is chosen for execution at time t.

In addition to the implementations CHOOSEREQUEST-G and CHOOSEREQUEST-LO that do not use stochastic knowledge, the authors propose different implementations of the function CHOOSEREQUEST which make use of stochastic knowledge. All of these implementations use a concept denoted as Multiple Scenario Approach (MSA) which is based on the MPA (see Sect. 4.4). In MSA-based approaches, pro-active strategies are implemented by sampling a given probability distribution representing expected future request arrivals. By using this given stochastic knowledge, expected future requests called *sampled customers* are created (cf. Bent and van Hentenryck 2007). By sampling the distribution n times, n different realizations of an expected future scenario consisting of currently known and sampled customer requests are generated. These realizations are then used in the decision process. MSA-based implementations are similar to a sampling-based approach for online packet scheduling called *expectation* which was proposed by Chang et al. (2000). In this approach, for each sample scenario $s \in S$ the gain that is expected if request $r \in R$ is serviced next in scenario s is calculated by assuming that r is serviced next and then using the solution method to generate a solution for requests $R\backslash\{r\}$. For each request r, the sum of the expected gains of all scenarios is calculated and the request with the highest gain is chosen. The advantage is that this is likely to lead to a robust solution since different possible future outcomes are directly considered in the decision process. A drawback of the expectation approach is the time complexity since the solution method is required to be performed $|S| \cdot |R|$ times which can raise bottlenecks under time constraints which exist in real-time control approaches. In order to reduce the required computational effort, decisions in MSA-based implementations about which request is to be serviced next when a vehicle has finished servicing its current request are performed similar to the MPA which was described in Sect. 4.4. Specifically, while in the MPA a solution is constructed out of a set of stored solutions using a specific ranking function, in the MSA a set of sample scenarios S is created and a solution is generated for each of them. By utilizing a specific ranking function, the next request is selected using these $|S|$ generated solutions. The advantage is that only $|S|$ scenarios, instead of $|S| \cdot |R|$ scenarios, need to be solved which can be realized more easily in time-restricted real-time systems. Multiple-vehicle routing problems are handled in MSAs by storing the decisions about the next request for each vehicle separately. In the first MSA-based implementations in Bent and van Hentenryck (2004b), the two ranking functions which have already been shown to be efficient in the MPA are used for selecting the next request. Specifically, the tested ranking functions distance and consensus, result in the MSA variants MSA^{d} and by MSA^{c}. In both MPA variants, the sampled customers are removed after generating the solutions for the sample scenarios providing flexibility for integrating future requests. Moreover, a real request is always selected by the ranking functions. Computational results show that each of the two tested ranking functions in the MSA significantly outperforms the corresponding MPA variant. Furthermore, the authors recommend the utilization of MSA^{c} instead of MSA^{d} since, especially for high degrees of dynamism as well as late arriving dynamic customers, MSA^{c} significantly reduces the number of unserviced customers compared to MSA^{d}.

In the working paper of Bent and van Hentenryck (2003), the authors evaluate the previously described MPA and MSA-based approaches on the problem setting considered by Larsen et al. (2002) (see Sect. 4.5.1). Computational results show that in the single-vehicle case, the utilization of the stochastic knowledge in the MSA can improve the solution quality compared to the original NN heuristic by Larsen when applied in scenarios with a larger delivery area. In the same paper, the authors extend the MSA^c approach to the MSA^c with *least commitment* (MSA-LC^c) that makes additional use of a waiting strategy. Specifically, if a sampled customer is chosen as the next requests in a vehicle's tour, the corresponding vehicle will wait at its current position until the sampled customer materializes or is removed from the first position in the tour, i.e., until a real request is assigned to this vehicle as its next request. Computational results are carried out in a multiple-vehicle setting with capacity constraints and hard time windows. The authors show that the MSA-LC^c implementation is able to significantly reduce the total travel distance since it postpones commitments to requests when future requests are expected to arrive. In contrast, when the minimization of unserviced customers is pursued, the best results are achieved by using the NN heuristic proposed by Larsen et al. (2002) as well as an MSA approach variant in which the NN heuristic is used for solving the individual sample scenarios.

In the work provided by Bent and van Hentenryck (2004c), the authors continue to study characteristics of the consensus ranking function. The authors mention that the consensus ranking function's main limitation is its elitism, i.e., only the best request is considered in a sample scenario while other suitable requests which may also provide robust solutions are neglected. Hence, the authors propose a modified version called quantitative consensus which calculates the consensus value by the sum of objective function values of the solved sample scenarios in which this request is scheduled next. Since this version of the consensus function does not reduce but rather even increases the observed elitism, the authors present two new ranking functions. The first one is a combination of the consensus and the expectation ranking function. In this function, first the consensus is applied in order to identify a set of promising requests. Afterwards, the expectation ranking function is applied to this set in order to achieve a more precise ranking among them. However, since the computational effort generated by this ranking function is still considerably high when compared to the consensus ranking function, the authors propose a second ranking function which is called *regret*. By quickly estimating negative impacts on other requests when a specific request is scheduled as the next request in a vehicle's tour, this ranking function provides additional ordering information among requests without having to solve additional scenarios. Details on this ranking function can also be found in Bent et al. (2005) and Bent and van Hentenryck (2007). Computational results show that the regret ranking function performs well compared to other evaluated ranking functions, especially in scenarios with a high degree of dynamism and tight time window restrictions.

In Bent and van Hentenryck (2007), the authors propose two extended MSA variants that make additional use of a waiting and a relocation strategy. Using the consensus or regret ranking function, the next request of a vehicle is identified while assigned sampled customers as next requests are regarded as a decision for the waiting action. If the waiting action is chosen by the ranking function, the vehicle waits at its current position. The relocation strategy is implemented by moving a vehicle to the next identified request by the ranking function regardless whether this request is a real or a sampled customer. In doing so, if the next assigned request is a sampled customer, the vehicle is relocated to sampled customer's position. Computational results on the consensus and regret ranking function reveal that the waiting and relocation strategy can significantly improve the attainable solution quality. Especially for problem instances that are highly dynamic and have many late customer requests, the regret ranking function with the relocation strategy achieves the highest improvements compared to the earlier MSA variants that do not make use of the waiting or relocating strategies. The authors conclude that both new presented strategies achieve the best performance for different objectives. Specifically, while the waiting strategy is beneficial when the minimization of total travel time is the main objective, the relocation strategy performs best if the primary aim is the minimization of unserviced customers within given hard time windows.

In all of the approaches described above, the existence and availability of a probability distribution that represents future request arrivals is assumed. In Bent and van Hentenryck (2005), the authors propose techniques to learn distributions either online from observed events during the transportation process or before the transportation process from available historical data. In order to deal with dynamic routing problems in which request arrival rates change during the planning horizon, the authors propose historical sampling. Specifically, request arrivals from the historical data are utilized as sampled customers in MSA-based approaches. First computational results using the MSA-LC approach are performed on historical sample sizes of 10 and 1,000 samples utilizing the consensus and regret ranking functions. The number of unserviced customers is compared to the instances in which the original distribution is available for sampling. From the obtained results, the authors conclude that historical sampling is an efficient approach for providing knowledge from historical data. For future research, they propose to analyze impacts of the quality of historical data on the solution quality as well as develop filtering techniques for identifying relevant parts of the historical data.

In the Master's thesis of Markus (2009), the author bases his research on the described findings of Bent and Van Hentenryck. The author proposes discretization of the considered service area so that sampled customers in the same subregion are regarded as the same for calculations performed by the ranking functions. This allows customer requests to be arbitrarily located within the considered service area. Furthermore, in the regret ranking function the author discusses which requests should be included in the calculations. For solving the sample scenarios, the author proposes a solution method based on an attribute-based hill climber metaheuristic. Computational experiments show that calculating the regret ranking function for the requests on the same tour only and including sampled requests in these calculations performs best. Furthermore, the author mentions different ways of performing the service activity at a sampled customer. For example, a sampled customer can be treated like a real request by traveling to the sampled customer, waiting for the beginning of its time window, and also simulating the service activity. Another possibility is to simply move to the sampled customer and directly drive to the next customer afterwards. As an outlook on future research, the author suggests several improvements for the algorithm and the model, e.g., considering additional sources of uncertainty.

Besides research work based related to OSCO, there are other approaches that use sampling-related techniques to utilize stochastic knowledge. Hvattum et al. (2006) consider a dynamic routing problem that is based on a real-world application in one of the leading distribution companies in Norway. The problem is modeled as a single-depot DVRP with Time Windows where a fleet of heterogeneous capacitated vehicles need to pick up goods at requests that arrive dynamically during the day. Each request has a specific demand and a hard time window. Uncertainty arises from dynamically arriving requests as well as from time windows and demand not known with certainty. In the approach, probability distributions for the mentioned sources of uncertainty that can be used as stochastic knowledge are assumed to be available and vehicle en-route diversion is not allowed. The considered hierarchical objective function consists of the topmost aim of minimizing the number of utilized vehicles and second the minimization of total travel distance. In order to model the problem, a two-stage stochastic formulation is presented that is similar to SVRP formulations. This two-stage model is then extended to a multistage stochastic model. The time horizon of the considered transportation process is divided into a predetermined number of intervals and in each interval, a new stage is generated in which the new requests that have arrived during the previous interval are integrated. Hence, new arriving requests are queued until the beginning of the next interval. The problem is solved by a proposed Dynamic Scenario Hedging Heuristic (DSHH) which gradually constructs the solution during the execution of the transportation process. Specifically, in each interval, the DSHH first generates a set of sample scenarios using the existing distributions. Each of these sample scenarios includes the currently known real requests and a set of expected future requests. In a second step, the sample scenarios are solved by a least cost insertion heuristic. After the tour plans for all sample scenarios have been generated, requests that are frequently scheduled in the generated tour plans within the upcoming time interval are identified and selected. This selection stops when one of the following two conditions is fulfilled: (a) all requests that have not yet been selected are only infrequently considered in the next time interval, i.e., below a predefined percentage in the solutions of the sample scenarios, or (b) a specific number of requests depending on the minimum number of vehicles used in the sample scenarios has been found. Afterwards, the sample scenarios that do not consider the selected requests in the next interval are solved again while enforcing the integration of these requests. At this time, the requests that are considered in the next time interval have been chosen. In the next step, the tour plan generation which determines the assignment and sequencing of the chosen requests to the vehicles is performed. For this purpose, a concept that is related to the previously described consensus ranking function is used. In the generated solutions of the sample scenarios, it is counted how often each request is assigned first to a specific vehicle and the most frequently identified request-vehicle pair is added to the generated tour plan. This selection process is continued until all of the chosen requests have been scheduled in the generated tour plan. In computational experiments, the performance of the proposed DSHH is compared to a myopic dynamic heuristic (MDH) that does not use stochastic knowledge and generates the tour plan using a local search heuristic in each interval. For reference purposes, an offline version of the problem that has access to all information before the execution of the transportation process is also considered. In this offline setting, the solution is solved using the same local search heuristic as in the MDH approach. In comparison to the MDH, the authors show that the DSHH is able to reduce travel distances up to 15 % while using a comparable number of vehicles. If more computational time is available by choosing longer intervals between solutions, the utilization of more samples can yield a better solution quality. The authors conclude that in a scenario with loose time windows (ten hours) using an interval length of two hours produces the best results in terms of travel distance. In contrast, in case of tighter time windows with a length of three to five hours, the DSHH produces longer travel distances. The authors presume this to happen due to the more elaborated local search utilized in MDH which can take advantage of finding good solutions in more constrained problems compared to the insertion heuristic used in DSHH. For future work, the authors suggest giving a detailed evaluation about why the DSHH tends to sometimes use an extra vehicle compared to the MDH.

In the following research work by Hvattum et al. (2007), the authors consider the same routing problem as the one in the work presented before. They address the previously mentioned issue of additionally used vehicles in the DSHH. They explain that choosing the request to be serviced next according to the decision in the majority of the sample scenario solutions may require the use of additional vehicles when a request which was only considered in the minority of the sample scenarios arrives. In order to deal with this issue, they extend the DSHH to a Branch-and-Regret Heuristic (BRH). Similar to the DSHH, samples in this approach are generated from given probability distributions. In comparison to the DSHH, a difference is that decisions produced by consulting the sample scenarios in the BRH are only implemented if it is unlikely that they will lead to suboptimal effects. The proposed BRH works as follows. After producing a solution for each sample scenario, the heuristic iteratively merges the individual sample scenario solutions by evaluating certain criteria on which to branch, for example if a request should be serviced during the next interval. This concept is similar to stochastic B&B (cf., e.g., Klein Haneveld and van der Vlerk 1999) except for the fact that only one path in the branching tree is considered by the BRH. DSHH and BRH differ mainly in two points. First, for each request, the DSHH decides whether the request will be serviced in the next interval based on a percentage threshold. In the BRH this type of decision is made by solving all sample scenarios twice; once with the request and once without the request. If the best decision is to service the request in the next interval, it is added to the set of chosen requests. Hence, in comparison to the DSHH, the threshold parameter is not required. According to the authors, this is advantageous because the value of this parameter can be significantly problem-dependent. Second, in order to choose the next request i for a vehicle k, the assignment proposed in the DSHH is evaluated in comparison to *alternative* assignments. The authors evaluate three different variants: (a) same request i on a different vehicle, (b) same vehicle k with another request, and (c) the setting of the previous two variants which can be found more frequently in the sample scenario solutions. In the computational experiments, the authors show that variant (c) performs best. Since each branching decision is evaluated against an alternative, more information about the regret of accepting or discarding a certain decision is provided which leads to a more efficient utilization of the vehicles which was initially pursued. This results in a lower number of required vehicles and thus a better solution quality. In further computational results, the BRH yields a vehicle utilization that is significantly better than the one achieved by the DSHH and almost comparable to the results achieved in the offline case. With regard to travel distance, the application of the BRH results in higher values than the DSHH, but still lower than the ones achieved by MDH. Furthermore, the authors evaluate the BRH strategy on the problem setting considered by Bent and Van Hentenryck which pursues a different objective function. In their comparison, BRH performs significantly better than DSHH but slightly worse than MSA^c. The authors mention that performance differences may result due to the fact that MSA^c directly integrates new arriving requests whereas BRH only considers new requests when a new interval starts so that time-critical decisions may be taken too late when time windows are tight. In addition, the authors extend their considered problem by allowing stochastic demands for known requests that only become known when the request is visited by the vehicle. This may lead to situations in which a request cannot be serviced due to a vehicle's capacity constraints. The BRH is extended by allowing requests to have different demands in the individual sample scenarios and demands are assumed to be uniformly distributed over a given interval. Computational results in which this BRH variant is compared to modified versions of the MDH approach which estimate request demands according to their minimum, average, and maximum possible value show that the BRH is able to improve the solution quality also in this more complex problem setting.

In the work of Ghiani et al. (2009), the authors consider a DPDP that is inspired by a real-world application in which an urban same-day courier service provider is considered. Requests arrive dynamically during the day and the objective consists of maximizing the customer service level by minimizing a non-decreasing convex customer cost function. This may be an inconvenience function, i.e., costs start directly after arrival of the request, or a penalty function in which costs arise when the request is not fulfilled within a given maximum time. In the problem it is assumed that a prescribed probability distribution that represents the request arrivals as well as their characteristics, e.g., origin and destination, is available. For this purpose, the considered service area is divided into 25×25 zones and the request arrival rates for all source-destination combinations of these zones are available as stochastic knowledge. Moreover, request arrival rates remain constant during the considered planning horizon. In order to increase flexibility for integrating dynamically arriving requests, a waiting strategy is implemented. For this purpose, the decision space is extended by decisions that define waiting times at each request location. In order to decide the length of waiting times at the request locations, values that increase in a given step size of, e.g., 5 minutes are evaluated. Furthermore, a limit on the maximum allowed total waiting time of each tour exists. Besides the waiting strategy, a relocation strategy for moving idle vehicles to waiting positions is applied. In the approach, the possible waiting positions are predefined as the stochastic medians of the service area. In order to solve the problem, two anticipatory heuristics are proposed that use the available stochastic knowledge about future demands by sampling the given request arrival distribution. In the first proposed approach, new arriving requests are integrated into the tour plan by an insertion heuristic that stores the best solutions for further examination. In this examination, different values for the waiting times at each request location as well as the waiting positions are evaluated resulting in a set of potentially good solutions. The evaluation of each good solution is carried out by individually generating a number of sample scenarios consisting of expected requests in the near future (denoted as sample horizon) for each good solution. The sampled requests are then inserted by a least cost insertion heuristic. Afterwards, the objective function value for each good solution is calculated as the average of the costs of all corresponding sample scenarios. The number of sample scenarios to be used for each good solution is determined using a simulation-based optimization method called fully sequential indifference zone selection. Starting with a fixed number of sample scenarios (10 in the experiments), confidence intervals of the objective function values of all good solutions are computed. Inferior solutions are directly discarded while additional sample scenarios are generated and solved in order to allow a precise discrimination between the remaining solutions. This process is continued until one solution remains in the set of good solutions. In the second proposed approach, a local search heuristic is applied which removes a request from a tour and reinserts it in another vehicle's tour. Expected changes in the objective function are calculated using the sampling method described above. In the computational experiments, two different types of scenarios are considered. The first scenario represents a standard courier service while the second one represents a situation with premium customers. In both scenarios, the objective is modeled by a linear lateness penalty function. While in the standard scenario this penalty begins 45 minutes after request arrival, in the premium scenario this penalty already starts after 20 minutes. Furthermore, in the standard scenario an arbitrary number of requests can be consolidated on a vehicle while the premium scenario allows only the transportation of one request per vehicle at a time. In order to obtain a similar vehicle utilization, the following instances were tested: 100 requests with 18 vehicles, 200 requests with 36 vehicles, and 300 customer requests with 54 vehicles. The anticipatory approaches are compared to a deterministic approach that does not use stochastic knowledge. In the computational results, the authors show that using a sample horizon with a length similar to the expected request response time for future requests achieves the best results. Compared to the deterministic approach, the anticipatory insertion heuristic achieves significant improvements of up to 69 % while reducing the vehicle utilization by 19 % from originally 60–70 % at the same time. The computational results also show that in the considered problem setting, the use of the more sophisticated anticipatory local search heuristic compared to the anticipatory insertion heuristic does not yield any significant additional improvements. For future research, the authors propose to evaluate the performance of the approach when stochastic knowledge is not given by prescribed probabilities but contains uncertainty.

In Ghiani et al. (2012), the authors compare two different methods for using stochastic knowledge in a dynamic routing application. These methods are inspired by a sample scenario based approach presented by Bent and van Hentenryck (2004b) as well as the approach utilizing anticipatory insertion proposed by Thomas (2007). The authors consider the single uncapacitated vehicle case in which the vehicle travels from a given source to a given destination while servicing a set of a priori known and a set of potential dynamically arriving requests. Requests have no time windows but the destination needs to be reached by a given time limit. Furthermore, all of the a priori known requests have to be serviced while dynamic requests may be rejected. The objective consists of maximizing the number of additionally serviced dynamically arriving customer requests while taking into account a latest time limit at the destination. For both approaches, it is assumed that a distribution is given for the arrival probability for each potential dynamic customer request. In the computational experiments, 192 instances with a total of 40 and 50 customers comprising a degree of dynamism of 25 % and 50 % are evaluated. With regard to potential dynamically arriving requests, different settings considering the arrival probabilities of these requests are evaluated. According to the computational results, the authors state that in case of a low degree of dynamism the anticipatory insertion performs better than the sample scenario approach whereas in case of a higher degree of dynamism the sample scenario approach performs better than the anticipatory algorithm but requiring larger computational times. The authors mention several aspects that could be tested in the future in order to assess the performance of both approaches under different settings, e.g., in the multiple-vehicle case or in presence of time windows.

Having described approaches for dynamic routing applications that provide an increased flexibility for dynamically arriving requests, in what follows, additional approaches covering various other characteristics that are important in the considered RDOPG applications are presented.

4.6 Related Objective Functions

As mentioned in Sect. 3.1, the objective pursued in the considered RDOPG applications consists of minimizing customer inconvenience in order to maximize quality of customer service. When dealing with such quality aspects, the objective function often contains some type of *penalty function* that makes solutions with poor service quality less attractive. As mentioned in Sects. 2.4.12.5 and 2.4.12.7, such a penalty function can contain fixed and variable components. Fixed lateness penalties can also be used to model DVRPs with hard time window constraints where customer requests are considered to be lost when they are not serviced within given time windows (cf. Bent and van Hentenryck 2004b). While fixed lateness penalties are completely determined by their parameter value, the definition of variable penalty costs also includes a definition regarding the *evolution* of the lateness-dependent penalty. In the marketing literature, two main approaches for modeling customer inconvenience depending on elapsed request response times can be found. While,

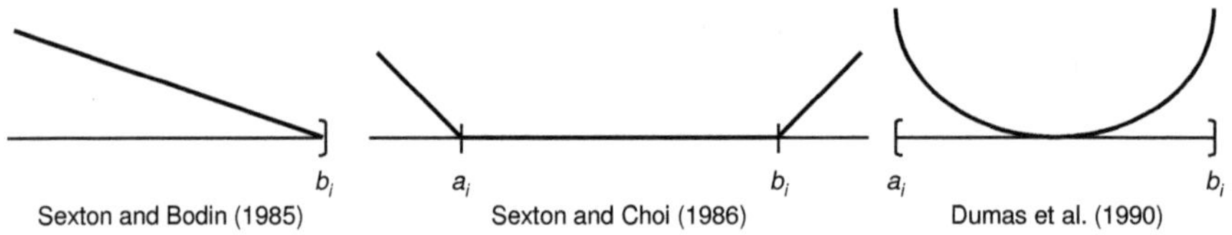

Fig. 4.4 Different inconvenience and penalty functions (cf. Desrosiers et al. 1995)

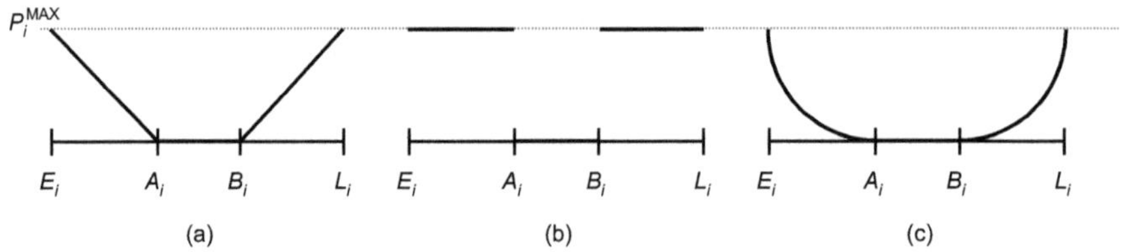

Fig. 4.5 Customer inconvenience functions considered by Fagerholt (2000, 2001)

among others, Davis and Maggard (1990) propose that customer inconvenience increases linearly with resulting response time, other authors such as Kristensen et al. (1992) derive a quadratic dependency.

In our definition, a customer quality-related objective function is called an *inconvenience function* if penalties directly occur with the arrival of a request in the system while in a *lateness function*, penalties only occur when service has not occurred within specified time windows. Moreover, depending on the application, quality-related objectives can also be used to penalize the unwanted case in which a request is serviced too early, i.e., before the start of its time window. In what follows, routing problem-related approaches known in the literature that consider inconvenience and lateness functions are presented.

In the routing literature of approaches for static routing problems, different types of inconvenience and penalty functions can be found. For example in DARPs, considered service quality consists of time differences experienced by customers that demand to be picked up at a specified time point (cf. Cordeau et al. 2007a). Hence, passengers should neither be picked up too early nor too late. Furthermore, the length of the ride time experienced by passengers is another quality criteria. Desrosiers et al. (1995) mention three different types of penalty functions (see Fig. 4.4). The first one described by Sexton and Bodin (1985) denotes a linear decreasing inconvenience function with a hard time window at time point b_i. The second function describes a lateness function proposed by Sexton and Choi (1986). Here, too early or too late service, i.e., service before a_i or after b_i, is allowed but linearly penalized. In the third function, an inconvenience function by Dumas et al. (1990) is presented that quadratically penalizes dispersion from a desired time point of service while also limiting this dispersion in both temporal directions by hard time windows.

Besides vehicle-based routing problems, aspects of the service quality are also considered in other routing applications, for example multi-ship PDPs (cf. Fagerholt

2000, 2001). In this static problem, outer and inner time windows exist. Specifically, the outer time window of request i defined by $[E_i, L_i]$ can be seen as a hard time window before or after which no service is possible and, instead, high penalty costs denoted by P_i^{MAX} occur. As illustrated in Fig. 4.5, the inner time window $[A_i, B_i]$ represents a soft time window so that service that occurs too early or too late results in an either (a) linear, (b) fixed, or (c) quadratic penalty.

In dynamic routing problems in which new requests arrive during the execution of the transportation process, a linear customer inconvenience function is used among others in the described approaches of Psaraftis (1980), Bertsimas and van Ryzin (1991, 1993a, 1993b), Papastavrou (1996), in the approaches of Gendreau et al. (1999a), Swihart and Papastavrou (1999), Richter (2005), Attanasio et al. (2007), Schmid (2012), and in van de Klundert and Wormer (2010). Moreover, a linear lateness function is used in the work of Larsen (2000), Yang et al. (2004), Ichoua et al. (2006), Branchini et al. (2009), and Ghiani et al. (2009). In the work of Krumke et al., the authors also consider lateness functions in a DTRP encountered by the German Automobile Club ADAC. For example, in Krumke et al. (2002), a linear lateness function is considered whereas in other publications, e.g., in Westphal and Krumke (2008), a quadratic lateness function is used in order to penalize longer lateness more. The authors formulate the problem as a set partitioning model that is solved by a column generation approach. Although the considered problem is strongly NP-hard, the authors show that due to a low request per vehicle ratio in tested real-world instances, limited tour lengths of vehicles are often present in optimal solutions of the considered instances. Hence, the evaluated real-world problem instances can be solved exactly or with a very small gap compared to a proposed lower bound within a couple of minutes. Moreover, the authors report that for all tested instances the optimal objective function value is at most 0.494 % above the lower bound so that it can be concluded that the proposed lower bound is quite tight and hence of high quality. Test instances of 106 requests and 84 vehicles up to 775 requests and 211 vehicles are tested and the performance of the proposed exact solution method is evaluated against different variants of a GA provided by an industrial partner. Since, as mentioned above, the computational time required to solve the problem instances which occur during the transportation process is very important in real-time environments, the authors test the achieved solution quality after computational times of 5, 15, 30, 60, and 120 seconds. The computational results show that the proposed exact solution method is able to perform significantly better than the GA in terms of solution quality but also with regard to required computational times. For all of the tested instances, the exact method provides solutions with an objective function value of at most 1 % above the optimal objective function value or also often directly an optimal solution within 5 seconds. In the bigger test instances, the evaluated GAs still have a gap of more than 5 % after 600 seconds computational time. In Westphal and Krumke (2008), the authors improve the approach for the previously presented real-world problem by proposing a more sophisticated pruning scheme for the search tree used in the column generation approach. Computational experiments show the effectiveness of the new applied scheme. Specifically, compared to the old pruning method, the new pruning approach requires only 38.2 % of the computational time and explores only 24 % of the nodes in the search tree.

Besides customer quality of service criteria, driver inconvenience aspects are considered in the objective function in the work of Savelsbergh (1992) and Ascheuer et al. (2000) by minimizing the completion time of vehicle tours (see also Sect. 2.4.12.4).

4.7 Other Relevant Factors for RDOPG Applications

In this section, we deal with other relevant factors that are existent in the considered RDOPG applications in this book. First, approaches that evaluate the impact of allowing *en-route diversion of vehicles* are presented. This is followed by a description of approaches which utilize a *road network* to map vehicle movements more realistic. In the next section, approaches which describe *simulator* concepts which can be used to simulate dynamic transportation processes are illustrated. Furthermore, research work describing *real-time control* approaches are introduced. Since the RDOPG applications are carried out in an urban area, *city logistics-related decision support systems and communication technologies* are also presented.

4.7.1 En-Route Diversion of Vehicles

As mentioned in Sect. 2.6.3, the utilization of technical equipment makes it possible to locate vehicles and to track their movements in real-time. Hence, when a new request arrives close to the route that a vehicle is currently traveling to its next assigned request, the application of vehicle en-route diversion can redirect the vehicle to service this new request first. As a result, the overall service quality is increased if an objective considering the minimization of customer inconvenience is pursued. Furthermore, allowing vehicle en-route diversion can also be advantageous when the objective of minimizing total travel distance is considered since it allows for the reduction of required detours. In Regan et al. (1995, 1998), different vehicle en-route diversion strategies are investigated and positive effects on system performance in truckload routing applications are shown. Besides, vehicle en-route diversion is only considered in a few approaches, e.g., in Ichoua et al. (2000), Richter (2005), Attanasio et al. (2007), Branchini et al. (2009), and van de Klundert and Wormer (2010). In Ichoua et al. (2000), a long-haul courier service routing problem is considered. The objective consists of minimizing the weighted sum of travel distance and customer lateness. In order to adapt the tour plans, the Parallel Tabu Search metaheuristic of Gendreau et al. (1999a) is used and adapted. Compared to the original approach in which no vehicle en-route diversion is allowed, computational experiments show that the solution quality can be significantly improved by using vehicle en-route diversion. Moreover, as mentioned in Sect. 4.2, Richter (2005) concludes that allowing vehicle en-route diversion can significantly reduce customer inconvenience and therefore increase quality of service.

4.7.2 Road Network

As can be concluded from the literature review, only a few of the described proactive real-time control approaches make use of a road network to map vehicle movements and to calculate travel distances between relocations, such as Bell et al. (1983), Fleischmann et al. (2004b), Tarantilis et al. (2004), Thomas and White III (2004), and Schmid (2012). The utilization of a road network is important for determining accurate travel times from current vehicle positions to request locations by taking the existing road infrastructure into account. The utilization of a real road network in practical applications is especially useful when vehicle en-route diversion is allowed. When considering transportation problems in real road networks that cover a big service area, improvements with regard to the required computational time compared to the Dijkstra algorithm can be found among others in Bauer et al. (2008) and Schultes (2008).

4.7.3 Simulator

In order to empirically assess the performance of dynamic real-time control approaches, mathematical models are used to represent the system, i.e., the considered transportation process, in terms of logical and quantitative relationships. In general, "a system is defined to be a collection of entities, e.g., people or machines, that act together toward the accomplishment of some logical end" (Schmidt and Taylor 1970). Since a system changes over time, its current *state* is defined to be a "collection of variables necessary to describe a system at a particular time, relative to the objectives of a study" (Law and Kelton 2000, p. 3). Simulation-based methods are often used if mathematical models are too complex to be studied analytically. Using simulation allows for "numerically exercising the model for the inputs in question to see how they affect the output measures of performance" (Law and Kelton 2000, p. 5).

Basically, for modeling the advance of the simulation time in the considered system, two types of discrete simulators exist: (a) simulators in which the simulation time is incremented by a fixed amount of time and (b) simulators in which the simulation time is directly advanced to the time point of the next relevant event (cf. Larson and Odoni 1981, p. 590). In approaches of the latter type, the "[simulation] clock is advanced only to those instants in time when certain important (for the simulated system) events take place. Such events might be the arrival or departure of customers in a queueing system, or the occurrence of a fire alarm in the simulation of a fire department" (Larson and Odoni 1981, p. 590). According to the authors, this is also known as *advancing the clock simulation*. The used term *discrete* represents the fact that the state variables of the simulated system only change their values at specific discrete time points. These time points are determined by *events* that do not consume any time (cf. Biethahn et al. 1999, p. 52, Law and Kelton 2000,

p. 6). Hence, by using events, all state changes occurring in the considered dynamic system are covered.

In dynamic routing problems, applying a discrete event-based simulator is often appropriate since the relevant system state changes whenever an event occurs. The advantage of event-based simulators is that no decision has to be taken about *time precision* as is the case in simulators which increase the simulation time in fixed time intervals. Hence, in event-based simulators, the simulation time is always directly set to the point in time of the next event while time points at which no relevant event occurs are skipped so that less computational time is required for simulating the system itself (cf. Larsen 2000, p. 111).

A discrete event-based simulator needs to distinguish between *endogenous* and *exogenous* events during the simulation (cf. Mihram 1972, p. 219 et seq.). The first type of events are generated by the system, e.g., when a vehicle has finished servicing its current request and requires further orders. These events may trigger other activities to be performed. For example, in approaches which iteratively generate the tour plan by calculating the next request for a vehicle when this is demanded, an activity is required at the occurrence of such an event. The second type of events is represented by exogenous events which come from outside into the system, such as the arrival of a new request which are required to be integrated into the transportation process by an appropriate adaptation activity. Furthermore, when vehicle en-route diversion is allowed and a real road network is used, the simulator framework is more complex since vehicle movements need to be modeled explicitly in order to enable a determination of their current position at any point in time. More information about simulation environments can be found among others in Regan et al. (1998), Larsen (2000), Ichoua et al. (2000), Gendreau et al. (2006), and Barceló et al. (2007). More information about the discrete event-based simulator which is developed to simulate the considered RDOPG applications in this book will be presented in Sect. 8.1.

4.7.4 Real-Time Control

As mentioned in Sect. 2.5, in contrast to static routing problems that only consider the initial tour plan generation before the execution of the transportation process, dynamic routing problems are required to be controlled in real-time in order to monitor the execution of the transportation process and to adapt the tour plan when necessary. Giaglis et al. (2004) mention that, once a tour plan is generated, classic planning systems for static VRPs do not take into account the possibility of changes in the relevant information that may require an adaptation of the constructed tour plan. Hence, the only option for these classic systems is to discard the current tour plan in execution and generate a completely new one. This procedure is expected to raise problems in real-world applications since vehicles may get completely different tours assigned to them than were assigned before. Furthermore, it may not be possible to consider information about current vehicle positions and statuses as well as already loaded goods in such systems. Therefore, the application of real-time control approaches that are able adapt the current tour plan in order to integrate changes in the relevant information caused by dynamic events is advantageous.

An important aspect in real-world routing applications is that while tour plans are adapted according to changes in the relevant information, the transportation process continues during this adaptation. Since most adaptation methods require an amount of time that cannot be considered as zero time[1] anymore, one important aspect of real-time control approaches is that they need to take the concurrency of tour plan execution and tour plan adaptation into account. Hence, since utilized adaptation methods usually consider individual problem instances that do not change during the adaptation, it has to be ensured that decisions which are taken during the adaptation can still be made after the adaptation has been completed. This is, for example, accounted for in Ichoua et al. (2000) which utilizes the Parallel Tabu Search metaheuristic by Gendreau et al. (1999a). In the proposed real-time control approach the current system state at time point t is anticipated to its state in δt time units. Incompatible tour plans in the adaptive memory are then removed and the remaining time is used for the improvement method. The authors state that the choice of the length of δt can be of high importance. Specifically, while a large value of δt may miss adaptation possibilities, a too small value of δt may prohibit the improvement method from unfolding its full potential. The authors evaluate different strategies for setting δt, in particular for setting δt to a fixed value or for dynamically determining δt according to the system situation at t. Moreover, Hvattum et al. (2006) and van de Klundert and Wormer (2010) consider the research question about choosing the length for the time intervals in which tour plans should be adapted. As can be seen from the different results in these research work, this value is highly application-dependent. The utilized value in the real-time control approaches of this book will be described in Sect. 8.2.1.

Another real-time control approach that considers the concurrency of tour plan execution and tour plan adaptation can be found in Bock (2010). The author considers a real-time control concept that pre-simulates the system situation which exists at the beginning of each adaptation activity for a specified time period denoted as the anticipation horizon. After the pre-simulation, the remaining time of the anticipation horizon is used to adapt the tour plan based on the problem data of the anticipated system state. Since the real-time control approaches developed in this book are based on this real-time control concept, more information about it will be described in Chap. 5. Moreover, when uncertain travel times are considered, real-time control approaches can be extended by methods for estimating vehicle arrival times proposed by Cathey and Dailey (2003) and Jula et al. (2008).

In Zeimpekis and Giaglis (2005), the authors propose a general architecture of real-time control approaches applicable for DVRPs (see Fig. 4.6). It can be described as follows. In the monitoring state, the current system state which includes

[1]Zero time defines a time period during which the system state of the considered dynamic process can be assumed to remain constant. Note that the length of this time period is highly application-dependent—in the considered RDOPG applications its length is one second at most.

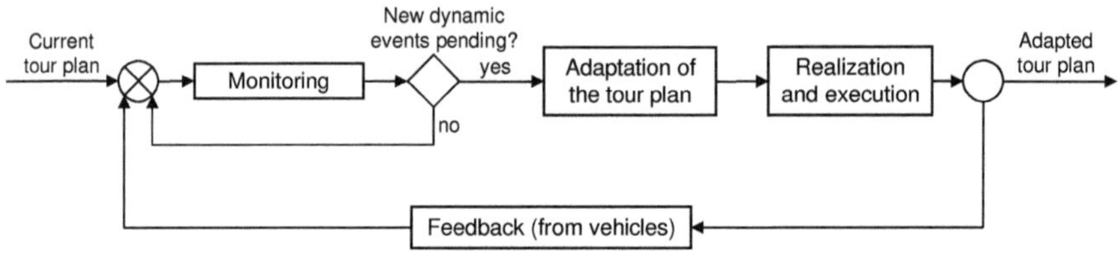

Fig. 4.6 General architecture of real-time control approaches (cf. Zeimpekis and Giaglis 2005)

the current vehicle positions and their status as well as the arrival of newly arrived dynamic events. When dynamic events that require an adaptation of the tour plan occur, the solution method that adapts the tour plan with regard to the occurred changes in the relevant information is invoked. In this step, also measures for considering the concurrency of tour plan execution and tour plan adaptation, e.g., the pre-simulation of the tour plan, are considered. After the execution of the solution method has been completed, the adapted tour plan is given into execution. Feedback can be provided from the vehicles after the changes in the tour plan have been communicated to them. Besides the acknowledgment of the transmitted changes, such a feedback can also consist of various other elements. For example, if the approach allows the rejection of requests, drivers may reject assigned newly requests. Furthermore, it may happen that vehicles cannot execute the adapted tour plan as planned, for example when vehicles have a different position than has been anticipated during the adaptation. In such cases, another invocation of the adaptation procedure can be triggered.

4.7.5 City Logistics-Related Decision Support Systems and Communication Technologies

Since the considered RDOPG applications can be considered as transportation problems in urban regions, research on city logistics reflects an important factor for increasing the practicability of the proposed real-time control approaches. Larson and Odoni (1981) provide detailed information about foundations of appropriate mathematical models as well as suitable scientific methods for handling city logistic problems. Furthermore, the use of technological advances in routing applications has been widely researched. This is advantageous in order to realize a real-time control system at a central dispatching center, the current location of the vehicles needs to be automatically and accurately updated and transmitted to the dispatching center in real-time as mentioned in Sect. 2.6.3. The use of a bidirectional communication system allows the automatic transmission of information between the vehicles and the dispatching center. Several authors have conducted research on technical systems that are appropriate for this purpose. Among others, in Giaglis et al. (2004) and Zeimpekis and Giaglis (2005), several information about available technical infrastructure, e.g., positioning systems such as GPS and data transfer systems such as

GSM and GPRS, are presented. Furthermore, a possible technical implementation of a real-time system is presented. The application of GPS technology in transportation problems is also investigated in Mintsis et al. (2004). In addition, the use of information about expected changes in travel times is advantageous. In order to obtain such information, floating car data can be processed (cf. Ehmke et al. 2009, 2010). Furthermore, traffic sensors which are commonly installed at highways and major roads allow for an automatic identification of traffic jams (cf. Ghiani et al. 2003 and Attanasio et al. 2007).

In Tarantilis et al. (2004), a decision support system for an open DVRP is proposed that incorporates a real road network from a geographical information system (GIS) and GPS-assisted vehicle location techniques in order to assist the dispatcher during the execution of the tour plan process. By using the implemented decision support system, dynamic events arising from urgent dynamic customer requests, request cancellations, or vehicle damages are automatically handled and the tour plan is dynamically adapted according to occurred dynamic events. For handling city logistics with vehicle delays and vehicle breakdowns, Zeimpekis et al. (2005) propose a dynamic fleet management system that monitors the location of vehicles in real-time and recommends re-routing activities.

In the work of Barceló et al. (2007), a framework for modeling city logistics applications is developed. This framework includes several data connectors in order to retrieve necessary data as well as transportation-specific components such as a simulator and a routing optimization module and a graphical user interface. The framework is successfully applied in case studies conducted in two Italian cities.

4.8 Conclusion of the Literature Review

According to the presented literature review, it can be observed that there are a large variety of pro-active approaches in the literature which integrate various strategies for providing an increased flexibility for the integration of dynamically arriving requests during the execution of the transportation process. As can be seen from the provided computational results, the presented approaches are able to efficiently utilize available stochastic knowledge so that the solution quality can be significantly improved compared to using deterministic real-time control approaches. However, with regard to applying the presented approaches in RDOPG applications, different issues arise that need to be additionally considered. First, the considered objective in the RDOPG applications dealing with the minimization of customer inconvenience has only been assessed in a few dynamic pro-active real-time routing approaches (see Sect. 4.6). As mentioned in Sect. 3.1, in the considered RDOPG applications, both a linear and a quadratic customer inconvenience are evaluated in this book in order to evaluate the impacts of different customer inconvenience functions on the attainable solution quality. Furthermore, additional prohibitive penalty costs occur if a defined maximum response time is not fulfilled which also prevents an indefinite deferment of customer requests (see aspect 8 of Psaraftis' approach presented in Sect. 2.5.2.2). Furthermore, it enables the modeling of routing problems in which the maximization of the number of dynamically arriving requests within their hard time windows is pursued. As can be seen in the described literature review, this is an often considered scenario. Second, since the RDOPG applications represent inner-city problems, the utilization of a real road network is advisable. Note that the utilization of a road network is only considered in a few approaches in the literature while many approaches assume Euclidean connections between the relevant locations. Moreover, as reported among others by Richter (2005) and van de Klundert and Wormer (2010), using vehicle en-route diversion is likely to allow a significant improvement with regard to solution quality when customer quality-related aspects are considered. Hence, vehicle en-route diversion should be implemented in routing applications in which a real road network is used and the attainable improvements should also be assessed. To the best of our knowledge, in the literature, this has only been performed in Ferrucci (2006) and Attanasio et al. (2007). Furthermore, allowing vehicle en-route diversion may increase driver inconvenience due to the fact that drivers are obliged to quickly react to occurring changes in their tour while they are traveling to their next assigned request. We are not aware that this issue has been dealt with yet in dynamic routing problems. In order to deal with this research question, we propose a new approach that additionally considers driver inconvenience aspects during the execution of the transportation process.

In pro-active approaches existing in the literature, another issue often arises when utilizing stochastic knowledge. Specifically, while developed pro-active real-time control approaches offer sophisticated methods for integrating available stochastic knowledge into the transportation process in order to efficiently adapt the tour plan according to expected future requests, in these approaches it is often assumed that available stochastic knowledge is given by prescribed probability distributions. Furthermore, it is often assumed that the dynamic requests exactly arrive according to these prescribed probability distributions. Note that these are strong assumptions which are often not present in real-world applications. However, since past request information is available in many real-world applications, the development of forecasting methods which are able to automatically extract stochastic knowledge from past request information is advisable. In the literature, only few approaches that present such forecasting methods can be found, e.g., in Bent and van Hentenryck (2005), Attanasio et al. (2007), and Schmid (2012) but these approaches deal with applications which have significant differences to the considered RDOPG applications. Moreover, Bent and van Hentenryck (2005) and Ghiani et al. (2009) suggest that future research should analyze the impact of the quality of the available past request information on the performance of pro-active real-time control approaches. This research question is addressed in this book by determining the *structural quality* of a request data set according to its *structural distortion* and *structural diversity* (see also Chap. 8). Specifically, it is advisable to assess the quality of available past request information with regard to characteristics of this information that are assumed in the forecasting methods and in the pro-active real-time control approach. For example, if request arrivals based on Poisson processes which are constant from day to day are assumed to exist, it is advisable to verify whether the available past request information is likely to fulfill this assumption. The forecasting method for generating stochastic knowledge from past request information proposed in this book in

Chap. 6 checks whether the assumed characteristics are likely to exist in the available past request information. For this purpose, the structural distortion of a request data set is determined. Computational results on the efficiency of the proposed approach for identifying structural distortion in given request data sets are provided in Sect. 8.6. Besides evaluating the structural distortion of a request data set, the structural diversity of a request data set is defined and its impact on the attainable solution quality is evaluated. For this purpose, request data sets comprising different levels of exploitable information for generating stochastic knowledge are designed (see Sect. 8.2.4) and two approaches for determining the structural diversity of given request data sets are proposed. The first approach is a qualitative assessment which makes use of a visual analysis (see Sect. 8.8). The second approach denoted as the degree of structural diversity (*dosd*) is a quantitative approach for determining the structural quality of given request data sets. It is proposed in Sect. 8.9 and its efficiency is shown by using the computational results which have been attained on the evaluated request data sets.

Two other aspects are connected with the generation of dynamically arriving requests that remain unclear in some pro-active approaches. The first aspect is about whether dynamically arriving requests appear at arbitrary locations in the considered service area or if they can only appear at few predefined locations which is similar to the modeling of uncertain customers in classic SVRPs (see Sect. 2.5.3.2.1). The second aspect concerns changes in the probability distributions of request arrivals, in particular changes in request arrival rates during the planing horizon. In order to classify and compare differences in the structural diversity of individual considered problem settings, it is advantageous to know whether the utilized distributions for generating dynamic events such as newly arriving requests remain constant or whether they change and, if so, in which dimensions, during the planning horizon of the considered transportation process. Note that this does not only apply to global changes of request arrival rates which occur during the considered planning horizon. In particular, it is especially important to take into account changes in the request arrival rates which occur in individual subregions of the considered service area since, as will be shown, these changes have a significant impact on the structural quality of request data sets. The proposed *dosd* presented in Sect. 8.9 explicitly considers such characteristics of given request data sets.

This book contributes new approaches for dealing with the mentioned aspects. The proposed approaches and computational studies provide information on defining the structural quality of arbitrary request data sets and the correlation between their structural quality and attainable improvements. These contributions provide promising starting points for new avenues of further research which will be discussed in Sect. 9.2.

Chapter 5
A New Deterministic Real-Time Control Approach for RDOPG Applications

In this chapter, the deterministic real-time control approach is introduced. It is used to coordinate the transportation process of the considered RDOPG applications, which were presented in Chap. 3, in real-time during its execution. First, a *general categorization of different control concepts in logistic and production systems* is described followed by the *deterministic real-time control approach*. It solves the dynamic problem by considering consecutive static snapshots, i.e., problem instances which represent the current system state of the transportation process during its execution. In order to represent these problem instances in the deterministic real-time control approach, the *mathematical model of individual problem instances* is presented next. Moreover, a *characterization of the deterministic real-time control approach* according to the general categorization introduced in Sect. 5.1 is given. Finally, with regard to the considered layers of transportation processes described in Sect. 1.2, the *utilization of the real-time control approach in RDOPG applications* is presented.

5.1 General Categorization of Different Control Concepts in Logistic and Production Systems

Control concepts applied to coordinate dynamic processes in logistic and production systems share common characteristics. According to Bock (2004, p. 67 et seq.), such control concepts can be classified according to the following five attributes:

1. **Adaptation-frequency:** This attribute specifies the points in time at which an adaptation is performed.
    - Disturbance-oriented concepts: The adaptation is started after the occurrence of a disturbance, i.e., a dynamic event.
    - Event-oriented concepts: Specific predefined events start the adaptation.
    - Periodic concepts: The adaptation is performed in predefined but longer time intervals.

F. Ferrucci, *Pro-active Dynamic Vehicle Routing*, Contributions to Management Science, DOI 10.1007/978-3-642-33472-6_5, © Springer-Verlag Berlin Heidelberg 2013

- Continuous concepts: The adaptation is continuously performed taking into account almost all or, in the ideal case, all decisions that have not yet been implemented.

2. **Adaptation-synchrony:** Using this attribute, the continuation of the process during its adaptation is defined.

    - Planning priority: It is assumed that the considered process is stopped during the adaptation. By this assumption, it is expected that the system state present at the beginning of the adaptation still exists after the adaptation has been performed. Since processes usually cannot be stopped at arbitrary points in time during their execution, this assumption is usually only valid for adaptation methods which can guarantee adaptation of the process in a time period which can be still considered as zero time (see also Sect. 4.7.4).
    - Process priority: It is accounted for that decisions are executed in the process during its adaptation. Hence, certain parts of the plan currently in execution are not able to be changed during the adaptation. This needs to be considered in the real-time control concept, for example by anticipating the system state that is assumed to exist after the adaptation stops and by using this anticipated system state in the adaptation method.
    - Simultaneity of process execution and plan adaptation: In this constellation, plan adaptations are carried out parallel to the process execution. Note that this is an ideal setting in which all decisions which are currently not in execution can be adapted.

3. **Adaptation-technique:** This attribute defines the methods that are used to perform the adaptation.

    - Priority rules: The plan is adapted by greedy approaches which can usually be carried out in zero time.
    - Improvement procedures: Plan adaptation methods such as local search or metaheuristics are applied. Note that due to their complexity, the utilization of such methods usually cannot be performed in zero time.

4. **Adaptation-range:** This attribute represents the range of decisions that can be changed in the plan during its adaptation.

    - Complete adaptation range: All decisions that have not yet been performed in the plan at the beginning of the adaptation can be changed.
    - Usable range: A reduction of the adaptation range by the time period which is required to transmit and realize plan adaptations in the process is considered. Hence, such decisions cannot be changed.
    - Time-limit depending range: A reduction of the adaptation range by a specified time limit is performed. This time limit may consider the time period utilized for performing the adaptation process and transmitting changes. With this setting, the concurrency between process adaptation and process execution can be considered.
    - Selected range: Predefined subsets of decisions are excluded from adaptation. For example, all decisions that are made within the next hour are excluded.

5. **Adaptation-duration:** This attribute defines the computational time which is available for the applied adaptation method.
   - Time limit based: The available time is defined by a time limit. Hence, it is known in advance when the adaptation will be complete.
   - Zero time: The available computational time is defined by a time period that can be interpreted as zero time, i.e., no noticeable changes occur in the considered process during the adaptation.
   - Infinite time: The adaptation method is continuously applied. Note that the time points in this setting at which performed adaptations are applied in the considered process are required to be specified.

5.2 The Deterministic Real-Time Control Approach

In this section the deterministic real-time control approach which is used to solve the problem represented by RDOPG applications is proposed. As mentioned in Sect. 3.1, in the considered problem, a set of $\mathcal{R}$ customer requests $R = \{1, \ldots, \mathcal{R}\}$ which dynamically occur during the day need to be serviced by a set of $\mathcal{K}$ homogeneous vehicles $K = \{1, \ldots, \mathcal{K}\}$ which are originally located at one central depot. During the execution of the transportation process, the current time is denoted by τ. When the execution of the transportation process starts, it holds that $\tau = 0$. Let r_i^{a} denote the arrival time of request i in the system. Note that this arrival time is not known to the real-time control approach in advance. The set of requests that are known in the real-time control approach at time point τ is then defined by $R_\tau = \{i \in R | r_i^{\mathrm{a}} \leq \tau\}$. Since the real-time control approach only utilizes information currently available while not making use of information about expected future requests but provides additional tour plan flexibility by utilizing a waiting strategy, it is classified as a deterministic real-time control approach (see Sect. 2.5.3.2). In Chap. 6, it will be extended to a pro-active real-time control approach which utilizes stochastic knowledge.

In what follows, the *applied real-time control concept* is described first. Afterwards, the *update handling* procedure consisting of activities which are carried out in each step of the real-time control approach are presented. Moreover, the *vehicle scheduling strategy* as well as the *idle vehicle waiting strategy* which are utilized in this deterministic real-time control approach are introduced.

5.2.1 The Applied Real-Time Control Concept

Since the considered RDOPG applications are dynamic, the existing system situation changes during the day. Hence, the tour plan in execution needs to be adapted according to incoming dynamic events. As mentioned in Chap. 3, dynamism arises in the RDOPG from newly arriving requests. In order to integrate these dynamically arriving requests, the tour plan currently in execution is coordinated during the execution of the transportation process by the applied real-time control concept. However, since the considered problem is NP-hard in the strong sense (see Sect. 3.2.2), sophisticated tour plan adaptations cannot be carried out by zero time heuristics. Therefore, appropriate real-time control approaches which consider relevant practical aspects of dynamic routing problems have to modify the tour plan simultaneous to its execution, i.e., while the vehicles are continuing to perform assigned activities.

In order to fulfill this requirement, the proposed real-time control approach implements a concept related to the one described by Bock (2010). In this concept, the real-time control of the considered transportation process during its execution is conducted by solving a series of individual static *problem instances*. They arise from system snapshots of the ongoing transportation process that are generated at intervals of t^{a} time units each during the transportation process. In what follows, these intervals are denoted as *anticipation horizons*. Note that each static problem instance represents the current system state of the ADFSVRPSTW as described in Sect. 3.2.1. In order to simplify the following explanations, we designate the tour plan that is currently in execution as the *relevant tour plan*. It is denoted as P_τ^{r}, where the index τ denotes the current time at the beginning of the considered anticipation horizon. Analogously, P_τ^{t} defines a *theoretical tour plan* that is optimized during this anticipation horizon utilizing a specified solution method. Additionally, P_τ^{bt} denotes the best theoretical tour plan found during the performed optimization.

At the beginning of each anticipation horizon, the corresponding static problem instance is generated. In order to consider the concurrency of the tour plan adaptation and tour plan execution, the static problem instance at τ is derived from the *future system state* which is expected at the end of the current anticipation horizon. This future system state is attained by pre-simulating P_τ^{r} for t^{a} time units. Consequently, after applying the solution method, P_τ^{t} and P_τ^{bt} are directly implementable at the end of the respective anticipation horizon. In order to enable a detailed mapping of system snapshots, each corresponding static problem instance comprises all currently known unserviced requests as well as positions and statuses of vehicles. Since dynamically arriving requests are buffered throughout each anticipation horizon, this problem instance additionally comprises a list of these newly arrived requests. After the pre-simulation that has consumed t_τ^{calc} time units is completed, the remaining time of the current anticipation horizon denoted by $t^{\mathrm{a}} - t_\tau^{\mathrm{calc}}$ is available for optimizing P_τ^{t} by the solution method. Clearly, since P_τ^{t} contains only decisions that are taken after the elapse of the current anticipation horizon, all decisions that will be made by P_τ^{r} in the current anticipation horizon become unchangeable at its beginning and thus are executed in the system as planned. Therefore, the assignment of a request to a vehicle becomes fixed if service at this request starts according to P_τ^{r} before $\tau^+ = \tau + t^{\mathrm{a}}$. Since no uncertainties in travel times or service times are considered in the RDOPG applications, in what follows, we assume that the anticipated system state is in agreement with the actual system state at τ^+. By optimizing P_τ^{t} throughout the considered anticipation horizon, P_τ^{bt} is stored for a possible subsequent replacement of P_τ^{r}. Note that due to the utilization of the presented pre-simulation concept, all existing tour plans are directly applicable at τ^+ and are therefore feasibly interchangeable.

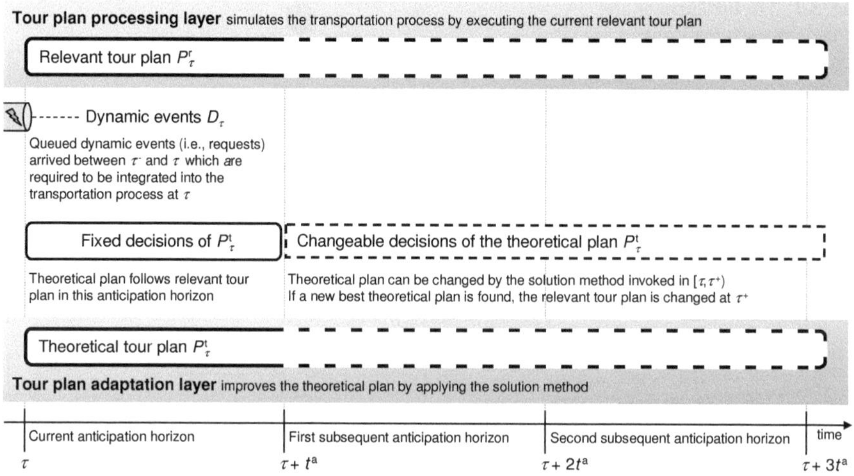

Fig. 5.1 The utilized real-time control concept adapting the transportation process at τ

The applied real-time control concept is illustrated in Fig. 5.1. The length of the anticipation horizon t^a can be chosen arbitrarily but is constant during the considered transportation process. Depending on the considered applications, a trade-off may arise here: While longer anticipation horizons may allow a more comprehensive adaptation by the applied solution method, shorter anticipation horizons can more directly react to newly arriving requests. In what follows, the applied update handling which is performed in each anticipation horizon is presented in detail.

5.2.2 Update Handling

In what follows, we describe in detail how tour plans are adapted and selected in each anticipation horizon of the previously described real-time control concept. In each anticipation horizon, the applied update handling comprises the following four steps (see also Fig. 5.2):

1. As mentioned above, since requests that have arrived in the previous anticipation horizon have only been buffered, all requests of the set $D_\tau = \{i \in R_\tau | r_i^a > \tau^-\}$ with $\tau^- = \tau - t^a$ need to be integrated into the three tour plans P_τ^r, P_τ^t, and P_τ^{bt}. Therefore, for all new requests in D_τ, travel times from and to all other active relevant locations (including current vehicle's positions) are calculated using Dijkstra's algorithm (cf. Dijkstra 1959). This calculated information is used to update the digraph G (see Sect. 5.3.1) with travel time information of the customer requests in D_τ. After these calculations, the pending requests are subsequently inserted in a sequence of non-decreasing arrival times into all three tour plans using a least cost insertion heuristic. Since this least cost insertion can be per-

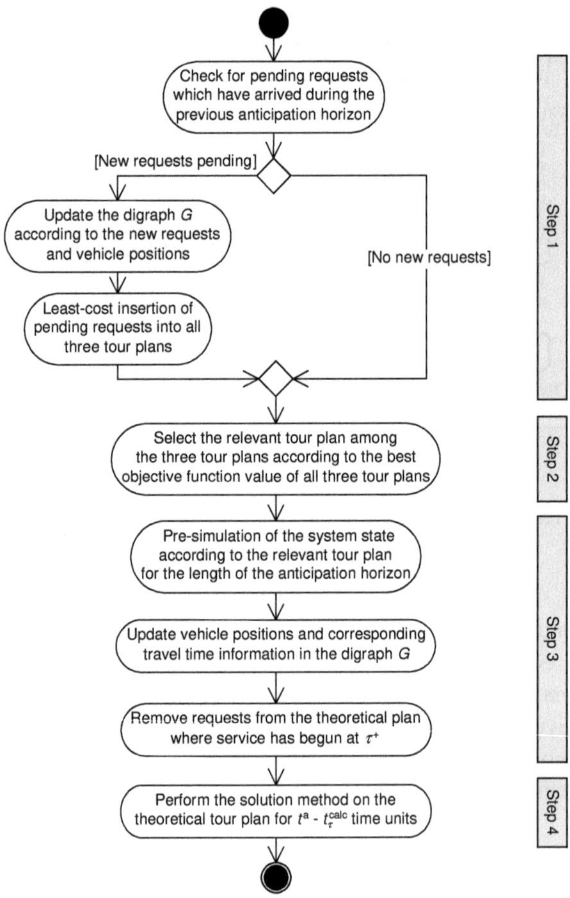

Fig. 5.2 The update handling of the proposed real-time control approach

formed in zero time, we obtain the set $\tilde{P}_\tau = \{\tilde{P}^{\mathrm{r}}_\tau, \tilde{P}^{\mathrm{t}}_\tau, \tilde{P}^{\mathrm{bt}}_\tau\}$ of modified tour plans that are directly applicable in the transportation process at τ.

2. The tour plan of the set $\tilde{P}_\tau$ with the best objective function value is selected as tour plan $P^{\mathrm{r}*}_\tau$. This tour plan becomes the new relevant tour plan and therefore controls the transportation process until time point τ^+. Note that we do not select $\tilde{P}^{\mathrm{bt}}_\tau$ directly as $P^{\mathrm{r}*}_\tau$ because this would not take into account effects of inserting the pending requests of set D_τ on the objective function value of all three tour plans in the set $\tilde{P}_\tau$. Since this selection can also be performed in zero time, $P^{\mathrm{r}*}_\tau$ remains directly applicable at τ.
3. Based on $P^{\mathrm{r}*}_\tau$, the system state at time point τ^+ is anticipated. This state is computed by pre-simulating all vehicle activities up to τ^+ according to $P^{\mathrm{r}*}_\tau$. This updates the vehicle positions and their availability times, i.e., the time points at which vehicles can react to new orders (see also Sect. 5.3.2.2). Apart from the resulting anticipated static problem instance at τ^+, the removal of all requests, where service has begun during the pre-simulation, leads to the relevant tour

plan $P^{\mathrm{r}}_{\tau+}$ applicable at time point τ^+. Clearly, in order to adapt the theoretical tour plan so that it is also applicable at time point τ^+, those requests are also removed from $\tilde{P}^{\mathrm{t}}_{\tau}$. The resulting theoretical tour plan is defined as $P^{\mathrm{t}*}_{\tau+}$ and is used as the initial solution for the solution method. The solution method is a Tabu Search metaheuristic which will be described in Chap. 7.

Note that by resuming the search process of the Tabu Search procedure from the solution examined last in the previous anticipation horizon, a larger diversification is attained. This is not the case when the search process is always started using the solution that is currently in execution. However, in case new requests have arrived (i.e., $D_\tau \neq \emptyset$) or in case requests have been serviced in the previous anticipation horizon, the Tabu Search procedure starts its search directly on a solution that is equal to $P^{\mathrm{r}}_{\tau+}$ using intensifying neighborhood operators. This methodology is applied since changes in the system state justify the direct application of intensifying neighborhood operators on the tour plan $P^{\mathrm{r}}_{\tau+}$.

4. Based on the tour plan $P^{\mathrm{r}}_{\tau+}$, all travel times from the anticipated vehicle positions at τ^+ to all other relevant locations are updated using Dijkstra's algorithm. Note that only vehicles need to be updated that have changed the current road in the road network during this anticipation horizon. Otherwise, only their availability time (see Sect. 5.3.2.2) at the next street crossing needs to be updated. Similar to step 1, the digraph G is updated according to this calculated travel time information that will be used in the solution method after the update handling.

At the end of the four steps, the update handling is completed. Note that one further important aspect for practical applications should be mentioned. Especially when a large road network is used, time-consuming calculations, due to Dijkstra's algorithm in step 4 of the update handling, may occur. As a consequence, the described update handling may consume an amount of time that cannot be considered as zero time in all situations. Hence, the time used for anticipating the new system situation at τ is measured by t^{calc}_τ. After the update handling has been completed at time point $\tau + t^{\mathrm{calc}}_\tau$, the solution method is executed on $P^{\mathrm{t}*}_{\tau+}$. Note that the computational time that is available for the solution method is not equal to t^{a} since the time needed for the update handling needs to be considered in order to stay in synchronization with the transportation process that is simultaneously executed. Hence, as mentioned above, the computational time available to the solution method is given by $t^{\mathrm{a}} - t^{\mathrm{calc}}_\tau$.

After the solution method has been executed, analogous to the previous anticipation horizon, the best solution that has been found during the application of the solution method is stored as $P^{\mathrm{bt}}_{\tau+}$ and the tour plan examined last is stored as $P^{\mathrm{t}}_{\tau+}$. These tour plans are utilized in the next anticipation horizon in order to integrate new requests and to select the relevant tour plan $P^{\mathrm{r}*}_{\tau+}$ that will be executed during the next anticipation horizon.

5.2.3 *Vehicle Scheduling Strategy*

According to our definition, a vehicle scheduling strategy determines when a vehicle will leave its last service request location in order to travel to its next assigned request location. The vehicle scheduling strategy that is used in this deterministic real-time control approach is straightforward. Specifically, since customer inconvenience values directly start when requests arrive in the system according to the objective function, vehicles are scheduled to directly drive to their next assigned request location when they have finished servicing the current request. Clearly, this only applies when a vehicle currently has further requests assigned to its tour. Otherwise, the idle vehicle waiting strategy is utilized as described in what follows.

5.2.4 *Idle Vehicle Waiting Strategy*

Since new requests arrive during the day, we utilize the following waiting strategy for idle vehicles in order to increase flexibility for integrating future requests in the tour plan. It works as follows. If a vehicle has no further requests assigned to its tour during the day, it waits at its last serviced request location for new requests. Preliminary computational experiments show that this strategy is superior compared to an immediate return of the vehicles to the depot. In other words, in the considered RDOPG applications, computational results showed that it is more advantageous to keep the vehicles at positions where past requests have arrived than directly sending the vehicles back to the depot.

5.3 The Mathematical Model of Individual Problem Instances

In this section, the *road network and the derived digraph* that models information regarding the relevant locations of the transportation process are described. After that, the *deterministic dynamic mathematical model* that is used to represent each of the individual problem instances in the proposed real-time control approach is introduced.

5.3.1 *Road Network and the Derived Digraph*

As mentioned in Sect. 3.1.1, a real road network is utilized in order to enable the real-time control approach to consider more realistic vehicle movements in the road network. The utilized road network $\tilde{R}$ is defined by $\tilde{R} = (\tilde{N}, \tilde{A}, \tilde{t})$ with $\tilde{N}$ and $\tilde{A}$ denoting its nodes and arcs, respectively. While the nodes represent crossings and road endpoints, the arcs represent the roads. The value $\tilde{t}(\tilde{a})$ of function $\tilde{t} : \tilde{A} \to \mathbb{R}$

describes the travel time in the road network required on an arc $\tilde{a} \in \tilde{A}$. In order to model the current situation in the transportation network at τ, the digraph G that maps all relevant locations at τ, i.e., request locations, the depot, and current vehicle positions, is used. It represents an abstraction layer from the road network $\tilde{R}$ and is used to provide the information that is required during tour plan adaptation to the solution method. Specifically, since in the solution method only the total travel times between relevant locations are required in order to calculate arrival times at relevant locations and thereby resulting objective function values, no information about the exact routes that vehicles will take is required to be provided to the solution method. As mentioned above, the set of known requests at time point τ is defined by R_τ whereas the set of unserviced requests at τ is defined by $R_\tau^{\mathrm{U}} = \{1, \ldots, \mathcal{R}_\tau^{\mathrm{U}}\}$, i.e., all requests $i \in R_\tau$ where no service has begun up to τ. Furthermore, $K = \{1, \ldots, \mathcal{K}\}$ represents the set of available homogeneous vehicles.

The digraph is given by $G = (N, A)$. It models the relevant locations, i.e., the depot location, request locations, and vehicle locations, of the system state at τ. The set $N = \{0, \ldots, \mathcal{R}_\tau^{\mathrm{U}} + \mathcal{K}\}$ is the associated set of nodes and $A = (N \times N)$ the corresponding set of arcs. Node 0 represents the depot which is located at node $D^{\mathrm{p}} \in \tilde{N}$ and nodes $1, \ldots, \mathcal{R}_\tau^{\mathrm{U}}$ correspond to the set of unserviced request locations R_τ^{U} at τ. Similar to the depot, it holds that each request $i \in R_\tau^{\mathrm{U}}$ is located at $r_i^{\mathrm{p}} \in \tilde{N}$, i.e., at a crossing or at a road endpoint in the road network. The start location of vehicle $k \in K$ is represented by node $v_k^+ = (\mathcal{R}_\tau^{\mathrm{U}} + k)$ of the digraph. This node corresponds to the location of vehicle k at τ in the road network defined by node $v_k^{\mathrm{p}} \in \tilde{N}$. Note that if vehicle k is not at a node $\tilde{n} \in \tilde{N}$ at τ but on an arc $\tilde{a} \in \tilde{A}$, v_k^{p} represents the node of the road network that vehicle k will reach next. Furthermore, $v_k^{\delta t}$ represents the *availability time* of vehicle k when it is available at its start location v_k^+. This availability time is a time duration relative to τ. Hence, this value is larger than 0 if the vehicle is busy with traveling on a road or with servicing a request at τ (see also Sect. 5.3.2.2). In the considered RDOPG applications, the end location of vehicle k denoted as v_k^- is always the depot that is represented by node 0 of digraph G. We define $K^+ = \bigcup_{k \in K} v_k^+$ and $K^- = \bigcup_{k \in K} v_k^-$. A non-negative value t_{ij} is associated with each arc $(i, j) \in A$. It represents the time needed for a vehicle to travel from location i to location j in the road network $\tilde{R}$ on the shortest path.[1] With regard to the considered RDOPG applications, shortest paths in terms of least travel times are sought. Specifically, with $p_{ij} = (p_{ij1}, p_{ij2}, \ldots, p_{ijn}) \subset \tilde{A}$ representing the sequence of arcs in $\tilde{R}$ that describe the shortest path from node i to node j, it holds that $t_{ij} = \sum_{a \in p_{ij}} \tilde{t}(a)$. Note that in practical applications, the shortest path routing information given by p_{ij} can be used in order to route a vehicle to the location of its next assigned customer request. Specifically, further information regarding traffic jams and current travel times could be integrated into the shortest path calculations in the dispatching center and then transmitted to the vehicle's

[1] The shortest path is defined as the path in a weighted graph connecting two given nodes i and j with the property that the sum of the weights of all the edges is minimized over all such paths (cf. Sedgewick 1983, p. 413).

on-board unit (see also Sect. 2.6.3). In the conducted empirical studies, the simulation component utilizes this information to simulate vehicle movements in the road network (see also Sect. 8.1).

In the proposed real-time control approach, two different types of shortest path calculations are required. Specifically, while highly detailed information about the shortest paths is required for routing vehicles in the road network, the solution method only needs information about total travel times between the relevant locations as mentioned above. In these calculations, the geographical positions of the relevant locations in the road network are given by D^{p}, r_i^{p} $(i \in R_\tau^{\mathrm{U}})$, and v_k^{p} $(k \in K)$. Studies performed in Ferrucci (2006, p. 86 et seq.) show that it is beneficial to use Single-Point Shortest-Path (SPSP) calculations for the first type of calculations and Single-Source Shortest-Path (SSSP) as well as Single-Source Shortest-Path Reverse (SSSP_R) calculations for the second type of calculations. Specifically, SPSP, also denoted as one-to-one shortest paths, calculates the shortest path for a given source-destination pair while SSSP calculates the shortest paths starting from a given source to a set of destinations. Moreover, since travel time information is required between the relevant locations in both travel directions in the solution method and travel time information may be direction-dependent due to one-way streets, SSSP_R is utilized in order to calculate the shortest paths starting from a given set of sources to one destination. More information on shortest path calculations can be found among others in Ahuja et al. (1993, p. 93 et seq.) while details on efficient SPSP and SSSP implementations in road networks can be found in Zhan (1997), Zhan and Noon (1998), and Zhan and Noon (2000). Further sophisticated shortest path calculation methods which are suitable for calculating shortest paths in road networks are proposed, among others, by Bauer et al. (2008) and Schultes (2008).

5.3.2 *The Deterministic Dynamic Mathematical Model*

In this section, the mathematical model for representing the individual static problem instances generated from the system snapshot at time τ is introduced. Note that the problem data of the model is derived from the pre-simulated system state at τ^+ in order to consider the concurrency of tour plan execution and tour plan adaptation. The solution of this model determines the tours of the vehicles in order to service the currently pending requests so that the value of the considered objective function is minimized. Since the generated vehicle tours are applied at τ^+, they start at the respective anticipated vehicle positions and take into consideration the anticipated vehicle status and availability times at this point in time. According to the classification of Powell (see Sect. 2.5.2.3), the presented model is *dynamic* since it is adapted according to the current system state during the execution of the transportation process. Moreover, the model is *deterministic* since no stochastic knowledge about future requests is integrated and the model is only solved on the basis of the current information that is known with certainty at τ. This model will be extended to a stochastic dynamic model by integrating stochastic knowledge in Chap. 6. In the following, we consider the deterministic dynamic version of the model and its problem data in more detail.

5.3.2.1 Modeling the Requests

In the RDOPG applications, customer requests have soft time windows as well as variable and fixed penalty costs. Each unserviced customer request $i \in R^{\mathrm{U}}_{\tau+}$ is unambiguously defined by the tuple $r_i = (r^{\mathrm{a}}_i, r^{\mathrm{e}}_i, r^{\mathrm{l}}_i, r^{\mathrm{s}}_i, r^{\mathrm{w}}_i, r^{\mathrm{p}}_i)$. In this tuple, r^{a}_i denotes the arrival time of request i in the system whereas r^{e}_i and r^{l}_i define the earliest and latest valid start of service without fixed penalty costs denoted by R^{pen} occurring. Hence, the time period between r^{e}_i and r^{l}_i represents the maximum allowed response time as introduced in the problem description in Sect. 3.1. Since in the RDOPG applications, each customer request is directly waiting for service when it arrives in the system, its time window simultaneously opens upon its arrival, i.e., r^{e}_i is equal to the arrival time r^{a}_i. Note that both these parameters may have different values for dummy customer requests which will be introduced in Chap. 6. The customer inconvenience is modeled by variable penalty costs which start at r^{e}_i. The time required to service a request i at its corresponding customer location is defined by the service time denoted by parameter r^{s}_i. As mentioned in Sect. 3.1, since the maximum allowed response time as well as the service time are assumed to be equal for all requests in the considered example application given by the subsequent delivery of newspapers, we define $\forall i \in R^{\mathrm{U}}_{\tau+}: r^{\mathrm{s}}_i = R^{\mathrm{st}}, r^{\mathrm{l}}_i = r^{\mathrm{e}}_i + R^{\mathrm{mrt}}$. Parameter r^{w}_i determines the weight factor with which the request is considered in the objective function. Finally, r^{p}_i determines the aforementioned location of the request in the road network. Since all requests are located at a node in the road network, it holds that $r^{\mathrm{p}}_i \in \tilde{N}\ \forall i \in R^{\mathrm{U}}_{\tau+}$.

5.3.2.2 Modeling the Vehicles

Each vehicle $k \in K$ is modeled by the tuple $v_k = (v^+_k, v^-_k, v^{\mathrm{p}}_k, v^{\delta t}_k)$. As mentioned above, parameter v^+_k defines the start location of vehicle k in the digraph represented by node $(\mathcal{R}^{\mathrm{U}}_{\tau+} + k)$ whereas v^-_k defines the end location which is equal to the depot location for all vehicles. Parameter $v^{\mathrm{p}}_k \in \tilde{N}$ defines the node in the road network which corresponds to the start location of vehicle k. Moreover, relative to τ^+, parameter $v^{\delta t}_k$ determines the availability time of vehicle k at v^{p}_k. In what follows, we introduce how three vehicle statuses that are possible at τ^+ are transformed into different values of the parameters v^{p}_k and $v^{\delta t}_k$:

1. **Vehicle *k* is busy and drives to its next request:** As mentioned in the problem description in Sect. 3.1.4, the first request may be reassigned due to the allowed vehicle en-route diversion. Therefore, the current destination of a vehicle may be changed. However, note that such a change is only possible when the vehicle has reached a node in the road network, i.e., a road junction or road end point $\tilde{n} \in \tilde{N}$.

Thus, v_k^{p} represents the node in the road network that the vehicle will reach next. Moreover, $\tau^+ + v_k^{\delta t}$ is defined as the point in time when vehicle k arrives at node v_k^{p}.

2. **Vehicle k is busy and services a request:** Assuming that vehicle k is servicing request i at its location $r_i^{\mathrm{p}} \in N$ at time point τ^+ and let x be the remaining service time of request i at τ^+. Then, it holds that $v_k^{\mathrm{p}} = r_i^{\mathrm{p}}$ and $v_k^{\delta t} = x$ since the vehicle needs to complete servicing request i before it is available to fulfill further orders.
3. **Vehicle k is idle and has no assigned requests or it waits unused at the depot:** As mentioned in Sect. 5.2.4, the vehicle waits at the location of its last serviced request i or at the depot, respectively. Since vehicle k is immediately available for fulfilling further orders, it holds that $v_k^{\delta t} = 0$ and $v_k^{\mathrm{p}} = r_i^{\mathrm{p}}$ where $i \in R_{\tau+}^{\mathrm{U}}$ represents the request i that vehicle k has serviced last or $v_k^{\mathrm{p}} = D^{\mathrm{p}}$, respectively.

5.3.2.3 Model Formulation

In this section, we describe the model which is utilized to represent the individual problem instances which occur during the execution of the transportation process. In the model, binary decision variables of set $X = \{x_{ki}^{\mathrm{s}}\}$ ($k \in K$, $i \in R_{\tau+}^{\mathrm{U}}$) $\cup \{x_{kij}^{\mathrm{d}}\}$ ($k \in K, i \in \{R_{\tau+}^{\mathrm{U}} \cup K^+\}, j \in \{R_{\tau+}^{\mathrm{U}} \cup K^-\}$, $i \neq j$) define request-vehicle assignments as well as vehicle travel activities which form the tours of the vehicles. Specifically, for each vehicle $k \in K$, x_{ki}^{s} is 1 if and only if vehicle k services request $i \in R_{\tau+}^{\mathrm{U}}$ while x_{kij}^{d} is 1 if vehicle k directly drives from the relevant location i to the relevant location j and 0 otherwise. Note that this formulation of binary decision variables relates to the three-index vehicle flow formulation presented for the CVRP in Sect. 2.3.2. The variable set $Y = \{y_i^{\mathrm{a}}\} \cup \{y_i^{\mathrm{s}}\}$ determines the time assignments of the requests. For each request $i \in R_{\tau+}^{\mathrm{U}}$, y_i^{a} denotes the arrival time of the assigned vehicle at request i and y_i^{s} the start time of service at request i. Note that the variable set Y depends on set X so that the solution of the model is completely defined by set X.

Based on these definitions, each static problem instance occurring in the transportation process of the considered RDOPG applications can be modeled as follows:

$$\forall i \in R_{\tau+}^{\mathrm{U}}: \quad \sum_{k \in K} x_{ki}^{\mathrm{s}} = 1 \tag{5.1}$$

$$\forall k \in K, i \in R_{\tau+}^{\mathrm{U}}: \quad \sum_{j \in \{v_k^+ \cup R_{\tau+}^{\mathrm{U}}\}} x_{kji}^{\mathrm{d}} = \sum_{j \in \{R_{\tau+}^{\mathrm{U}} \cup v_k^-\}} x_{kij}^{\mathrm{d}} = x_{ki}^{\mathrm{s}} \tag{5.2}$$

$$\forall k \in K: \quad \sum_{j \in \{R_{\tau+}^{\mathrm{U}} \cup v_k^-\}} x_{kv_k^+ j}^{\mathrm{d}} = \sum_{i \in \{v_k^+ \cup R_{\tau+}^{\mathrm{U}}\}} x_{kiv_k^-}^{\mathrm{d}} = 1 \tag{5.3}$$

$$\forall i \in R_{\tau+}^{\mathrm{U}}: \quad y_i^{\mathrm{s}} = \max\left(r_i^{\mathrm{e}}, y_i^{\mathrm{a}}\right) \tag{5.4}$$

$$\forall i, j \in R^{\mathrm{U}}_{\tau+}, k \in K: \quad x^{\mathrm{d}}_{kij} = 1 \Rightarrow y^{\mathrm{a}}_j = y^{\mathrm{s}}_i + r^{\mathrm{s}}_i + t_{ij} \tag{5.5}$$

$$\forall j \in R^{\mathrm{U}}_{\tau+}, k \in K: \quad x^{\mathrm{d}}_{kv^{+}_{k}j} = 1 \Rightarrow y^{\mathrm{a}}_j = \tau^{+} + v^{\delta t}_k + t_{v^{+}_{k}j} \tag{5.6}$$

In this model, constraints (5.1) ensure an unambiguous assignment of requests to vehicles. Constraints (5.2) make sure that a vehicle correctly visits all of its assigned requests. Constraints (5.3) guarantee that a vehicle starts and ends its tour at its defined start and end location. Constraints (5.4) derive the start time of service at each request location while restrictions (5.5) define the arrival time of each vehicle at the request locations of its tour. Constraints (5.6) determine the arrival time of each vehicle at the first request location assigned to its tour. Besides the required travel time, this arrival time also depends on the availability time of vehicle k at its start location v^{p}_k. As mentioned above, depending on the current state of vehicle k at τ^{+}, this availability time is greater than or equal to τ^{+}.

In what follows, detailed information about modeling the objective functions utilized in the considered RDOPG applications is provided.

5.3.2.4 Objective Function

As introduced in Sect. 3.1, the objective function pursues the minimization of the total customer inconvenience. This total customer inconvenience consists of variable and fixed inconvenience costs and is operationalized by a function of request response times (see Sect. 2.4.12.7) as well as fixed lateness costs. As defined above, the point in time when service begins at request i is defined as y^{s}_i. We denote the response time of each request $i \in R^{\mathrm{U}}_{\tau+}$ as t^{res}_i defined by $t^{\mathrm{res}}_i = y^{\mathrm{s}}_i - r^{\mathrm{e}}_i$. Since each request $i \in R^{\mathrm{U}}_{\tau+}$ has to be serviced with certainty, each of these requests obtains the same weight factor r^{w}_i by which it is weighted in the objective function. For these requests that already have arrived in the system, this weight factor is set to 1. When exceeding the predefined maximum allowed response time R^{mrt}, a prohibitive increase of the customer inconvenience occurs due to the penalty costs that are defined by R^{pen}. Utilizing the specified response time t^{res}_i, the objective function that is pursued in the model of the considered RDOPG applications is defined by

$$\min z = \sum_{i \in R^{\mathrm{U}}_{\tau+}} r^{\mathrm{w}}_i \cdot \Big(\underbrace{F\big(t^{\mathrm{res}}_i\big)}_{\substack{\text{Variable customer} \\ \text{inconvenience}}} + \underbrace{R^{\mathrm{pen}} \cdot \Theta\big(t^{\mathrm{res}}_i - R^{\mathrm{mrt}}\big)}_{\substack{\text{Lateness penalty costs} \\ \text{(i.e., lateness inconvenience)}}} \Big)$$

with the function $\Theta(x) = 1$ if $x > 0$ and 0 otherwise. As can be seen, the primary objective is to avoid a late service of customer requests, i.e., all customers should be serviced within their maximum allowed response time. Furthermore, the secondary objective is to minimize the total variable customer inconvenience which depends on the required request response times. As mentioned in Sect. 3.1, the effects of a linear and a quadratic customer inconvenience increase are assessed in the RDOPG applications. Therefore, F is defined either as a linear or as a quadratic customer inconvenience function.

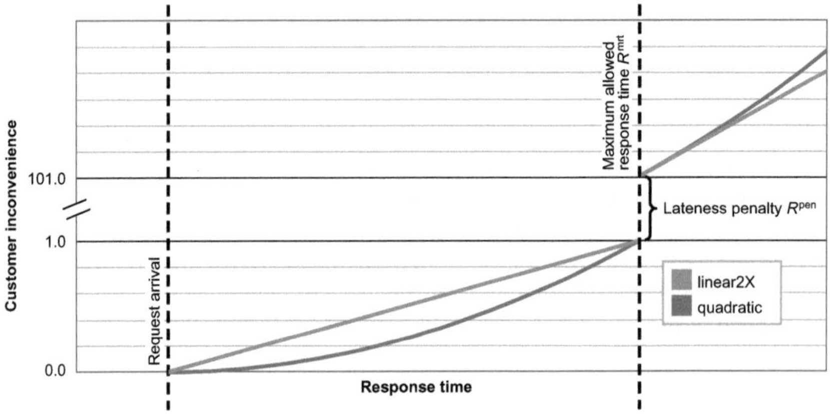

Fig. 5.3 Illustration of the linear2X and quadratic customer inconvenience

As illustrated in Fig. 5.3, both customer inconvenience functions have the same customer inconvenience contribution of 1.0 at the maximum allowed response time at R^{mrt} whereas further increase, i.e., a violation of this maximum allowed response time, are penalized by R^{pen}. The prohibitive effect of this penalization is realized by the significantly higher value of these lateness penalty costs in comparison to the customer inconvenience values. The presented objective functions are used in the real-time control approach for coordinating the transportation process during its execution. In the computational results presented in Chap. 8, the lateness penalty costs are separately considered by specifying the number of late customer requests. Hence, the presented improvements which were attained refer to the variable customer inconvenience of the objective function.

5.3.2.4.1 Linear Customer Inconvenience (linear2X)

In the linear case, F is defined by:

$$F\left(t_i^{\mathrm{res}}\right) = \frac{1}{R^{\mathrm{mrt}}} \cdot \left(\min\left(t_i^{\mathrm{res}}, R^{\mathrm{mrt}}\right) + 2 \cdot \max\left(0, t_i^{\mathrm{res}} - R^{\mathrm{mrt}}\right)\right)$$

Note that in this function the steepness is doubled after the maximum allowed response time R^{mrt} in order to still consider those late requests with high priority. Hence, in the computational results, this type of evaluated customer inconvenience is denoted by *linear2X*.

5.3.2.4.2 Quadratic Customer Inconvenience (quadratic)

In this case, the function F is simply defined as $F(t_i^{\mathrm{res}}) = (\frac{t_i^{\mathrm{res}}}{R^{\mathrm{mrt}}})^2$. Since the steepness of the quadratic function increases implicitly, no adaptation at R^{mrt} is required.

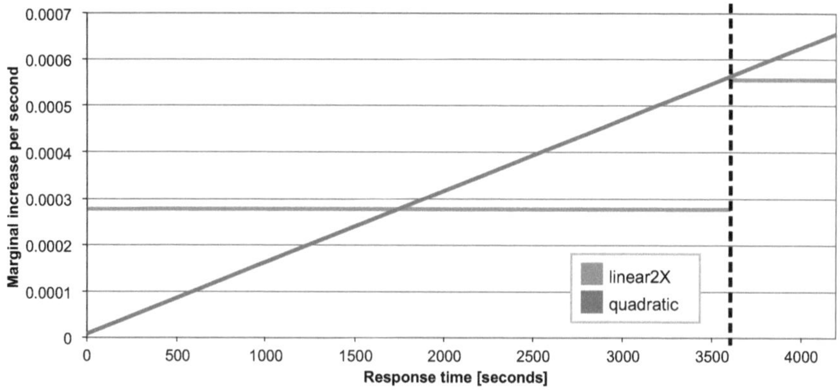

Fig. 5.4 Marginal increase in variable customer inconvenience using a maximum request response time of one hour

In the computational results, this type of customer inconvenience is denoted as *quadratic*.

In Fig. 5.4, the marginal increase in variable customer inconvenience, i.e., without the lateness penalty costs R^{pen}, in both evaluated customer inconvenience functions is illustrated. In this figure, the explicitly defined increase of the marginal increase of the linear2X customer inconvenience function for requests serviced after the maximum allowed request response time as well as the implicit increase of the quadratic customer inconvenience function can be observed.

5.4 Characterization of the Deterministic Real-Time Control Approach

According to the general classification of real-time control concepts which has been introduced in Sect. 5.1, the proposed real-time control concept can be characterized as follows:

1. The *adaptation-frequency* is implemented by using a *continuous concept*. Specifically, in time intervals of equal length which is defined by the anticipation horizon t^{a}, the transportation process is adapted by integrating intermediately arriving dynamic events into consecutive problem instances. Hence, the complete time which is available during the execution of the transportation process is used for applying the solution method.
2. The *adaptation-synchrony* is implemented by a *process priority* since the transportation process cannot be interrupted while the tour plan adaptations are performed by the Tabu Search metaheuristic. Since this proposed solution method consumes an amount of computational time which cannot be considered as zero time, changes in the transportation process which occur during the adap-

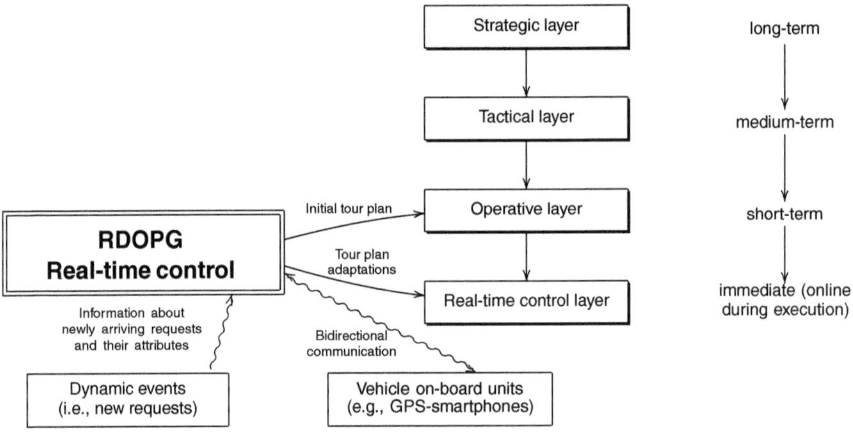

Fig. 5.5 Integration of the real-time control approach into RDOPG applications

tation need to be considered. In order to consider such changes, the adaptation range (see below) is set accordingly.

3. The applied *adaptation-technique* is an *improvement method* implemented in the form of a developed stage-based Tabu Search metaheuristic which will be presented in Chap. 7.
4. The *adaptation-range* is implemented as a *time-limit dependent range*. Specifically, since decisions which are made in the transportation process during the adaptation of the tour plan cannot be changed after the adaptation, they are excluded from the available adaptation range. The length of this range is defined by the length of the anticipation horizon t^{a}.
5. The *adaptation-duration* is realized by a *time-limit* based concept. Its length is individually defined for each problem instance by the remaining length of the anticipation horizon after the pre-simulation has been performed, i.e., by $t^{\mathrm{a}} - t_{\tau}^{\mathrm{calc}}$.

5.5 Utilization of the Real-Time Control Approach in RDOPG Applications

In Sect. 1.2, the individual layers of transportation processes were introduced. In Fig. 5.5, the integration of the proposed real-time control approach into the considered RDOPG applications is illustrated. As was described before, the proposed real-time control approach for the RDOPG applications is applied to the operative and real-time control layer of the transportation process. In accordance with the example architecture of a real-time control system which was presented in Fig. 2.10, the RDOPG real-time control is connected to two systems. The first system provides information about new dynamic events (new requests) while the second one establishes the bidirectional communication with the vehicles. With regard to the considered layers, the real-time control approach supplies one initial tour plan before the execution of the transportation process (operative layer) and repeatedly provides adaptations to the relevant tour plan during the execution of the transportation process (real-time control layer).

5.6 Summary

In this chapter, the real-time control approach which is utilized for coordinating the transportation process of RDOPG applications was described. Based on a deterministic real-time control approach proposed by Bock (2010), the proposed real-time control approach solves the transportation process during its execution by dividing the dynamic problem in individual problem instances. A new problem instance is generated in time intervals of a specified length defined by the anticipation horizon. New requests which occur during the transportation process are buffered until the generation of the next problem instance. Hence, each problem instance can be seen as a static problem instance which is adapted by the developed solution method which will be presented in Chap. 7. Since the solution method consumes a time period which cannot be considered as zero time for adapting the tour plan and since the transportation process continues its execution during the adaptation, the concurrency of transportation process execution and adaptation has to be considered. For this purpose, before the adaptation, the state of each static problem instance at the end of the current anticipation horizon is anticipated so that the calculated solution is directly applicable in the transportation process after the adaptation is complete. The update handling which is used for integrating newly arrived requests and generating the individual static problem instances was presented in detail. By making use of an idle vehicle waiting strategy, idle vehicles wait for further arriving requests at their last serviced request location instead of directly traveling back to the depot. Moreover, vehicles with pending requests directly drive to their next assigned request. Furthermore, the mathematical model used to represent the individual problem instances in the real-time control approach was formulated including constraints which need to be regarded in the considered RDOPG applications. In this mathematical formulation, both evaluated objective functions comprising a linear and a quadratic evolution of customer inconvenience were introduced. After the description of the utilized real-time control approach, it was categorized according to the classification of real-time control approaches proposed in Bock (2004, p. 67 et seq.) which was introduced at the beginning of the chapter. Since the proposed real-time control approach does not make use of stochastic knowledge, it is a deterministic real-time control approach and therefore the utilized mathematical model is a deterministic dynamic model. The proposed deterministic real-time control approach will be extended to a pro-active real-time control approach in the next chapter.

Chapter 6
A New Forecasting Approach for Generating Stochastic Knowledge from Past Request Information and Utilizing the Stochastic Knowledge

The real-time control approach presented in the previous chapter utilizes a deterministic dynamic model which has been introduced in Sect. 5.3.2. Apart from the idle vehicle waiting strategy (see Sect. 5.2.4) no methods for providing an increased flexibility for integrating dynamically arriving requests are utilized in this real-time control approach. Hence, according to the classification in Sect. 2.5.3.2.2, this real-time control approach is a deterministic real-time control approach which only adapts the tour plan according to problem data known with certainty. Specifically, although past request information is available (see Sect. 3.1) out of which stochastic knowledge can be generated, this deterministic real-time control approach does not make use of it. Therefore, each time a static problem instance is solved, the performed adaptations do not take into account the fact that *further changes* in the relevant information are likely to occur during the execution of the transportation process. Consequently, tour plans generated by the deterministic real-time control approach could in many cases be improved in order to service additionally arriving requests more efficiently, i.e., with lower response times. This is especially true when newly arriving requests are located far away from current vehicle positions which happens when previous requests have only arrived in other regions of the considered service area. This may occur if changing request arrival rates in the individual regions during the planning horizon exist in the considered application.

In order to take advantage of the available past request information, we propose new forecasting methods for generating and integrating stochastic knowledge into the deterministic real-time control approach in this chapter. By using these methods, the stochastic knowledge is generated from the available past request information, i.e., from request arrivals observed on elapsed working days. The generated stochastic knowledge is integrated into the real-time control approach in order to implement the three pro-active strategies mentioned in Sect. 4.3:

- **Relocation strategy:** By using this strategy, vehicles can be relocated to subregions, in which future requests are likely to arrive, at suitable time points and can be kept there for a specified amount of time. Such promising regions in the service area are automatically derived from the available request information by

F. Ferrucci, *Pro-active Dynamic Vehicle Routing*, Contributions to Management Science,
DOI 10.1007/978-3-642-33472-6_6,

the proposed forecasting methods. Note that keeping vehicles in promising subregions after relocation activities have been performed also involves the use of a waiting strategy which is described next.

- **Waiting strategy:** When a vehicle has been relocated to a promising subregion or is already located in such an area, a decision has to be made about how long this vehicle will be kept there in order to wait for expected future requests. For this purpose, a maximum value of this amount of time is defined by the generated stochastic knowledge. Furthermore, the real-time control approach can dynamically determine to keep vehicles for a shorter amount of time in these subregions according to the current request situation during the transportation process. Comparable to the idle vehicle waiting strategy applied in the deterministic real-time control approach (see Sect. 5.2.4), idle vehicles with no assigned relocation activities wait at their last serviced request location.
- **Request assignment strategy:** In addition to the previous two strategies which are primarily applied to improve the utilization of idle vehicles, the available stochastic knowledge is also directly integrated into the tour plan adaptation process. Specifically, since request assignments to vehicles affect future positions of the vehicles, resulting effects on the response times and customer inconvenience values of expected future requests are anticipated and directly considered during tour plan adaptations.

According to our classification in Sect. 2.5.3.2.2, using these pro-active strategies extends the deterministic real-time control approach to a *pro-active real-time control approach*. The general aim of the pro-active real-time control approach is to enable earlier service of expected future requests by efficiently utilizing the available fleet of vehicles according to the expected requests. This is especially beneficial in the considered RDOPG applications in which the pursued objective solely consists of minimizing customer inconvenience. Therefore, an earlier service of pending requests is likely to increase the solution quality significantly.

In order to enable relocation and waiting activities of vehicles according to expected future requests, their temporal and spatial occurrence has to be forecasted. For this purpose, we propose a new forecasting method for generating stochastic knowledge from the available past request information. The integration of the stochastic knowledge into the real-time control approach supports the generation of tour plans with an increased flexibility for integrating expected future requests which offers possibilities to increase the solution quality of generated tour plans. By using the stochastic knowledge during the adaptation of the individual problem instances, the deterministic dynamic model (see Sect. 5.3.2.3) is extended to a stochastic dynamic model.

In what follows, the proposed approaches for generating and integrating stochastic knowledge into the real-time control approach are described. We first describe the general ideas that are used for *integrating stochastic knowledge into the proposed real-time control approach*. Since the stochastic knowledge is integrated into the pro-active real-time control approach by utilizing dummy customers, we present how the *generation of dummy customers using segment-based clustering* is carried out by analyzing the available past request information. Finally, methods for *handling dummy customers in the pro-active real-time control approach* during the execution of the transportation process are presented.

6.1 Integrating Stochastic Knowledge into the Proposed Real-Time Control Approach

In the deterministic dynamic model described in Sect. 5.3.2, the set of unserviced requests at τ^+ denoted by $R^{\mathrm{U}}_{\tau+}$ only comprises requests known with certainty so that no expected future requests are considered. In order to integrate stochastic knowledge about expected future requests, the model is extended as follows. In addition to the mentioned requests that are known with certainty, the set $R^{\mathrm{U}}_{\tau+}$ also contains *dummy customer requests*, or in short *dummy customers*, that additionally have to be considered during tour plan adaptations and the execution of the transportation process. As mentioned in Chap. 4, dummy customer-related approaches have been used among others by Ichoua et al. (2006) and Bent and van Hentenryck (2007). In contrast to many of these approaches, our forecasting approach generates dummy customers by the application of a sophisticated offline procedure before the execution of the transportation process. This procedure provides stochastic knowledge of customizable quality about expected future requests. Since this stochastic knowledge defines the temporal and spatial position of the generated dummy customers, only one scenario comprising these high-quality dummy customers has to be solved during the time-critical adaptations of the transportation process. Note that this is a significant advantage compared to many other approaches which require solving numerous sample scenarios simultaneous to the execution of the transportation process. Consequently, in order to provide useful decision support, these approaches have to generate their stochastic knowledge in real-time under high time pressure which may limit their practical applicability and efficiency. Additionally, our offline approach allows for extending the proposed forecasting methods to more complex procedures which may require more computational time that is available before the execution of the transportation process.

As mentioned above, all dummy customers are generated in an offline step so that they are known before the execution of the transportation process. Therefore, all dummy customers are directly considered in the initial tour plan generation together with the real requests that are already known before the execution of the transportation process. Dummy customers and real requests are similarly handled by the real-time control approach except for the following four aspects:

- **Arrival time and vehicle scheduling:** In contrast to real requests, whose time window directly begins upon their arrival in the system, time windows of dummy customers open at a specified time point derived from the past request information. Specifically, the start of the time window relates to the expected arrival time of future requests that are represented by the respective dummy customer. This also requires an extended vehicle scheduling strategy which is described in Sect. 6.3.2.

- **Location and service time:** In contrast to real requests that have a specific location in the road network, a dummy customer guides a vehicle to the *spatial barycenter* of a promising subregion denoted as a request-likely area. As mentioned in Sect. 3.1.3, a request-likely area represents a temporally and spatially defined area in which future requests are expected. Since the locations of requests that are expected to arrive in a request-likely area are presumably different from its spatial barycenter, the travel time which the assigned vehicle requires from the barycenter to the location of the expected future requests needs to be considered. Hence, dummy customers provide information about the estimated total amount of time that a vehicle requires within a request-likely area. For each request-likely area, this amount of time is individually determined by the total sum of travel and service times required to fulfill all expected future requests represented by the respective dummy customer of the request-likely area. The service time of the corresponding dummy customer is set to this calculated time period.

  Note that since the service time represents the estimated amount of time that a vehicle will require for servicing the expected requests in a request-likely area, the integration of dummy customers into the transportation process makes it possible to have a more foresighted tour plan generation. Specifically, the service time of a dummy customer acts as a buffer that increases the robustness of generated tour plans which may lead to better results. Due to this increased robustness of the tour plans, substantial tour modifications under high time pressure become less likely and therefore tour plans with a high solution quality can be more frequently implemented as planned. Moreover, actively guiding vehicles to request-likely areas before real requests arrive there enables an earlier service of these requests. Therefore, the utilization of the generated stochastic knowledge provides the desired increased flexibility for integrating expected future requests.
- **Uncertainty of occurrence:** As mentioned in Sect. 5.3.2.4, each request i that is known with certainty is considered in the objective function with a weight factor of $r_i^{\mathrm{w}} = 1$. In contrast, a dummy customer i represents the expected arrival of future requests in the corresponding request-likely area with some uncertainty. Therefore, it is considered in the objective function with a weight factor $r_i^{\mathrm{w}} < 1$ which depends on the probability that at least one real request will actually arrive in the assigned request-likely area. As an advantage, this enables the pro-active real-time control approach to decide, depending on the current pending request situation, if and for how long each vehicle should be kept in a request-likely area. With regard to the considered objective function, this flexible vehicle utilization is more advantageous compared to defining fixed rules.
- **Parameter updating:** In contrast to known requests that are removed from the system after service at their request location is completed, dummy customers represent place-holders for expected future requests. Since dummy customers are only assigned to a request-likely area which is temporally restricted, dummy customers need to be updated during the transportation process. Specifically, as time passes by during the execution of the transportation process, the probability that further requests will arrive in the corresponding request-likely area of a dummy customer decreases. Hence, the dummy customer parameters representing the expected service time and the weight factor need to be dynamically updated during

the execution of the transportation process. Furthermore, when the probability that more requests will occur in the assigned request-likely area is too low, the dummy customer has to be removed from the system. The parameter updating and removal of dummy customers is discussed in Sect. 6.3.1.

6.2 Generation of Dummy Customers Using Segment-Based Clustering

In this section, we propose a new forecasting procedure for generating stochastic knowledge which is integrated into the pro-active real-time control by using dummy customers. As mentioned in Sect. 3.1.3, this stochastic knowledge is solely generated by analyzing past request information that has been observed on a specified number of previous working days which are assumed to be representative of the next working day, i.e., the next planning horizon. In what follows, the number of past working days that is utilized to generate the stochastic knowledge is denoted by n^{f}.

In the proposed forecasting procedure, no information about underlying request distributions and arrival rates is explicitly given. It is assumed that request arrival rates follow certain patterns which may depend on the region as well as on the time of the day. Specifically, it is assumed that dynamically arriving requests can be modeled by a set of unknown non-homogeneous time-space Poisson processes. In order to describe non-homogeneous time-space Poisson processes, we introduce the following definitions:

- **Counting process:** A stochastic process $\{N(t), t \geq 0\}$ is denoted as a counting process if $N(t)$ denotes the number of events which have occurred by time point t (cf. Ross 2010, p. 312 et seq.).
- **Poisson process:** A counting process is a Poisson process with rate parameter λ if the following three criteria are fulfilled (cf. Ross 2010, p. 313):

  (a) $N(0) = 0$.
  (b) For an arbitrary time interval of length t', the number of events occurring in this time interval is Poisson distributed with mean $\lambda t'$. Due to the properties of the Poisson distribution, it holds that the expectancy value is $E(N(t')) = \lambda t'$. The parameter λ is also denoted as *rate parameter* since it represents the number of expected events, i.e., request arrivals, per time unit.
  (c) Increments are independent, i.e., the number of events which occur in disjoint time periods are independent. Hence, in Poisson processes, request arrivals are independently and uniformly distributed in time (cf. Larson and Odoni 1981, p. 46 et seq. and p. 75).

- **Time-space Poisson processes:** The previously described Poisson process is also denoted as a one-dimensional Poisson process in time t. A two-dimensional (spatial) Poisson process is defined by a two-dimensional random variable $X \in \mathbb{R}^2$

on a considered area A with $N(B)$ representing the number of events in a subarea $B \subset A$. A three-dimensional Poisson process in which events have an occurrence time as well as a two-dimensional point location (i.e., $X \in \mathbb{R}^3$) is denoted as a *time-space Poisson process* (cf. Larson and Odoni 1981, p. 196).

- **Homogeneous and non-homogeneous Poisson processes:** A Poisson process is *homogeneous* if its rate parameter stays constant. In contrast, it is called *non-homogeneous* if the rate parameter comprises changes in one or more dimensions of the random variable X (cf. Larson and Odoni 1981, p. 193).

In our application, each of the unknown non-homogeneous time-space Poisson processes has a temporally and spatially varying rate parameter assigned to them. This may result in time- and space-varying request arrival rates during the considered planning horizon. Moreover, the rate parameters of the Poisson processes are also unknown but they are assumed to remain constant from day to day. Note that the proposed forecasting approach is also applicable if the specified assumptions are not given. However in this case, a reduced efficiency of the forecasting approach is likely. Therefore, we develop methods which check specific quality criteria of the generated stochastic knowledge and integrate them into our forecasting approach.

In what follows, the proposed forecasting approach for generating stochastic knowledge is described in detail. First, we first describe the *segment-based cluster generation* which is used to generate all valid clusters, i.e., those that fulfill defined quality criteria. After all valid clusters have been generated, we describe how the *cluster selection and dummy customer generation* is performed.

6.2.1 Segment-Based Cluster Generation

In order to derive reliable information about expected future requests, the developed forecasting methods determine appropriate time-space Poisson processes by analyzing the available past request information. For this purpose, the temporal and spatial area of the considered service area during the considered planning horizon is divided into *subareas* which are further temporally divided into *segments*. According to Fig. 6.1, each segment $s \in S$ has the same predetermined quadratic spatial ($DC^{se} \in \mathbb{R}^2$) as well as a temporal ($DC^{te} \in \mathbb{R}$) extension. Note that all segments are disjoint and cover the entire considered service area and planning horizon. For each segment $s \in S$, expected request arrivals in s are modeled by a non-homogeneous time-space Poisson process p_s. As mentioned above, it is assumed that request arrival rates remain constant from day to day. Therefore, the corresponding rate parameter $\lambda(p_s)$ is calculated by the average number of past request arrivals in s over the last n^f working days. Clearly, since p_s represents the expected request arrivals in segment s each day, p_s is only considered during segment s. Hence, the time unit of $\lambda(p_s)$ is given by DC^{te}.

Preliminary computational experiments indicated that defining the spatial and temporal extension of a segment by $DC^{se} \times DC^{te} =$ (2.5 kilometers $\times$ 2.5 kilometers) $\times$ 1 minute provides a reasonable setting for the considered RDOPG application (see also Sect. 8.2.1). The division of the considered service area into

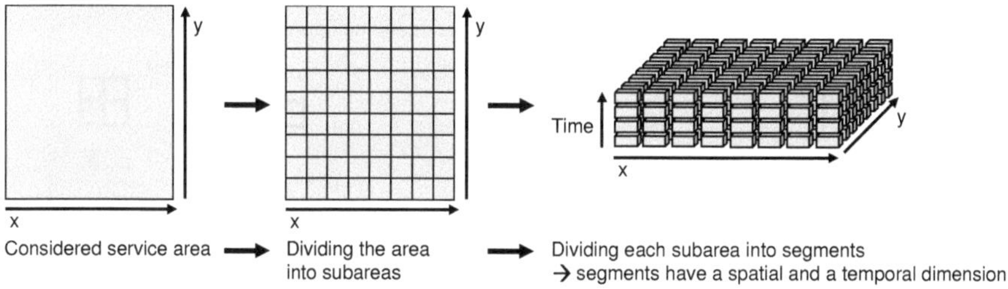

Fig. 6.1 Dividing the service area into subareas and segments

segments with such a highly detailed time resolution allows for a precise temporal mapping of request arrival rates but usually leads to low individual segment rate parameter values $\lambda(p_s)$, $s \in S$. In order to allow more reliable forecasts of expected future requests, adjacent segments are combined into *clusters* modeled by *compound Poisson processes*. Specifically, n non-homogeneous time-space Poisson processes p_s, $s \in S' \subset S$ with $|S'| = n$ are combined into a new compound Poisson process $p^*_{S'} = \bigcup_{s \in S'} p_s$ with rate parameter $\lambda(p^*_{S'}) = \sum_{s \in S'} \lambda(p_s)$. Note that $\lambda(p^*_{S'})$ represents the total number of expected requests in the segments $s \in S'$ on one day. Hence, the time unit of $\lambda(p^*_{S'})$ is defined by $T' \cdot DC^{\mathrm{te}}$ with T' denoting the temporal height of $p^*_{S'}$, i.e., the number of consecutive temporal levels from which segments $s \in S'$ are combined in $p^*_{S'}$. This combination of adjacent segments is performed for the purpose of $p^*_{S'}$ reaching a rate parameter which is sufficiently high, thereby justifying the assumption that real requests will actually arrive within the generated cluster. Such a cluster represents the aforementioned request-likely area.

In order to generate clusters with a specified minimum forecasting precision and a minimum reliability of the occurrence of at least one real request in the cluster during the execution of the transportation process, adjacent segments are assigned to a cluster c_{temp} so that the following three requirements are met:

(i) **Appropriate shape and extension:** Segments are combined so that the cluster c_{temp} is a three-dimensional convex area which represents the aforementioned request-likely area. The assigned segments do not exceed a predefined maximum spatial and temporal extension DC^{mse} and DC^{mte}, respectively.

(ii) **Minimum occurrence probability:** The sum of the rate parameter values of the segments $s \in S_{c_{temp}}$ which are assigned to c_{temp} reaches a defined minimum score $DC^{\min\lambda}$, i.e., $\lambda(c_{temp}) = \lambda(p^*_{S_{c_{temp}}}) \geq DC^{\min\lambda}$.

(iii) **Suitable request distribution and Poisson quality:** Both these quality aspects are described later on.

Preliminary tests have revealed that clusters of the desired quality defined by the three requirements above cannot be generated by applying the well-known k-means algorithm (cf. MacQueen 1967). Specifically, k-means requires the number of clusters that shall be generated. However, for generating stochastic knowledge in our application, this is a restrictive and inappropriate requirement since the number of

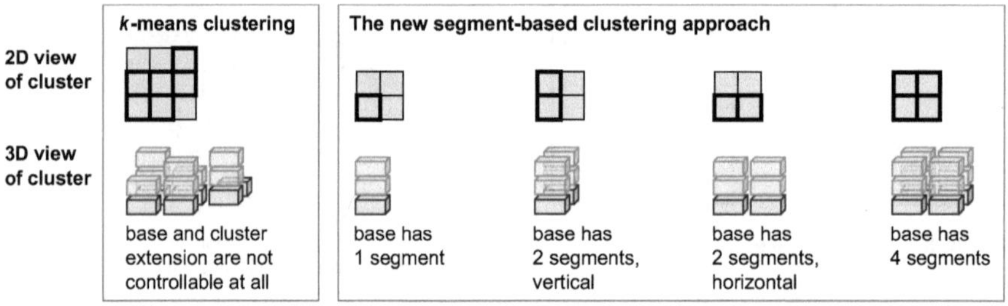

Fig. 6.2 The k-means clustering algorithm and the new proposed cluster generation approach with different convex bases

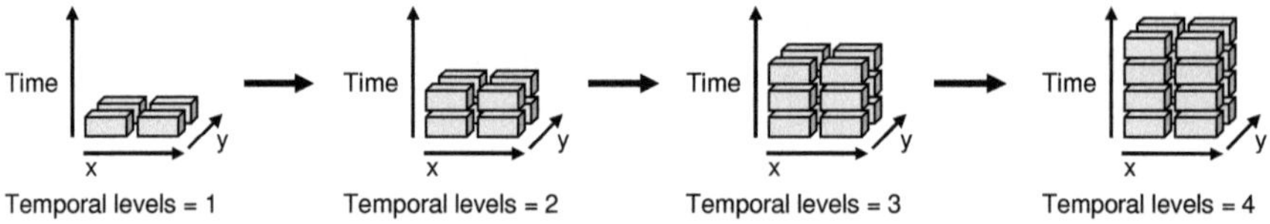

Fig. 6.3 Steps of the cluster generation approach generating one valid cluster using a base consisting of four segments

clusters that can be generated depends on the used clustering method itself and the structural quality of the available past request information. Moreover, according to requirement (i), a generated cluster is only useful if it represents a convex area and its spatial and temporal extensions do not exceed certain limits. This requirement cannot be controlled in the k-means algorithm. Additionally, clusters generated by k-means are rarely compact and frequently irregular (see Fig. 6.2). Therefore, we have developed a new cluster generation method that does not have the above-mentioned disadvantages.

Our proposed clustering approach works in two phases. It starts by generating all valid clusters that meet the three requirements mentioned above. In the second phase, the maximum set of non-overlapping clusters is determined. In order to generate all valid clusters in the first phase, a new cluster is iteratively generated by starting from all possible *bases*. Such a base comprises either 1, 2, or 4 segments that are adjacently located at the same temporal level (see again Fig. 6.2) and together form a convex cluster c_{temp} with a temporal height of one segment. Note that only bases are considered whose sum of assigned rate parameters is larger than 0.

Note that the definition of these bases provides a compact spatial cluster shape that guarantees the fulfillment of the convexity and the maximum spatial extension in requirement (i). In order to simultaneously fulfill requirement (ii), segments located at the successive temporal levels are iteratively inserted into c_{temp} (see Fig. 6.3). This is repeated until the calculated rate parameter $\lambda(c_{temp})$ is at least $DC^{\min\lambda}$ or the temporal height of c_{temp} exceeds DC^{mte}. In the latter case, c_{temp} is discarded, since it violates the maximum temporal extension in requirement (i). Otherwise, if c_{temp} fulfills both requirements, the additional quality criteria of requirement (iii) are checked by conducting the following two steps:

1. As mentioned above, real requests which arrive in a cluster are likely to have a location which is different from the corresponding dummy customer location. Hence, a vehicle currently positioned at the dummy customer location is required to perform a certain detour which consumes additional time in order to reach the real request. If the expected detour is too large, positive effects of using a dummy customer are likely to disappear. Therefore, the quality of clusters with regard to the expected travel time for such detour activities is assessed. For this purpose, a suitable location for the dummy customer in c_{temp} is derived. The spatial barycenter of the past requests that have been observed within c_{temp} over the last n^{f} working days is calculated and mapped on the road network by determining the closest located node $n'_{c_{temp}}$. Since this node may be positioned within an area consisting of roads with a low speed limit, it is relocated to a node that fulfills the following two requirements: First, it is located within a maximum predefined travel time radius DC^{radiusTT} of $n'_{c_{temp}}$. Second, in order to provide a reasonable road connection, it is located at a road of a minimum desired road class where lower road classes denote roads with higher speed limits. This node called $n_{c_{temp}}$ is determined by conducting a Dijkstra search that starts at $n'_{c_{temp}}$ and is limited by the parameter DC^{radiusTT}. Note that such a node may not exist in the considered radius. In this case, the closest node with minimum road class (i.e., largest speed limit) is determined. The node $n_{c_{temp}}$ represents the potential location of the dummy customer assigned to c_{temp}. After $n_{c_{temp}}$ is calculated, the average travel time $c_{temp}^{\mathrm{avgTT}}$ from $n_{c_{temp}}$ to all past requests observed within c_{temp} over the last n^{f} working days is calculated. If $c_{temp}^{\mathrm{avgTT}}$ exceeds a predefined maximum average travel time DC^{maxAvgTT}, c_{temp} is discarded since the estimated detour in c_{temp} is too large to provide a reliable forecast for an efficient positioning of the assigned vehicle.
2. As mentioned above, we assume in our forecasting method that request arrivals can be modeled by non-homogeneous time-space Poisson processes whose rate parameters are constant from day to day. In order to satisfy this assumption to a degree which is practically reasonable,[1] the quality of c_{temp} is analyzed. Specifically, the three requirements of Poisson processes (see Sect. 6.2) are considered for each cluster c_{temp} as follows. While requirement (a) is straightforward, requirement (b) is explicitly checked by applying Pearson's Chi-Square Goodness-of-Fit test (cf. Kvam and Vidakovic 2007, p. 155 et seq.). Using this test it is checked whether there exist significant deviations between the number of actual customer request arrivals in c_{temp} over the considered n^{f} working days and those that are expected according to the Poisson distribution with rate parameter $\lambda(c_{temp})$.

   In general, the Chi-Square Goodness-of-Fit test is applied to test the null hypothesis $H_0 : F_X(x) = F_0(x)$ where $F_X(x)$ is the observed sample $X =$

[1] As noted by Larson and Odoni (1981, p. 194), almost any practical system has a non-perfect degree of conformity with the postulates of the Poisson process. Hence, a trade-off between accuracy and insights possible by utilizing the Poisson model has to be found.

$\{X_1, X_2, \ldots, X_n\}$ and $F_0(x)$ represents the distribution to be tested. The domain of this distribution is partitioned into r non-overlapping intervals $I_1, I_2, \ldots, I_r$. Moreover, $n_1, n_2, \ldots, n_r$ represents the number of samples that have been observed in the corresponding interval with $\sum_{i=1}^{r} n_i = n$. Furthermore, np_i is the expected value of interval I_i according to the distribution to be tested under H_0. The Chi-Square test statistic for the sample X is then defined as follows:

$$X^2 = \sum_{i=1}^{r} \frac{(n_i - np_i)^2}{np_i}$$

The X^2 test statistic is asymptotically χ^2 distributed with $f = r - k - 1$ degrees of freedom where k represents the number of parameters for the distribution under H_0 that were estimated by using sample X (cf. Kvam and Vidakovic 2007, p. 156). Let $\chi_f^2(x)$ be the cumulative distribution function of the χ^2 distribution with f degrees of freedom and value x. After calculating the X^2 test statistic of the sample X, it is decided whether the null hypothesis is to be rejected or not. For this purpose, a type I error α which is application-dependent is defined. If it holds that $\chi_r^2(X^2) > 1 - \alpha$, the null hypothesis is rejected.

In our forecasting approach, the Chi-Square Goodness-of-Fit test is applied in order to test the null hypothesis H_0: the number of daily request arrivals observed in c_{temp} over the last n^{f} working days represented by the sample X are Poisson distributed with rate parameter $\lambda = \lambda(c_{temp})$. Since we estimated the rate parameter λ by using the sample X, k is 1 and consequently $f = r - 2$. We set $n = n^{\mathrm{f}}$ to the number of analyzed working days and accordingly $X_1, \ldots, X_{n^{\mathrm{f}}}$ represents the number of observed request arrivals in c_{temp} over the past n^{f} working days. Let $PS_\lambda(x)$ denote the probability mass function of the Poisson distribution with rate parameter λ and value x. According to the estimated rate parameter λ, we generate l intervals $I_1, I_2, \ldots, I_l$ with np_i denoting the expected number of days on which exactly $i - 1$ requests arrive in c_{temp}, i.e., $np_i = n \cdot PS_{\lambda(c_{temp})}(i - 1)$. The interval I_l represents the expected number of days on which at least $l - 1$ requests arrive in c_{temp}. Consequently, it holds that $\sum_{i=1}^{l} p_i = 1$. According to the past request information used in the cluster generation process, n_i represents the number of days out of the considered n^{f} working days on which exactly $i - 1$ requests have been observed in c_{temp}.

As mentioned above, the X^2 test statistic is asymptotically χ^2 distributed. In order to use this property also for practical sample sizes, different authors mention that certain data conditions should be met (cf., e.g., Yates et al. 1999, p. 734 and Kvam and Vidakovic 2007, p. 156). Specifically, it is recommended that at least 80 % of all intervals $I_1, \ldots, I_l$ have an expected value of $np_i \geq 5$ while all of the intervals fulfill $np_i \geq 1$. In order to fulfill these conditions, we pool the first and last intervals until both requirements are fulfilled. After this pooling has been performed, we obtain r intervals $I_1, \ldots, I_r$ with correspondingly adapted values of np_i and n_i. After calculating the X^2 test statistic for c_{temp} denoted as $X^2_{c_{temp}}$, the null hypothesis is rejected and the cluster is discarded if $\chi^2_{r-2}(X^2_{c_{temp}}) > 1-\alpha$. Clearly, smaller values of $\chi^2_{r-2}(X^2_{c_{temp}})$ indicate a higher conformity of observed request arrivals in c_{temp} with those expected by a Poisson distribution having rate parameter $\lambda(c_{temp})$. Hence, we define the *Poisson quality* of c_{temp} by $PQ(c_{temp}) = 1 - \chi^2_{r-2}(X^2_{c_{temp}})$. Since cluster c_{temp} is accepted if $PQ(c_{temp}) \geq \alpha$ we interpret α as the minimum demanded Poisson quality for valid clusters. For this reason, setting $\alpha = 0.05$ which is often proposed in the literature (cf., e.g., Israel 2008, p. xxii, Sapsford and Jupp 2006, p. 217) might be a value that is too low for our application since our aim is to accept only clusters whose observed request arrival rates possess a Poisson distribution to a certain extent. In Sect. 8.6, we will evaluate the impacts of different values for parameter α on the generated dummy customers.

Finally, requirement (c) of Poisson processes can be analogously checked by using a Chi-Square Goodness-of-Fit test and testing the request arrival times which were observed in c_{temp} over the last n^{f} working days for uniformity. However, due to the small temporal extensions of generated clusters, we refrain from performing this test and assume that request arrivals are independent of each other. This is because the required accuracy in request arrival times is frequently not fulfilled by practical request data sets and therefore would lead to an unfavorable rejection of clusters.

If both steps have been successfully completed, the cluster c_{temp} fulfills the demanded quality criteria and is therefore added to the set of valid clusters denoted by C. Furthermore, request parameter values of the corresponding dummy customer request of c_{temp} are calculated. The dummy customer request i belonging to c_{temp} gets the weight r_i^{w} defined by:

$$r_i^{\mathrm{w}} = P(X \geq 1) = 1 - PS_{\lambda(c_{temp})}(0)$$

This value corresponds to the probability that at least one request will arrive in c_{temp}. Moreover, the service time r_i^{s} of dummy customer i is defined by:

$$r_i^{\mathrm{s}} = \left(R^{\mathrm{st}} + c_{temp}^{\mathrm{avgTT}}\right) \cdot \lambda(c_{temp})$$

As mentioned above, r_i^{s} is set to the sum of service and travel times that the assigned vehicle is estimated to require for servicing all expected requests in c_{temp}.

Note that for each base (see Fig. 6.2) on which valid clusters can be generated, our clustering method generates the valid cluster with the smallest temporal height. Since generating such precise clusters allows for producing two temporally small clusters instead of a large one, two dummy customers can also be created. This is advantageous since it increases the flexibility of the pro-active real-time control approach by allowing to assign the integrated stochastic knowledge to different vehicles.

6.2.2 Cluster Selection and Dummy Customer Generation

After generating all valid clusters, the clusters which are integrated as stochastic knowledge into the transportation process have to be selected. In order to integrate as much stochastic knowledge as possible, it is pursued to select the maximum number of clusters that do not overlap from the set C. For this purpose, the following Mixed Integer Programming (MIP) problem is solved. The set $S_C \subseteq S$ comprises all segments which are assigned to clusters of the set C. The binary parameter c_{is} is set to 1 if and only if cluster $i \in C$ contains segment $s \in S_C$. Moreover, parameter c_i^{startTL} denotes the temporal level at which cluster c_i starts. The binary decision variable x_i is 1 if and only if cluster $i \in C$ is selected, 0 otherwise. Furthermore, parameter M is a big number. Using these definitions, the cluster selection procedure is formulated as follows:

$$\max z = \sum_{i \in C} \left(M \cdot x_i - c_i^{\text{startTL}} \cdot x_i \right) \quad \text{s.t.} \tag{6.1}$$

$$\forall s \in S_C\colon \quad \sum_{i \in C} c_{is} \cdot x_i \leq 1 \tag{6.2}$$

$$\sum_{i \in C} x_i \leq C^{\max} \tag{6.3}$$

The primary goal of the hierarchical objective function (6.1) is the maximization of the number of selected clusters. Since this also maximizes the number of dummy customers that are integrated in the tour plan, it ensures that the maximum stochastic knowledge generatable by our forecasting approach is utilized. The secondary goal is to start clusters at the earliest point in time in order to support an earlier service of the requests that are expected in the identified request-likely areas. This also supports the pursued aim of guiding the vehicles into request-likely areas at the right time. Constraints (6.2) ensure that all segments $s \in S_C$ are assigned to, at most, one cluster $i \in C$. Constraint (6.3) imposes a theoretical upper bound on the number of clusters that can be selected. Note that this restriction was defined for technical reasons; since it allows the MIP solver to generate tighter bounds during the examination process which enables a faster solution process of the considered cluster selection problem. For the computational experiments, this bound is initially set to:

$$C^{\max} = \left\lfloor \frac{\sum_{s \in S_C} \lambda(p_s)}{DC^{\min\lambda}} \right\rfloor$$

After solving the cluster selection problem, for each selected cluster a dummy customer with the parameter settings defined above is added to the set of initial requests. These dummy customer requests are integrated into the initial tour plan which is constructed before the execution of the transportation process. This is described in Sect. 7.4.

6.3 Handling Dummy Customers in the Pro-Active Real-Time Control Approach

As mentioned in Sect. 6.1, request-likely areas are only present for a specific time period during the execution of the transportation process. Hence, the proposed utilization of stochastic knowledge in the real-time control approach requires the updating of specific dummy customer parameters and appropriate actions during the transportation process. In what follows, we describe in detail the performed *dynamic parameter updating during the transportation process*. This is followed by the presentation of the *extended vehicle scheduling strategy with regard to dummy customers*. Specifically, if a dummy customer is the next request in the tour of a vehicle, it is advantageous to determine its departure according to the start of the dummy customer's time window. Finally, an *extended vehicle waiting strategy using dummy customer information* is proposed.

6.3.1 Dynamic Parameter Updating During the Transportation Process

As stated above, a dummy customer i belonging to cluster c_i guides a vehicle to a request-likely area whose spatial and temporal dimensions are equal to the ones of the corresponding cluster c_i at the appropriate time. Analogous to the definition of c_i^{startTL} presented above, we define c_i^{endTL} as the temporal level at which cluster c_i ends. Consequently, the start and end time of cluster c_i denoted as c_i^{start} and c_i^{end} in the transportation process is defined by $c_i^{\text{start}} = c_i^{\text{startTL}} \cdot DC^{\text{te}}$ and $c_i^{\text{end}} = (c_i^{\text{endTL}} + 1) \cdot DC^{\text{te}}$. From the moment the time point of the anticipated system situation τ^+ reaches c_i^{start} during the execution of the transportation process, the values of the parameters of dummy customer i continuously change. This is because the number of segments, which are assigned to the remaining part of cluster c_i that is still in the future at τ^+, diminishes so that the rate parameter value of the remaining assigned compound Poisson process decreases with τ^+. Therefore, whenever a new problem instance which represents the anticipated system situation at τ^+ is generated at τ, the remaining rate parameter value of each cluster c_i is updated to $\lambda(c_i, \tau^+)$. In particular, this value represents the rate parameter values of the Poisson processes p_s of segments that are assigned to c_i and are still in the future at τ^+. Moreover, the remaining rate values of segments, within whose time dimension the time point τ^+ is currently situated, are linearly approximated. Clearly, for each cluster with $c_i^{\text{start}} \geq \tau^+$, it holds that $\lambda(c_i, \tau^+) = \lambda(c_i)$. Moreover, for each cluster c_i we define the start time of the remaining part of the cluster c_i at τ^+ as $c_i^{\text{start}}(\tau^+) = \max(c_i^{\text{start}}, \tau^+)$. Based on the anticipated system state at τ^+ and the resulting rate parameter $\lambda(c_i, \tau^+)$, the updating of the three parameters $r_i^{\text{e}}, r_i^{\text{s}}$, and r_i^{w} of each dummy customer i is conducted as follows:

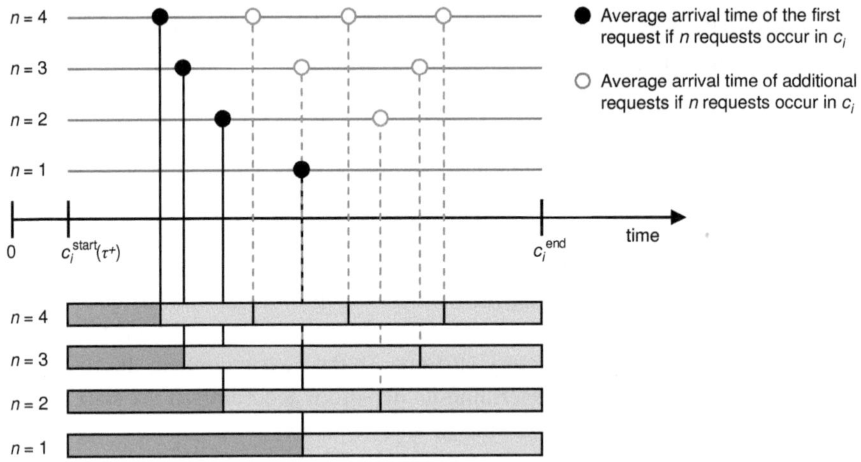

Fig. 6.4 Average arrival time of the first request when n request arrivals occur in cluster c_i

- **Beginning of the dummy customer time window (r_i^e):** As mentioned above, requests represented by dummy customer i are only expected and thus do not occur with certainty. Hence, the opening of time window r_i^e is set to the expected arrival time of the first request in the remaining part of cluster c_i which is still in the future at τ^+ under the assumption that at least one more request will occur in this remaining part of c_i. Clearly, this value is always in the future. The expected arrival time of the first request is updated as follows:

$$r_i^e(\tau^+) = c_i^{start}(\tau^+) + (c_i^{end} - c_i^{start}(\tau^+)) \cdot \frac{\sum_{n=1}^{\infty} PS_{\lambda(c_i,\tau^+)}(n) \cdot \frac{1}{n+1}}{1 - PS_{\lambda(c_i,\tau^+)}(0)} \tag{6.4}$$

The expected arrival time of the first request in cluster c_i is derived by determining the average arrival time of the first request in all scenarios with $1, 2, \ldots, \infty$ requests that arrive in the remaining time period of cluster c_i. The calculation of this average arrival time is described in what follows. Due to the assumed properties of Poisson processes (see Sect. 6.2), request arrivals are on average uniformly distributed over the time period of cluster c_i. Hence, as illustrated in Fig. 6.4, when $n > 0$ requests arrive in cluster c_i, the remaining time period of cluster c_i is divided into $n + 1$ intervals of equal length so that the average arrival time of the first request, relative to the beginning of cluster c_i, is $\frac{1}{n+1} \cdot (c_i^{end} - c_i^{start}(\tau^+))$. This time point is multiplied by the probability $PS_{\lambda(c_i,\tau^+)}(n)$ which represents the probability that exactly n requests will arrive in cluster c_i. The resulting values for all $n \in \{1, 2, \ldots, \infty\}$ are summed up and divided by the probability that at least one request will arrive in cluster c_i. By adding the current start time point $c_i^{start}(\tau^+)$ of cluster c_i, the expected arrival time of the first request in cluster c_i is determined (see (6.4)).

Since request arrivals are independent of each other in Poisson processes, no requests which have already arrived in c_i need to be considered. Thus, due to the impact of $r_i^{\mathrm{e}}(\tau^+)$ on the objective function value, generated tour plans do not prefer to visit dummy customer i earlier than the expected occurrence of the next request within cluster c_i if other requests can be visited before.

Note that updating r_i^{e} in the described way has an additional advantage. Consider the case in which r_i^{e} is not updated and remains at a value defined by $r_i^{\mathrm{e}} = c_i^{\mathrm{start}}$. In this case, although a vehicle is already waiting at the dummy customer, the corresponding dummy customer nevertheless contributes a customer inconvenience value to the objective function. This is an unwanted scenario since it is not taken into consideration whether or not a vehicle has arrived at the dummy customer location in time. In contrast, by updating r_i^{e} in the way described above, it is positively acknowledged when a vehicle arrives at the location of dummy customer i in time by omitting a contribution of a customer inconvenience value to the objective function value of the generated tour plan.

- **Service time** (r_i^{s})**:** As mentioned above, the service time r_i^{s} of dummy customer i represents the time period which a vehicle is estimated to require in the corresponding request-likely area. This time period is represented by the sum of average travel and service time for fulfilling all expected future requests. Since the number of expected future requests in c_i at τ^+ is given by $\lambda(c_i, \tau^+)$, the service time is updated to:

$$r_i^{\mathrm{s}}(\tau^+) = \left(R^{\mathrm{st}} + c_i^{\mathrm{avgTT}}\right) \cdot \lambda(c_i, \tau^+)$$

- **Weight factor** (r_i^{w})**:** For each dummy customer i, the weight factor r_i^{w} is updated according to the current remaining probability at τ^+ that at least one more request will arrive in cluster c_i. Therefore, for each dummy customer i, this probability can be calculated by:

$$r_i^{\mathrm{w}}(\tau^+) = P(X \geq 1) = 1 - PS_{\lambda(c_i,\tau^+)}(0)$$

Furthermore, since dummy customers are placeholders for expected future real requests, dummy customers are not serviced. Hence, the time point when a dummy customer is removed from the tour plan has to be defined. In our approach, a dummy customer is removed from the set of unserviced requests $R_{\tau^+}^{\mathrm{U}}$ when it becomes unlikely that further requests will occur in the considered request-likely area defined by cluster c_i. This time point denoted as $t_{c_i}^{\mathrm{rem}}$ is defined by the time when $\lambda(c_i, \tau^+)$ becomes smaller than a predefined minimum threshold $DC^{\lambda\mathrm{rem}}$. Note that since $t_{c_i}^{\mathrm{rem}}$ is independent of dynamic events, it can be calculated offline before tour plan execution. Hence, the point in time when each dummy customer is removed from the tour plan during the transportation process is exactly defined before the execution of the transportation process.

6.3.2 Extended Vehicle Scheduling Strategy with Regard to Dummy Customers

During the execution of the transportation process, situations may occur in which a dummy customer is scheduled as the next request of a vehicle tour but its service is planned in the far future due to a much later opening of its time window. Using the vehicle scheduling strategy presented in Sect. 5.2.3, the vehicle would nevertheless immediately drive to the dummy customer i and wait there for a much longer time than is intended by the dummy customer. In contrast to a scenario where the current transportation process execution time τ is after the time point c_i^{start} and therefore placing an idle vehicle in the corresponding request-likely area is reasonable, this could lead to unwanted scenarios. For instance, vehicles are guided to remotely located areas much too early and are therefore not directly available for intermediately arriving requests in more central regions.

In order to deal with this issue, an extended vehicle scheduling strategy is applied. Specifically, if a vehicle's next request is a dummy customer i, the vehicle waits at its current position as long as the difference between the current transportation execution time to c_i^{start} is larger than the travel time that is needed to reach the dummy customer i. This is performed because a visit to dummy customer i might also be useful in the interval $[c_i^{\text{start}}, r_i^{\text{e}})$ since requests are also expected to occur with a certain probability before the expected arrival time of the next request in the cluster c_i denoted by r_i^{e}.

6.3.3 Extended Vehicle Waiting Strategy Using Dummy Customer Information

As mentioned before, dummy customers are used in order to guide vehicles to request-likely areas and keep them there for a specified amount of time. The length of this time period is decided by the pro-active real-time control process itself. Specifically, a vehicle waits at a dummy customer until another request is assigned to its tour as the first request. This happens when another request is assigned before the dummy customer to the vehicle's tour or when the dummy customer is removed from the tour plan. By allowing the pro-active real-time control approach to dynamically decide about the length of this waiting period, the waiting strategy can be utilized in a flexible manner according to changing situations during the execution of the transportation process.

In order to enable a vehicle to wait at the location of a dummy customer for future customer requests to arrive, an additional vehicle status has to be defined in the mathematical model introduced in Sect. 5.3.2. This is due to the fact that while real requests can always be serviced immediately, the start of the time window of a dummy customer is always in the future as described in Sect. 6.3.1. In this constellation, the vehicle simply waits at the dummy customer in order to be prepared for quickly servicing future arriving requests in the corresponding request-likely area. The additional vehicle status can be handled analogous to the third status presented in Sect. 5.3.2.2. Specifically, if, at τ^+, vehicle k is currently waiting at a dummy customer location, the vehicle location v_k^{p} is set to the location of the dummy customer and the vehicle availability time $v_k^{\delta t}$ is set to 0 since vehicle k is directly available for servicing other requests.

6.4 Summary

In this chapter, a new forecasting approach was presented in order to extend the deterministic real-time control approach to a pro-active real-time control approach. The proposed forecasting approach makes it possible to generate stochastic knowledge out of the available past request information. After its generation, the stochastic knowledge is integrated into the real-time control approach by using dummy customers. In contrast to real requests, dummy customers are used to guide vehicles into request-likely areas in which future requests are expected to occur with a high probability according to the available stochastic knowledge. The generation of stochastic knowledge is performed by partitioning the considered service area into subareas and segments. For each of these segments, request arrivals are modeled by a time-space Poisson process whose rate parameter is calculated by analyzing the number of request arrivals in the respective segment on the considered past working days. Afterwards, dummy customers are created by aggregating segments to clusters that fulfill certain quality criteria, for example that one or more requests actually arrive in place of a dummy customer with a minimum specified probability. In order to check the assumption that the number of request arrivals in the segments can be modeled by Poisson processes, the likelihood that past request arrivals which are observed in each generated cluster follow a Poisson distribution is evaluated. If this is too unlikely, the cluster is discarded. In order to generate as many valid dummy customers as possible, an exact two-step solution method was presented. In the first step, all valid clusters are generated while in the second step the maximum number of non-overlapping clusters is selected by making use of an MIP solver. Moreover, the methods by which the deterministic real-time control approach is extended in order to integrate and handle dummy customers were described. This includes how relevant parameters of dummy customers are updated during the execution of the transportation process and how it is decided when dummy customers are to be removed from further consideration. Furthermore, the extension of the deterministic real-time control approach included the introduction of an extended vehicle scheduling strategy. This strategy makes a vehicle wait at its current position if its next assigned request is a dummy customer and if its departure at the current time in the transportation process would otherwise make the vehicle arrive earlier than desired at the dummy customer's location.

Chapter 7
The Proposed Tabu Search Solution Method

In this chapter, the developed solution method which is utilized to solve individual problem instances that occur during the execution of the transportation process is presented. This solution method is a specific stage-based Tabu Search metaheuristic. In order to efficiently integrate intensification and diversification aspects into the enumeration process, the Tabu Search procedure switches between different stages that control the currently applied neighborhood operator. In each iteration, depending on the current stage, a neighborhood operator to be applied is deterministically selected. If a defined number of iterations has been performed without finding a new best solution, the Tabu Search algorithm switches into the next higher stage in order to continue the search with a different neighborhood operator. Neighborhood operators on lower stages intensify the search in direct vicinity of the current solution, whereas higher stages apply more diversifying neighborhood operators. If a new best solution is found, the search returns to the first stage.

As mentioned in Sect. 5.2.1, a static problem instance that is adapted during a run of the solution method is not modified by dynamic requests that intermediately occur in the transportation process since these events are buffered until the generation of the next static problem instance. Therefore, each adaptation of the solution method is similar to solving a static routing problem. In order to support the approach to enable a continuous optimization over static problem instances, an appropriate tabu list structure is developed. In the static problem instance generated at τ, the Tabu Search solution method is applied for $t^{\mathrm{a}} - t_{\tau}^{\mathrm{calc}}$ time units. After the execution of the solution method, the tour plans $P_{\tau+}^{\mathrm{bt}}$ and $P_{\tau+}^{\mathrm{t}}$, i.e., the best and the last tour plan found during the search process, are generated. Hence, since the computational time that is available for the solution method is temporally restricted, different concepts for efficiently computing the objective function are utilized in order to speed up the solution process.

In this chapter, an introduction to Tabu Search by describing *general concepts of the implemented solution method* is given. Specifically, basic concepts of this metaheuristic are explained as well as the way a tour plan is represented in the developed solution method. Moreover, it is illustrated how a tour plan is modified by the application of the neighborhood operators and how the objective function value

F. Ferrucci, *Pro-active Dynamic Vehicle Routing*, Contributions to Management Science,
DOI 10.1007/978-3-642-33472-6_7, © Springer-Verlag Berlin Heidelberg 2013

is efficiently updated according to performed modifications. After that, the individual *neighborhood operators* that are applied in the proposed Tabu Search solution method are presented in detail. Furthermore, the *implemented tabu list* for identifying tabu-active solutions is described. Moreover, the *construction of the initial tour plan* is illustrated. In the following section, the *stage-based neighborhood operator selection scheme* that controls the application of the individual neighborhood operators during the execution of the solution method is presented. Finally, the *integration of the solution method into the real-time control approaches* is explained.

7.1 General Concepts of the Implemented Solution Method

In this section we introduce the general concepts that are used in the proposed Tabu Search solution method. First, an introduction to the *Tabu Search metaheuristic* is given. Since Tabu Search is an iteration-based solution method, a *description of an iteration in the solution method* is given. This is followed by the *representation of a tour plan in the solution method* is and, finally, it is explained in which way an *efficient calculation of changes in the objective function value* is performed.

7.1.1 The Tabu Search Metaheuristic

In the real-time control approach that is used in this book to coordinate the transportation process of RDOPG applications, a stage-based Tabu Search metaheuristic (cf. Glover 1989 and Glover and Laguna 1998) is applied. The Tabu Search metaheuristic is an iteration-based heuristic improvement method which performs a neighborhood based search in the solution space. As mentioned in Sect. 2.3.5, the neighborhood of a neighborhood operator is defined as the set of solutions which can be generated from a current solution by applying this operator. In order to overcome local optima and to enable the metaheuristic to examine a broader part of the solution space, neighborhood operators are classified into *intensifying* and *diversifying* operators (cf. Glover and Laguna 1998, p. 7 et seq.):

- **Intensifying neighborhood operators:** Such operators often attempt to improve the current solution by applying small changes to the tour plan, e.g., only reinserting one customer to another location within the same tour. It is advisable to utilize such operators when the current solution is located in a part of the solution space where better solutions are expected to be attained by making slight changes in the current solution. Besides this, Glover and Laguna state that these operators are useful for exploring the neighborhood of elite solutions i.e., new best solutions, found during the search process. This is carried out in order to find local optima in the solution space that are close to the current solution. Moreover, these operators can also be used to combine parts of previously found elite solutions in order to generate a new best solution similar to adaptive memory techniques, e.g., as applied in Gendreau et al. (1999a). In doing so, previously visited parts of the solution space can also be re-examined in more detail.

- **Diversifying neighborhood operators:** Such operators attempt to improve the current solution by applying significant modifications to it. In so doing, guiding the search process to unvisited parts of the solution space is the pursued aim so that these parts are examined for other promising solutions. The utilization of diversifying neighborhood operators is advisable when intensifying neighborhood operators have not found a new best solution during a specific number of past iterations of the search process.

As a specific criterion of Tabu Search metaheuristics, information about the progress of the current search process is stored. Specifically, characteristics of solutions that have already been visited are stored in a list denoted as *tabu list*. Using this tabu list aims at avoiding the revisiting of solutions within a specified number of iterations during the solution process. For this purpose, visited solutions are marked as *tabu-active*. The utilization of the tabu list supports the search process in continuing to evaluate other parts of the solution space after a local optimum has been reached. In order to identify tabu-active solutions, characteristic attributes for determining the tabu-active state of solutions have to be defined. More information about the implemented attributes in the proposed Tabu Search metaheuristic is given in Sect. 7.3.

An iteration of the search process in a Tabu Search metaheuristic is generally carried out as follows. Its neighborhood is examined by applying a specific neighborhood operator starting from a current solution. After all solutions of this non-tabu active neighborhood have been evaluated, one neighbor is selected as the new solution according to specific criteria. The modifications that are applied to the current solution to obtain the new solution are called a *neighborhood move* or in short *move*. Note that in situations during the search process in which all neighbored solutions are worse than the current solution, the selected neighborhood move leads to a solution with an objective function value worse than the current solution. For example, this scenario happens when a local optimum is reached. In a Tabu Search metaheuristic, worse solutions are usually accepted in order to overcome these local optima. Specifically, the application of more diversifying neighborhood operators is likely to first result in worse solutions since major parts of the current solution are changed in order to guide the search process into a different part of the solution space. However, this is accepted since it is expected that this temporal deterioration leads to better solutions in subsequent iterations.

Since Tabu Search is an improvement metaheuristic, an initial solution has to be provided to it. In our proposed solution approach, the generation of the initial solution, i.e., the initial tour plan, is described in Sect. 7.4.

7.1.2 Description of an Iteration in the Solution Method

In each iteration, a neighborhood operator is selected and its neighborhood is evaluated. In the proposed solution method, this selection is deterministically performed by using a stage-based selection scheme which is described in Sect. 7.5. Once a neighborhood operator has been selected, it is applied to a given current solution (i.e., the current theoretical tour plan) by executing the following steps:

1. **Selecting the considered request set:** Depending on the chosen neighborhood operator, one or more unserviced requests of the set $R^{\mathrm{U}}_{\tau+}$ are selected. In what follows, the set of selected requests is denoted as R^*.
2. **Removing the considered request set:** The requests belonging to the set R^* are removed from the current solution.
3. **Temporarily reinserting the requests into the theoretical tour plan:** Each request of R^* is reinserted into the theoretical tour plan. The position at which each request is reinserted into the theoretical tour plan is determined by the neighborhood operator and can be either randomly chosen, exhaustive (i.e., successively evaluating all positions), or rule-based (defined by the neighborhood operator, e.g., at its least cost position). This set of temporarily performed reinsertion activities defines the currently evaluated neighborhood move. By this reinsertion, a tentative new solution is generated.
4. **Evaluating the tentative new solution:** In order to assess the quality of this new solution, the change in the objective function value denoted as the *move value* is calculated.
5. **Storing the best found neighborhood move:** If the current neighborhood move is non-tabu, it is compared to the currently known best neighborhood move that has been found during the ongoing neighborhood examination in this iteration. If the current neighborhood move is better than the best found move so far, it becomes the new best found move. Moreover, the best found move is stored as *escape move* which, regardless of whether it is tabu-active or not, is executed in case all evaluated neighborhood moves are tabu. Note that this only happens if the set $R^{\mathrm{U}}_{\tau+}$ comprises few requests so that all neighbored solutions of the current neighborhood operator have already been evaluated in the previous iterations of the current static problem.
6. **Evaluating further neighbors:** If the current neighborhood operator evaluates more than one reinsertion position of the requests of set R^*, further reinsertion positions are evaluated by continuing this procedure at step 3. Otherwise, the process continues with the next step.
7. **Restoring the initial solution:** After all reinsertion positions for the requests of set R^* have been evaluated, the considered requests are reinserted at their original positions so that the theoretical tour plan is now identical with the initial solution of this iteration.
8. **Selecting the next request set to be conducted:** If the current neighborhood operator considers the temporal reinsertion of different request sets R^*, the process continues with the next set at step 1, otherwise at step 9.
9. **Selecting and executing the best neighborhood move:** After all possible neighbored solutions have been evaluated, the best found neighborhood move (or, in case no non-tabu move has been found, the escape move, see step 5) is executed. This new generated solution now represents the current solution.

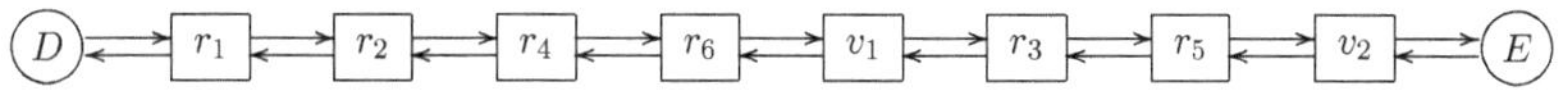

Fig. 7.1 Representation of a tour plan using a doubly linked list

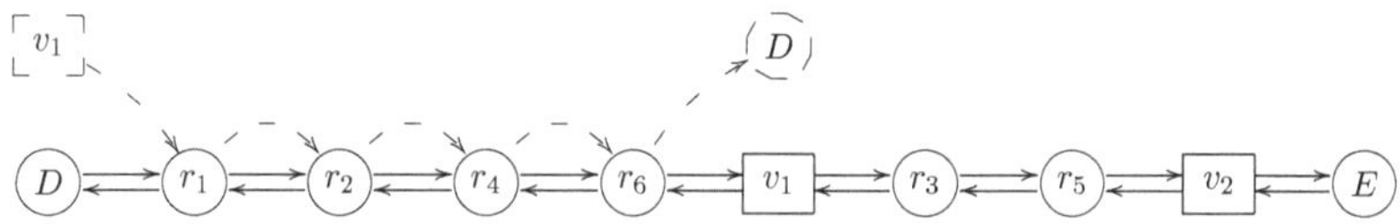

Fig. 7.2 Steps for determining the tour of vehicle 1

10. **Marking the current solution as tabu-active:** The tabu-active characteristics of the current solution are stored in the tabu list.
11. **Updating the best found solution:** If the new current solution possesses a better objective function value than the best found solution so far, it is replaced by the new current solution.

Here, the current iteration of the proposed Tabu Search metaheuristic is completed and the newly generated current solution is used as starting solution in the next iteration.

7.1.3 Representation of a Tour Plan in the Solution Method

In this section, an efficient data structure for modeling tour plans in the solution method is described (cf., e.g., Nanry and Wesley Barnes 2000). In this data structure, a tour plan is stored as a doubly linked list (see Fig. 7.1). The nodes of this list represent the relevant locations (i.e., requests and the depot) in the current situation of the transportation process and the currently utilized vehicles. The first node in this list is always the depot D and the last node is an artificial end tour plan element E. Moreover, each request i is modeled by node r_i and each vehicle j is modeled by node v_j. The tour of vehicle k is determined by all request nodes that are located to the left of vehicle k represented by vehicle node v_k in the doubly linked list up to the next vehicle node or the depot node, respectively. Each vehicle starts its tour at its current position determined by the vehicle node, services the assigned request nodes and lastly travels back to the depot node.

Note that unused vehicles waiting at the depot are not generally included in the tour plan and need to be inserted by a neighborhood operator. In the example given in Fig. 7.1, the tour of vehicle 1 is therefore described as $v_1 \rightarrow r_1 \rightarrow r_2 \rightarrow r_4 \rightarrow r_6 \rightarrow D$ and vehicle 2 has the tour $v_2 \rightarrow r_3 \rightarrow r_5 \rightarrow D$. In Fig. 7.2, the tour of vehicle 1 is illustrated.

7.1.4 Efficient Calculation of Changes in the Objective Function Value

As stated in step 4 in Sect. 7.1.2, the change in the objective function value, i.e., the move value, has to be evaluated for each neighbored solution within an iteration. Clearly, this evaluation can be realized by calculating the objective function value of the complete tentative tour plan. However, since this consumes computational time and the solution method is executed under a given time limit, it is advisable to determine the move value in more computational time-efficient ways. In what follows, we present two aspects which are utilized in the proposed solution method for efficiently calculating the move value. The first aspect consists of two important characteristics of the considered objective function value: (a) the objective function value of a tour plan is equal to the sum of all tours in the tour plan and (b) the objective function value of each tour is independent of the other tours. Hence, by storing the objective function values for each tour instead of storing only one value of the complete tour plan, the move value can be computed by only calculating differences in the objective function of tours which are modified by the current neighborhood move. In the second aspect, systematically examining the neighborhood of the current neighborhood operator enables the re-use of partial results of computed objective function values calculated in previously examined neighbored solutions. This reduces the computational effort of calculating changes in the objective function value (cf., e.g., Badeau et al. 1997). An example of systematically exploring the neighbored solutions is shown for the neighborhood operator REL which is described in Sect. 7.2.2.

7.2 Neighborhood Operators

In this section, the neighborhood operators that are applied in the proposed Tabu Search metaheuristic are presented. Note that all utilized neighborhood operators are node-exchanging operators, i.e., they move one or more requests in the tour plan to another location within the same vehicle tour or onto another vehicle tour. All utilized neighborhood operators only consider feasible tour plans which fulfill all restrictions of the mathematical model defined in Sect. 5.3.2.3.

First, intensifying neighborhood operators are described, i.e., *Within Tour Insertion (WTI)* and *Relocate (REL)* while neighborhood operators possess more diversifying characteristics, i.e., *MultiRelocate (MREL)*, *Large Neighborhood Search (LNS)*, and *Exchange Between Tours (XBT)* are presented afterwards. Besides describing the utilized neighborhood operators in detail, they are also presented in pseudo-code. In order to comply with the existing time limit, all neighborhood operators frequently check the remaining available computational time during their execution. Hence, our computational experiments show that the available computational time limit is always met by the proposed Tabu Search solution method. In what follows, we use the function $U(a, b)$ which returns a uniformly distributed continuous number from interval $[a, b]$ as well as the function $u(a, b)$ which returns a uniformly distributed discrete number from interval $[a, b]$.

7.2.1 Within Tour Insertion (WTI)

The WTI neighborhood operator removes a single request and reinserts it at its best position within the same tour. Specifically, for each tour the best performing local relocation that is not tabu-active is implemented if it leads to a better solution. In order to provide maximum intensification, this operator only allows request relocations that lead to a better solution. Therefore, within each application, for each tour of the tour plan, WTI removes a single request and reinserts it at its best position within the same tour if this leads to a solution improvement. This search is exhaustive, i.e., an evaluation is made for each request in the tour plan within one move while storing the best request relocation for each tour. After the evaluation, the neighborhood move consisting of all stored request relocations is applied to the tour plan unless this leads to a solution that is marked as tabu. In the latter case, the best found request relocations are applied successively to the tour plan in order of non-increasing objective function value improvements. After the application of each single request relocation, it is checked whether the current tour plan is tabu-active. If not, the current set of performed request relocations is stored as the current maximum implementable set $WTI^{\max}$ of request relocations that lead to a non-tabu solution. After evaluating all request relocations, the neighborhood move consisting of the first $WTI^{\max}$ stored tour moves is applied in order of non-increasing objective function value improvements to the tour plan. The pseudo-code of the WTI neighborhood operator is depicted in Algorithm 7.1.

Algorithm 7.1 Neighborhood operator Within Tour Insertion (WTI)

for each request $r \in R^{\mathrm{U}}_{\tau+}$ **do**
    Remove r from $P^{t*}_{\tau+}$
    Reinsert r at all positions within its original tour
    Store the best improving non-tabu tour move for each tour
end for
Sort best tour moves by their objective function value improvement (highest first)
Perform the combined neighborhood move consisting of the consecutive application of all stored best tour moves that provide an improvement
if combined neighborhood move is tabu-active **then**
    Perform first tour move. This move can be executed non-tabu, otherwise, it would not have been stored
    Perform improving best tour moves consecutively
    Store the highest number $WTI^{\max}$ of tour moves that lead to a non-tabu solution
    Perform the neighborhood move consisting of the $WTI^{\max}$ tour moves in consecutive order
end if

7.2.2 Relocate (REL)

The REL neighborhood operator works identically with the previously described WTI neighborhood operator except for the fact that REL evaluates all positions for request relocation that are on tours other than the one the selected request is currently assigned to. Hence, it also evaluates the benefits of additionally using currently idle and unused vehicles. Similar to the WTI neighborhood operator, REL also simultaneously performs multiple request relocations in one iteration, if possible. Since a neighborhood move of REL modifies two tours for each relocated request, both modified tours are marked as changed in order to exclude them from further request relocation operations in the current neighborhood move. Analogous to the neighborhood operator WTI, all stored request relocations are subsequently applied in order of non-increasing objective function value improvements. Potentially occurring solutions which are tabu-active are avoided in the same way as described in WTI. In order to allow an increased diversification compared to WTI, in case of non-existing improving request relocations in the current neighborhood, the best neighborhood move that leads to a non-tabu solution is applied, i.e., the one that generates the slightest cost increase. The pseudo-code of the REL neighborhood operator is depicted in Algorithm 7.2.

Algorithm 7.2 Neighborhood operator Relocate (REL)

for each request $r \in R^{\mathrm{U}}_{\tau+}$ **do**
  Remove r from $P^{t*}_{\tau+}$
  Reinsert r at all positions on tours other than the original tour
  Store the best improving non-tabu move for each tour from which r is removed. Furthermore, remember the second tour in this move on which r is relocated to
end for
Sort best tour moves by their objective function value improvement (highest first)
Perform best tour move regardless of improvement value. If further improving tour moves are available, perform the combined neighborhood consisting of consecutive application of improving best tour moves that only change tours which have not already been modified by previously applied tour moves
if combined neighborhood move is tabu-active **then**
  Perform first tour move. This move can be executed non-tabu, otherwise, it would not have been stored
  Perform improving best tour moves consecutively
  Store the highest number $REL^{\max}$ of tour moves that lead to a non-tabu solution
  Perform the neighborhood move consisting of the $REL^{\max}$ tour moves in consecutive order
end if

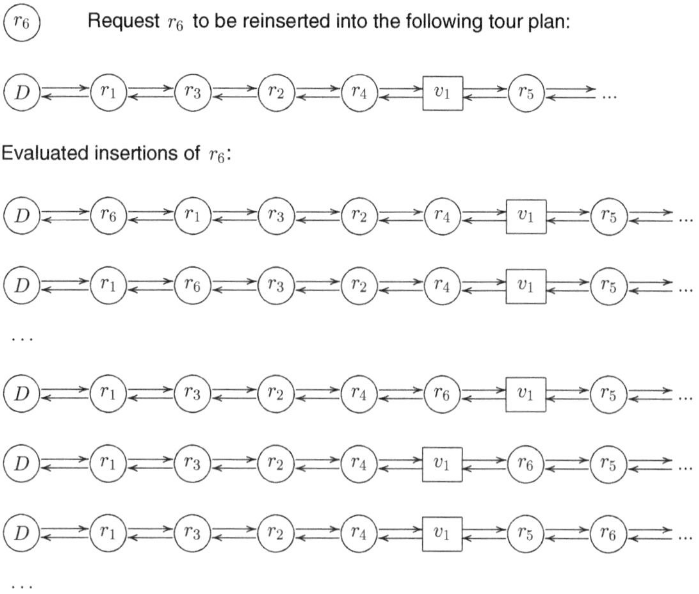

Fig. 7.3 The first steps of systematically reinserting request r_6 into the tour plan using the neighborhood operator REL

As described in Sect. 7.1.4, the evaluation of the current neighborhood can be performed in a time-efficient way which is described in what follows at the example of REL. For this purpose, let $c(t)$ denote the objective function value of tour t of a tour plan P. The origin tour from which request r will be removed is defined by t_r^{o} and t_r^{d} represents the destination tour to which r will be relocated to. With t_{r-}^{o} representing tour t_r^{o} after request r has been removed and t_{r+}^{d} defining t_r^{d} into which r is inserted at the next position to be evaluated, according to the used neighborhood operator. Using these definitions, the resulting move value can be calculated by $c(t_{r-}^{\mathrm{o}}) + c(t_{r+}^{\mathrm{d}}) - c(t_r^{\mathrm{o}}) - c(t_r^{\mathrm{d}})$. As can be seen, for a fixed origin and destination tour, $c(t_r^{\mathrm{d}})$, $c(t_r^{\mathrm{o}})$, and $c(t_{r-}^{\mathrm{o}})$ remain constant so that only $c(t_{r+}^{\mathrm{d}})$ changes depending on the reinsertion position of r. Moreover, by enumerating the insertion positions of the destination tour in ordered sequence, the computational effort required for calculating $c(t_{r+}^{\mathrm{d}})$ in each neighbored solution can be reduced. As depicted in Fig. 7.3 at the example of reinserting request r_6, the first part of tour $t_{r_6^+}^{\mathrm{d}}$ up to the current insertion position of r_6 remains constant and is extended by exactly one request in each step of the neighborhood evaluation. Hence, the objective function value of the first part of $t_{r_6^+}^{\mathrm{d}}$ up to r_6 can be calculated in constant time so that only the second part needs to be re-evaluated in each step. This efficient type of evaluation can be found in Badeau et al. (1997) in which also further techniques for approximating the objective function value of the second part of $t_{r_6^+}^{\mathrm{d}}$ in a time-efficient way are described. Since in the considered RDOPG applications tour lengths are quite short, these further approximation techniques are not implemented in the proposed approach.

7.2.3 MultiRelocate (MREL)

The neighborhood operator MREL simultaneously removes n requests from the tour plan where n is randomly drawn from interval $[MREL^{\text{reqMin}}, MREL^{\text{reqMax}}]$. Afterwards, the removed requests are successively reinserted at their least cost positions in the tour plan. Since a sub-procedure of the REL neighborhood operator is applied for this reinsertion, MREL also evaluates benefits of using additional idle and unused vehicles. Note that the sequence in which the removed requests are iteratively reinserted may have considerable effect on the resulting solution. Therefore, MREL considers all possible permutations for reinserting the removed requests into the tour plan. In order to make a time-efficient use of previously considered permutations, evaluated permutations are iteratively modified from last positions to first positions.

Clearly, due to the exponential number of possible request removals, an exhaustive examination of all existing subsets comprising n requests in an iteration is not reasonable. Therefore, the neighborhood operator focuses on requests that are currently assumed to be inefficiently positioned in the tour plan. These requests are identified by a high contribution value to the objective function value. This contribution is individually determined for each request r by calculating the objective function value improvement that occurs when only r is removed from the tour plan. Subsequently, all requests are ordered in a list denoted as R^{w} by non-increasing objective function contribution values. Since requests with highest contribution values are located at the beginning of this list, all requests with a later position than $n \cdot MREL^{\text{reqSelR}}$ are erased from R^{w}. Afterwards, n requests that are considered by MREL in the current evaluation are randomly chosen from R^{w} forming the set R^*.

Within one move, the described selection, removal, and reinsertion of requests is repeated c times, where c is randomly chosen from interval $[MREL^{\text{execMin}}, MREL^{\text{execMax}}]$. The best found non-tabu neighborhood move is stored and finally executed. The MREL neighborhood operator is described in pseudo-code in Algorithm 7.3.

7.2.4 Large Neighborhood Search (LNS)

In each iteration of the solution method in which the LNS neighborhood operator is applied, LNS iteratively evaluates the relocation of a chosen number of n requests by removing and subsequently reinserting them at their least cost positions in sequence of non-increasing objective function contribution values. Since LNS also uses a sub-procedure of REL for reinserting each individual request, the possibility

Algorithm 7.3 Neighborhood operator MultiRelocate (MREL)

$c \leftarrow u(MREL^{\text{execMin}}, MREL^{\text{execMax}})$
$n \leftarrow u(MREL^{\text{reqMin}}, MREL^{\text{reqMax}})$
Generate set R^{w}: Select the $n \cdot MREL^{\text{reqSelR}}$ worst requests from set $R^{\text{U}}_{\tau+}$, i.e., the set R^{w} contains the requests that contribute most to the objective function value
for $i \leftarrow 1$ to c **do**
    Generate set R^* by randomly selecting n requests out of R^{w}
    Remove each request $r \in R^*$ from $P^{t*}_{\tau+}$
    Reinsert all removed requests into the tour plan at least cost insertion positions. Perform this reinsertion consecutively in all possible permutations and store the best non-tabu neighborhood move
end for
Perform the best stored neighborhood move

of utilizing additional unused vehicles at the depot is also considered in this neighborhood operator. The selection and reinsertion process is repeated c times where c is randomly chosen from interval $[LNS^{\text{execMin}}, LNS^{\text{execMax}}]$.

Compared to the MREL operator, the LNS neighborhood operator provides a substantially higher variety of the number n of requests that are simultaneously considered in one move. Note that this has been shown to be efficiently applicable to VRPs (cf. Ropke and Pisinger 2006). In each LNS iteration, n is set to $\max(2, m \cdot |R^{\text{U}}_{\tau+}|)$, where m is randomly drawn from interval $[0, LNS^{\text{reqRemF}}]$, $LNS^{\text{reqRemF}} < 1$. These n requests are randomly chosen from list R^{w} that was introduced in MREL. Analogous to MREL, the size of the set R^{w} depends on n, in this case it holds that $|R^{\text{w}}| = \min(|R^{\text{U}}_{\tau+}|, n \cdot LNS^{\text{reqSelR}})$. Consequently, since n may be much larger than in MREL, it is computationally intractable to evaluate all existing $n!$ permutations for reinsertion. Therefore, LNS reinserts the selected requests in one particular sequence as mentioned above, that is, in sequence of non-increasing objective function contribution values. The pseudo-code of the LNS neighborhood operator is shown in Algorithm 7.4.

7.2.5 Exchange Between Tours (XBT)

In the neighborhood operator XBT, two requests which are located on different tours are exchanged so that their positions in both tours are preserved. Due to the fixed reinsertion positions, diversification is attained. Similar to the WRT neighborhood operator, this neighborhood operator works exhaustively, i.e., it considers all existing pairs of requests. After evaluating the possible combinations of two requests, the best found non-tabu move is implemented. The pseudo-code of this neighborhood operator is depicted in Algorithm 7.5.

Algorithm 7.4 Neighborhood operator Large Neighborhood Search (LNS)

$c \leftarrow u(LNS^{\text{execMin}}, LNS^{\text{execMax}})$
$m \leftarrow U(0, LNS^{\text{reqRemF}})$
$n \leftarrow u(2, m \cdot |R^{\text{U}}_{\tau+}|)$
Generate set R^{w}: Select the $\min(|R^{\text{U}}_{\tau+}|, n \cdot LNS^{\text{reqSelR}})$ worst requests from set $R^{\text{U}}_{\tau+}$, i.e., the set R^{w} contains the requests that contribute most to the objective function value
for $i \leftarrow 1$ to c **do**
  Generate set R^* by randomly selecting n requests out of R^{w}
  Remove each request $r \in R^*$ from $P^{t*}_{\tau+}$
  Reinsert all removed requests into the tour plan. Due to complexity reasons, do this only in one specific permutation. This permutation is determined by reinserting r^* of R^* first, i.e., the request that has the biggest contribution to the objective function out of r^* of R^*. Store the best non-tabu neighborhood move
end for
Perform the best stored neighborhood move

Algorithm 7.5 Neighborhood operator Exchange Between Tours (XBT)

for each request $r \in R^{\text{U}}_{\tau+}$ **do**
  for each request r' located on a tour other than r **do**
    Exchange r and r' while preserving their tour positions
    Store best non-tabu neighborhood move
  end for
end for
Perform the best stored neighborhood move

7.3 The Implemented Tabu List

As mentioned in Sect. 7.1.1, in order to avoid cyclical computations in the search process, Tabu Search approaches make use of a tabu list. For this purpose, the definition of the *tabu-active attribute* which is used to identify tabu-active solutions is an important task. Since solutions should be forbidden only if they have already been examined by the search process, the practicability of a tabu-active attribute depends on its ability to correctly identify solutions. Due to the exponential number of possible solutions, a unique identification cannot be reasonably implemented. Therefore, a reasonable compromise between uniqueness and computational effort has to be found.

In the literature, many authors use the objective function value of a solution as the tabu-active attribute (cf., e.g., Badeau et al. 1997 and Gendreau et al. 2006). However, our empirical studies for the considered RDOPG problems indicate that significantly different solutions may have the same objective function value. Specifically, tour plan constellations can occur in which several dummy customer requests i

can be reached by vehicles at or before r_i^e. Since dummy customer requests have no cost contribution to the objective function value before the time point r_i^e, significantly different tour plans might not be distinguishable by their objective function value. Consequently, these different tour plans could be falsely identified as equal and therefore erroneously considered tabu-active. Furthermore, in our real-time control approach, another reason speaks against using the objective function value as a tabu-active attribute. Specifically, since the considered static system state is being modified during the transportation process, the objective function values of all solutions also changes over different static problem instances. Therefore, using the objective function value as a tabu-active attribute does not allow for identification tour plans over different considered static problem instances.

In order to overcome these limitations, a *fingerprint* of the tour plan for realizing the tabu-active attribute is used. This fingerprint is based on the sequence of requests in the tour plan and their assignment to vehicles. This method allows the Tabu Search solution method to identify the same tour plans even between consecutive static problems. The fingerprint is implemented by a combination of the two quickly computable checksums CRC-32 (cf. Moon 2005, p. 147 et seq.) and Adler-32 (cf. Deutsch and Gailly 1996). For a given tour plan, these methods calculate checksum values by using the tours of the tour plan as input data, i.e., by considering the sequence of requests and their assignment to the vehicles. In these checksum calculations, each request is represented by a unique request ID and a value of the assigned vehicle which depends on its status. Specifically, if the vehicle has already been utilized, i.e., has left the depot, a unique vehicle ID is used. Otherwise, the same value for all unused vehicles waiting at the depot is used. Using this fingerprint, two tour plans P_1, P_2 are identified as identical if both described checksums are equal in the tour plans, i.e., $cs^{\text{crc32}}(P_1) = cs^{\text{crc32}}(P_2)$ and $cs^{\text{adler32}}(P_1) = cs^{\text{adler32}}(P_2)$. Note that in Stone (2001, p. 128 et seq.) and Stone et al. (2002), the authors state that the Adler-32 checksum has some drawbacks when it is applied to short input data. Specifically, since the Adler-32 checksum is basically calculated by the summation of individual values, in case of short input data, only small checksum values are usually generated. Hence, in problem instances with only a few requests, Adler-32 may have an increased tendency to produce equal checksums for different tour plans of short length. However, since CRC-32 does not possess this drawback and the combination of both checksums is always used to identify the tabu-active state of a solution, no disadvantages arise. The Adler-32 algorithm is used due to its fast computation time which additionally helps to reduce the false detection rate of different solutions in larger problem instances.

Besides the implementation of the tabu-active attribute, the tabu list requires the definition of the *tabu tenure* that determines the number of iterations for which tabu-active attributes are valid. Since preliminary empirical studies in the considered RDOPG applications reveal an almost unique identification of solutions by the applied fingerprint, a tabu-active solution is never withdrawn. This allows for identification of solutions over static problem instances. Clearly, since each change regarding the set of pending requests at time τ^+ defined by $R^U_{\tau^+}$ makes former and

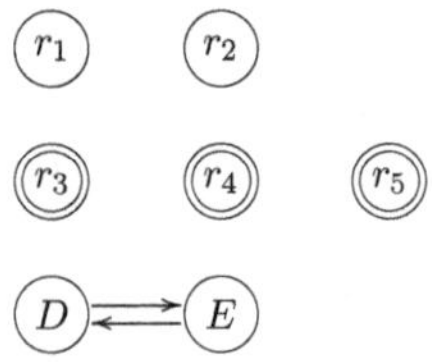

Fig. 7.4 Empty tour plan with two real requests and three dummy customer requests (denoted as r_3, r_4, and r_5) to be inserted by the construction heuristic

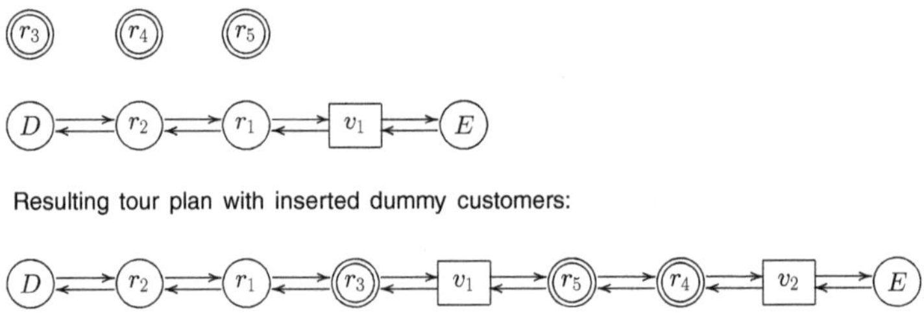

Fig. 7.5 Real requests inserted by the construction heuristic. If dummy customers are used, these are inserted afterwards

new solutions incomparable, the tabu list is reset whenever the pending request situation changes. Moreover, as mentioned in Sect. 5.2.2, the solution method returns to the first stage and is directly applied to the solution that is currently in execution in its pre-simulated state at τ^+.

7.4 Construction of the Initial Tour Plan

Since the Tabu Search metaheuristic is an improvement heuristic, an initial solution needs to be provided. Hence, before the real-time control of the considered transportation process starts, an initial tour plan is constructed at the beginning of the planning horizon.

In our approach, the initial tour plan is generated by an iterative insertion of all real requests that have arrived before the execution of the transportation process. These requests are inserted into the arising tour plan at their least cost position in sequence of non-decreasing arrival times. Note that for these requests it is assumed that their customer inconvenience starts at the beginning of the planning horizon. Afterwards, if stochastic knowledge is utilized, the additional dummy customer requests are inserted into the current tour plan at their least cost positions in sequence of non-decreasing starting times of their time window. In Figs. 7.4 and 7.5, the construction process is illustrated. Clearly, in both steps of the construction heuristic, the possible utilization of additional unused vehicles in the tour plan is evaluated.

Table 7.1 Stage-based selection scheme of the proposed Tabu Search solution method

Stage	Operator	Switch after number of iterations without new best solution
1	WTI	0 (only accept improvements)
2	REL	10
3	MREL	10
4	LNS	1,000
5	XBT	5

Table 7.2 Applied parameter values for the neighborhood operators MREL and LNS

Parameter	Value	Parameter	Value
$MREL^{\text{reqMin}}$	2	LNS^{reqRemF}	0.75
$MREL^{\text{reqMax}}$	3	LNS^{reqSelR}	1.5
$MREL^{\text{reqSelR}}$	3	LNS^{execMin}	1
$MREL^{\text{execMin}}$	10	LNS^{execMax}	2
$MREL^{\text{execMax}}$	20		

7.5 Stage-Based Neighborhood Operator Selection Scheme

In each iteration of the Tabu Search procedure, the neighborhood operator to be applied has to be chosen. In the proposed solution method, the neighborhood operator is chosen using a deterministic neighborhood operator selection scheme. This scheme has different stages where each stage consists of one of the neighborhood operators that have been introduced above. In each iteration, the neighborhood operator of the current stage is applied and the best found non-tabu move is executed. After a predefined number of performed iterations in which no new best known solution has been found, the stage is switched to the next one. If a new best known solution is found or if the last stage of the scheme has been completed by the search process, the search returns to the first stage and the current number of iterations without improvement is reset.

Preliminary tests have been conducted in order to determine reasonable parameter settings for the stages as well as for the neighborhood operators. Table 7.1 provides a detailed definition of the implemented stage-based selection scheme. Additionally, Table 7.2 shows information about chosen parameter values for the neighborhood operators MREL and LNS.

In order to combine intensification and diversification aspects in the enumeration process of the implemented Tabu Search procedure, each stage can be interpreted as representing a specific assumed distance of promising, still-unexplored solutions to the currently considered solution. Specifically, due to a large number of recent consecutively failed attempts at improving the currently best known solution, this distance is assumed to be significantly larger in higher stages than in lower stages. Hence, by switching to higher stages, the search process increasingly applies more globally-acting neighborhood operators that provide a higher diversification. Moreover, whenever a move leads to a new overall best known solution, the search process is intensified in its direct vicinity by switching back to the first stage.

7.6 Integration of the Solution Method into the Real-Time Control Approaches

As described in Sect. 5.2.2, the proposed Tabu Search solution method is invoked in both proposed real-time control approaches, i.e., in the deterministic and in the pro-active approach, after the theoretical tour plan has been adapted to the anticipated system situation at τ^+. Furthermore, the current execution status of the solution method, i.e., the current tabu list, the current stage, and the current number of iterations without improvement are stored at the end of the execution of the solution method in the current anticipation horizon so that this information can be used in the adaptation performed in the consecutive anticipation horizon. In so doing, the solution method is enabled to continue the search process in the same state over consecutive problem instances. If changes in the pending request situation have occurred, the tabu list is reset and the solution method starts on the first stage of the neighborhood operator selection scheme. This is performed since the changed request situation in the system makes it likely that the application of intensifying neighborhood operators quickly leads to improvements of the tour plan. After the available computational time for the solution method has elapsed, the solution method returns the best feasible tour plan that has been found during the application of the solution method as well as the last evaluated theoretical tour plan to the real-time control approach. In the update handling performed in the next anticipation horizon, the real-time control approach utilizes these generated tour plans in order to select the new relevant tour plan as described in Sect. 5.2.2.

7.7 Summary

This chapter introduced the solution method which is used to efficiently solve the individual static problem instances generated by the deterministic and pro-active real-time control approach during the execution of the transportation process. In order to cope with the tight computational time restrictions imposed by the real-time nature of the considered RDOPG applications, an elaborated Tabu Search metaheuristic solution method was proposed. It utilizes different neighborhood operators which are selected by a deterministic stage-based selection scheme. After describing the Tabu Search metaheuristic in general, the utilized neighborhood operators were presented in detail. Moreover, the applied tabu-active attribute in the tabu list was explained. In the proposed approach, a fingerprint consisting of two quickly computable checksums was used as the tabu-active attribute in order to allow an almost unique identification of individual solutions. This fingerprint also allows the identification of solutions across consecutive static problem instances. Furthermore, the generation of the initial tour plan and the stage-based selection scheme was presented. As mentioned above, this scheme is used to select the neighborhood operator which is to be applied in each iteration of the Tabu Search solution method. After a predefined number of iterations without finding a new best global solution, the current stage is switched to the next higher one. In doing so, neighborhood operators which provide a higher level of diversification are applied in the subsequent iterations. When a new best global solution is found, the stage is reset to the first one so that an intensifying neighborhood operator is applied in the next iterations. Moreover, the integration of the proposed Tabu Search solution method into both proposed real-time control approaches was described.

Chapter 8
Computational Results

In this chapter, we present the results of our detailed computational experiments conducted on the proposed real-time control approaches. The experiments are carried out in order to answer our three research questions posed in Sect. 1.3.1. Since no benchmark instances are available for the considered RDOPG applications, appropriate *test instances* were generated. Each test instance consists of a *problem instance* and a *setting*. A problem instance defines given problem characteristics such as the existing request data, customer inconvenience function, and the available vehicle fleet size. Hence, the problem instance represents the unchangeable system environment. In contrast, a setting comprises selectable parameters. Besides choosing the real-time control approach to be utilized for adapting the transportation process during its execution, this also includes defining appropriate parameters for it. Furthermore, if the pro-active real-time control approach is used, parameters for generating and integrating stochastic knowledge are defined in the setting.

Before presenting the results of the computational experiments, we first introduce the *discrete event-based simulator* which simulates the system environment of RDOPG applications including vehicle activities and dynamically arriving requests. This is followed by the description of different settings evaluated in our computational experiments consisting of *test environment, parameter values, and request data classes*. After that, the computational experiments are conducted in eight steps:

1. The *performance of the Tabu Search solution method* that is applied for solving individual static problem instances which occur during the transportation process is analyzed by comparing its performance to a lower bound of an exact solution procedure.
2. A detailed *performance evaluation of deterministic real-time control approaches* is provided. Specifically, the proposed deterministic real-time control approach is compared to a standardized least cost insertion procedure that integrates newly incoming requests. Moreover, a suitable time limit for the application of the Tabu Search metaheuristic in the deterministic real-time control approach is derived.
3. A detailed *performance evaluation of allowing vehicle en-route diversion* is provided. In this comparison, the deterministic real-time control approach is compared to a variant of this approach in which vehicle en-route diversion is for-

F. Ferrucci, *Pro-active Dynamic Vehicle Routing*, Contributions to Management Science, DOI 10.1007/978-3-642-33472-6_8,

bidden. Test results illustrating the positive effects of allowing vehicle en-route diversion on the solution quality are provided.
4. The efficiency of *measuring structural distortion using the Goodness-of-Fit test* used in our forecasting methods is evaluated. This evaluation is conducted on request data sets that have different levels of structural distortion. Moreover, a suitable value for the minimum demanded Poisson quality parameter for generating stochastic knowledge is derived.
5. Using the derived Poisson quality parameter value, *results of generating stochastic knowledge for* S^{REAL} *and* S^{GEN}, which is used in our main computational experiments, are presented.
6. Based on the results of the previous five evaluations, a comprehensive computational study regarding the *performance evaluation of the pro-active real-time control approach* is provided in the main computational experiments of this book in order to answer the first and second research question. Specifically, a detailed analysis of the conditions under which the generated stochastic knowledge can be reasonably utilized in the pro-active real-time control approach is carried out. This analysis is conducted by assessing the performance of the proposed pro-active real-time control approach in comparison to its deterministic counterpart, i.e., the deterministic real-time control approach which does not make any use of stochastic knowledge. In addition to an existing real-world request data set, test studies on designed request data sets are conducted.

   In order to analyze under which conditions the use of the proposed pro-active real-time control approach is recommendable, we identify crucial characteristics of the evaluated test instances. From the results of this study we conclude that the *structural quality* of the existing request data set is one of these crucial characteristics. Specifically, we propose that the structural quality of a request data set depends on its *structural distortion* and *structural diversity*:

   - **Structural distortion:** In the proposed forecasting approach it is assumed that request arrivals can be modeled by non-homogeneous time-space Poisson processes whose rate parameters remain constant from day to day (see Sect. 6.2). If this assumption is not fulfilled in the existing request data set, future requests are likely to occur with significant differences from what was forecasted. In this case, utilizing the stochastic knowledge generated by the proposed forecasting method is likely to lead to a worse solution quality. Hence, we state that it is important that the available request data has a *low structural distortion.* In our proposed forecasting approach the Chi-Square Goodness-of-Fit test is applied in order to measure the Poisson quality of a given request data set (see Sect. 6.2.1). We evaluate the efficiency of this test for identifying the structural distortion on request data sets with different levels of structural distortion in Sect. 8.6.
   - **Structural diversity:** In addition to requiring a low structural distortion, we classify the structural quality of request data sets according to the changes which occur in the request arrival rates in individual subregions during the considered planning horizon. We define that request data sets with nearly constant request arrival rates comprise a *low structural diversity* while request data sets

with significantly changing request arrival rates in the subregions have a *high structural diversity*. In order to analyze the effects of structural diversity on the attainable improvements, we generate test instances with designed levels of structural diversity (see Sect. 8.2.4.2) and evaluate the attainable improvements depending on the existing level of structural diversity. Moreover, we propose two methods for determining the structural diversity of given request data sets in Sects. 8.8.1 and 8.9.

According to the above criteria, a request data set is considered to have a high structural quality if it has a low structural distortion and a high structural diversity. Therefore, the structural quality of a request data set determines the value of the stochastic knowledge which can be generated out of this request data set. Consequently, a high structural quality indicates that high improvements of the solution quality can be achieved if the proposed pro-active real-time control approach is used compared with the results attained by its deterministic counterpart.

7. In the previous computational experiments, it was shown that high levels of structural diversity significantly increase the improvements attainable by the pro-active real-time control approach. Based on these insights, we propose a measure for determining the structural diversity of arbitrary request data sets. This measure is denoted as the *degree of structural diversity* and its significance is statistically proven.
8. In the third evaluation, the advantages of allowing vehicle en-route diversion in the considered RDOPG applications are shown. However, allowing vehicle en-route diversion may increase driver inconvenience since drivers are obliged to quickly react to occurring changes in their tour while they are traveling to their next assigned request. In order to deal with this issue, we additionally consider *driver inconvenience aspects in the real-time control approaches*. For this purpose, a method for limiting the number of performed vehicle en-route diversion activities is developed. The effects of this method on the solution quality and resulting driver inconvenience is assessed. Recommended parameter settings for achieving a suitable trade-off between the contradicting aims of driver inconvenience and customer inconvenience as well as parameter settings for prioritizing each of the aims are provided.

After these computational experiments, we conclude our evaluations by giving a *summary and answers to the three central research questions* which were posed in Sect. 1.3.1.

8.1 The Discrete Event-Based Simulator

In order to analyze the real-time control approaches under real-time conditions, all transportation processes of the considered RDOPG applications are carried out by an implemented discrete event-based simulator. In discrete event-based simulators, the simulation time is always set from the current time to the time point when the

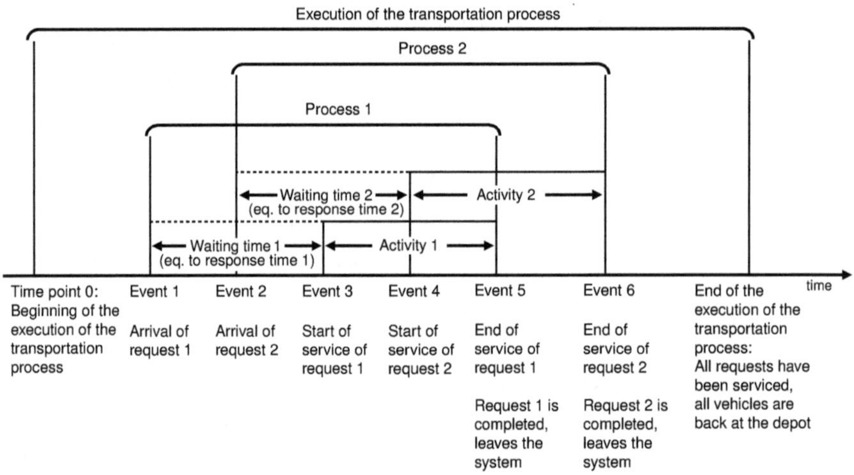

Fig. 8.1 Interrelation of event, process, and operation using the example of servicing two requests (cf. Biethahn et al. 1999, p. 52)

next event in the system takes place as mentioned in Sect. 4.7.3. In addition to conducting the computational experiments, the implemented simulator is also used for pre-simulating the system state at the beginning of each anticipation horizon. This pre-simulation concept was introduced in Sect. 5.2.1 in order to generate the anticipated system state at the end of the anticipation horizon so that the concurrency of the execution and adaptation of the transportation process is considered. By utilizing this pre-simulated system state, the static problem instance for the current anticipation horizon is generated (see Sect. 5.2.2). In the individual anticipation horizons conducted by the real-time control concept presented in Sect. 5.2.1, it is assumed that the real vehicle activities are in line with the activities that have been anticipated.

As mentioned in Sect. 4.7.3, a system consists of *entities* that act or interact together towards the accomplishment of some logical end (cf. Schmidt and Taylor 1970). Hence, in order to model a complex dynamic system such as RDOPG applications, it is necessary to identify all entities of the system. In the considered system, the entities consist of *requests* and *vehicles*. Dynamically arriving requests trigger so-called *exogenous events* that unexpectedly change the relevant information of the transportation process from outside the system since new requests arrive with uncertainty. In contrast, vehicle activities trigger *endogenous events* that modify the relevant information in ways that are generated by the system itself and can be expected with high certainty, such as request servicing activities performed by vehicles.

As depicted in Fig. 8.1, the executed transportation process can be subdivided into several individual *processes*. A process is defined by mutually dependent *events* and intermediate *operations*. A new process starts when a request arrives in the system. It consists of exactly two operations: *Waiting time* and *activity*. The waiting time operation directly starts upon the arrival of a new request. In the RDOPG applications, the length of the waiting time determines the response time and hence the resulting customer inconvenience. The activity operation starts when a vehicle reaches a customer request location. This operation is defined as an activity since the vehicle that services the request cannot perform other tasks during this activity. The length of this activity is defined by the service time of the request. In Fig. 8.1, the interrelation between the participating entities in the system is illustrated. In what follows, we describe the *simulation of requests* as well as the *simulation of vehicles*.

8.1.1 Simulation of the Requests

As mentioned in Sect. 3.1, dynamism in the considered RDOPG applications arises from newly arriving requests. The requests that arrive during the transportation process are read from a corresponding request data set file that contains all relevant request attributes, i.e., their arrival time and location. These request arrivals are stored in a list called *future event set* (FES). Since all request-related attributes are specified by the request data set, it is ensured that the same problem instance can be run several times using different settings so that different variants of the real-time control approaches can be compared with each other under the same conditions. During the execution of the transportation process, the request information is revealed to the simulator over time at the arrival times of the individual requests. The integration of the newly arrived requests is performed by the real-time control approach as described in Sect. 5.2.1.

In the computational experiments, relevant performance statistics observed during the simulation are recorded for each test instance in order to allow a performance analysis of the utilized setting under the given problem instance. Among these statistics, the request response times and the corresponding objective function values are determined. According to the objective function values of all requests, the solution quality for each test instance is determined by calculating the average customer inconvenience per request. This value is used as the main quality criterion in all computational experiments presented later. Since satisfaction of a maximum response time for all requests is also pursued, the number of late requests that violate this maximum response time is considered as another important performance indicator.

8.1.2 Simulation of the Vehicles

In each anticipation horizon, the vehicle activities are simulated by executing the current relevant tour plan. Specifically, at time point τ, after the relevant tour plan P_τ^{r} has been determined, each vehicle executes its tour according to P_τ^{r} for t^{a} time units. Depending on the current vehicle's status (see also Sect. 5.3.2.2), this execution comprises the following activities:

- **Traveling:** The vehicle travels in the road network to the next assigned request of its tour.
- **Servicing:** The vehicle services a request at the corresponding request location.
- **Waiting at a dummy customer request:** The vehicle waits at a dummy customer request location.
- **Waiting at the last request location:** The vehicle waits at the location of its last serviced request if the vehicle is currently idle since it has no further requests assigned.
- **Waiting at the depot:** The vehicle waits at the depot location if it is unused.

Within the time period $[\tau, \tau^+)$, it is possible that a vehicle performs more than one of these activities if the current activity requires less time than is available in the current anticipation horizon.

8.2 Test Environment, Parameter Values, and Request Data Classes

In this section, we describe settings and parameters which are evaluated in the computational experiments. First, we describe the *system environment and real-time control parameter values* under which the computational experiments are evaluated. Furthermore, utilized *tabu search solution method parameter values* are presented. Afterwards, appropriate *dummy customer request parameter values* that have been obtained in preliminary tests are described. This is followed by presenting the *request parameter values and request data classes* which are evaluated in the computational studies. Moreover, the *number of conducted computational experiments* is presented. Finally, the *performance evaluation using the additionally attainable improvement* is described.

8.2.1 System Environment and Real-Time Control Parameter Values

In order to analyze the performance of the different evaluated real-time control approach variants, all generated transportation processes are executed utilizing a detailed road network of Dortmund, a German medium-sized city. The road network is made available for our research by DDS Digital Data Services GmbH in Karlsruhe, Germany. The considered service region in this city comprises an area of 22.5 kilometers $\times$ 20 kilometers $=$ 450 square kilometers. Since only the urban area is considered, the utilized road network is quite small comprising about 17,300 nodes and 42,000 arcs. Hence, shortest path calculations can be performed within a short time, i.e., within a few milliseconds. In all experiments, the length of the anticipation horizon is set to $t^{\mathrm{a}} = 20$ seconds. Moreover, before each transportation process the initial generated tour plan is optimized by the Tabu Search solution method for 120 seconds.

The experiments are conducted with fleet sizes of 8, 10, and 12 vehicles. These fleet sizes have been empirically chosen from preliminary tests. They represent interesting constellations for the number of requests in the considered request sets. Specifically, while a fleet size of 8 vehicles represents a scenario with a very restricted set of available resources, the 12 vehicle setting provides significantly more flexibility for vehicle dispatching and pro-active strategies.

All proposed real-time control approaches and the segment-based forecasting approach were implemented using Borland Delphi while specific request generation and evaluation tools were developed using Microsoft Excel. The implemented applications comprise in total about 50,000 lines of code. The computational experiments were run on Intel Pentium D 2.8 GHz CPU (2.5 GB RAM) computers.

8.2.2 Tabu Search Solution Method Parameter Values

As illustrated in Sect. 7.2, some neighborhood operators in the proposed Tabu Search metaheuristic utilize random numbers during their application. For generating these random numbers, we use an extended version[1] of the random number generator proposed by Park and Miller (1988). In this extended version, an improved multiplicator (cf. Park et al. 1993) is applied and an individual stream of random numbers is provided for each randomly drawn parameter in all neighborhood operators of the Tabu Search method.

As mentioned in Sect. 5.2.2, static problem instances are adapted by the Tabu Search solution method which is applied for a given time period. Note that due to this time-limit based optimization, slightly imprecise operating system based measurements of the used CPU time may occur which may lead to a variation of the number of performed iterations in different replications of the same evaluated test instance. Consequently, this takes considerable effect on the random number generation process applied in some of the neighborhood operators of the Tabu Search solution method. Specifically, in different replications, those stochastic features caused by the real-time nature of the process can lead to varying performance results of the Tabu Search procedure (cf. Gendreau et al. 2006).

In order to reduce these unwanted effects, we use a variance reduction technique denoted as Common Random Numbers (CRN, cf. Law and Kelton 2000, p. 582 et seq.). By using this technique, all seeds of the random number streams used in a test instance are generated before the start of the simulation in an offline step. Therefore, in this offline step, a suitable upper bound on occurring static problem instances during the transportation process is determined. In order to derive suitable seed values for solving each static problem instance, we consecutively generate 10,000 random numbers of each utilized random number generator stream between two seed calculations. After each of these 10,000 random numbers, the resulting set of seed values is stored as initial seeds of the streams of random numbers in the corresponding static problem instance.

[1] Available at http://www.cs.wm.edu/~va/software/park/.

8.2.3 Dummy Customer Request Parameter Values

In order to obtain useful parameter values for generating and integrating dummy customer requests, extensive preliminary tests have been conducted. The obtained results indicate that the following parameter values for dummy customer generation and integration are reasonable:

- **Spatial and temporal precision:** According to results of preliminary computational experiments, the extension of each segment used in the segment-based clustering approach is set to $DC^{\text{se}} \times DC^{\text{te}} = (2.5 \text{ kilometers} \times 2.5 \text{ kilometers}) \times 1$ minute. Therefore, the considered service area of size 22.5 kilometers × 20 kilometers is divided into 9 × 8 subareas. Since the considered transportation process is controlled for four hours, we obtain the set S with altogether $|S| = 9 \cdot 8 \cdot 60 \cdot 4 = 17{,}280$ segments.
- **Range of past request information:** The number of past working days utilized in the generation of the stochastic knowledge is set to $n^{\text{f}} = 60$.
- **Maximum cluster extension:** The maximum spatial cluster extension DC^{mse} is set to 2 by 2 segments (i.e., 5 kilometers × 5 kilometers) while the maximum allowed temporal cluster height (DC^{mte}) is set to 15 segments (i.e., 15 minutes).
- **Dummy customer location:** For determining a reasonable location for the dummy customer, the maximum predefined travel time radius DC^{radiusTT} is set to 300 seconds whereas the maximum allowed average travel time within a cluster DC^{maxAvgTT} is set to 650 seconds.
- **Dummy customer prediction quality:** The following five values are evaluated for the minimum dummy customer quality $DC^{\text{min}\lambda}$: $DC^{\text{min}\lambda} \in \{1.0, 1.2, 1.5, 1.8, 2.0\}$. Consequently, in these cases, the expected occurrence probability of at least one request in an accepted cluster amounts to at least 63.21 %, 69.88 %, 77.69 %, 83.47 %, or 86.47 %, respectively.
- **Dummy customer removal:** Two different parameter values are evaluated for $DC^{\lambda\text{rem}}$, in particular $DC^{\lambda\text{rem}} \in \{0.50, 0.25\}$. Hence, a dummy customer is removed when the arrival probability of at least one more request in the assigned cluster becomes lower than 39.45 % or 22.12 %, respectively.

8.2.4 Request Parameter Values and Request Data Classes

The computational experiments are conducted on the following two request data classes: Real-world request data class S^{REAL} and generated request data class S^{GEN}. In addition to the equal request service time defined in Sect. 3.1 by $R^{\text{st}} = 60$ seconds, two further parameters are used for real requests as well as for dummy customer requests in both request data classes. Specifically, in all experiments, the maximum allowed request response time R^{mrt} is set to 3,600 seconds while the lateness penalty costs R^{pen}, that occur when R^{mrt} is exceeded, are set to 100. As mentioned in Sect. 5.3.2.4, since a request has a cost contribution of only 1.0 to the objective function value at a response time equal to R^{mrt}, the significantly higher lateness penalty costs R^{pen} are considered prohibitive costs. Clearly, due to the real-time nature of the considered problem and depending on the available fleet size, the arrival of additional dynamic requests may result in a scenario where some requests can only be serviced after the maximum allowed request response time R^{mrt}.

In what follows, we describe both evaluated request data classes in detail. First, the real-world request data class S^{REAL} is described followed by the generated request data class S^{GEN}. Note that all request data sets of both request data classes consist of 30 days. Consequently, in the computational experiments, each evaluation of a setting results in 30 test instances.

8.2.4.1 Real-World Request Data Class S^{REAL}

The request data class S^{REAL} comprises one request data set of 30 randomly chosen days out of over 400 weekdays and results from a case study of the subsequent delivery process carried out by a German newspaper publishing company. In this request data set, 150 requests per day have to be processed on average. The subsequent delivery of newspapers starts at 7 am. Since the majority of subsequent delivery requests occur before 11 am, the computational experiments consider all requests that arrive up to this point in time. Therefore, each simulated day consists of four hours during which new requests may occur. Clearly, for generating stochastic knowledge, for each of the 30 randomly chosen days of S^{REAL}, the $n^{\mathrm{f}} = 60$ past working days of the available real-world data are utilized.

Regarding request arrivals, an analysis of the provided data has shown that only a few requests arrive before 7 am so that more than 90 % of all requests arrive after 7 am (see Fig. 8.2). Therefore, according to Larsen et al. (2002) and as mentioned in Sect. 2.6.5, a high degree of dynamism can be identified for this request data set. Note that, due to the real-world nature of this request data set, underlying request distributions as well as other possibly existing non-deterministic structures are unknown in S^{REAL}. Hence, neither the structural distortion nor the structural diversity of S^{REAL} is known. Both criteria are evaluated by our proposed methods in this chapter.

8.2.4.2 Generated Request Data Class S^{GEN}

In contrast to S^{REAL}, request arrivals in the request data class S^{GEN} are designed according to the following characteristics. This request data class comprises requests that have been generated by time-space Poisson processes with time- and space-dependent rate parameters which remain constant from day to day. The generation is carried out in a manner similar to the procedure described in Ichoua et al. (2006). Specifically, the considered service area is subdivided into a set of $\mathcal{P}$ quadratic disjunct regions of equal size $a_1, \ldots, a_{\mathcal{P}}$ while the considered time

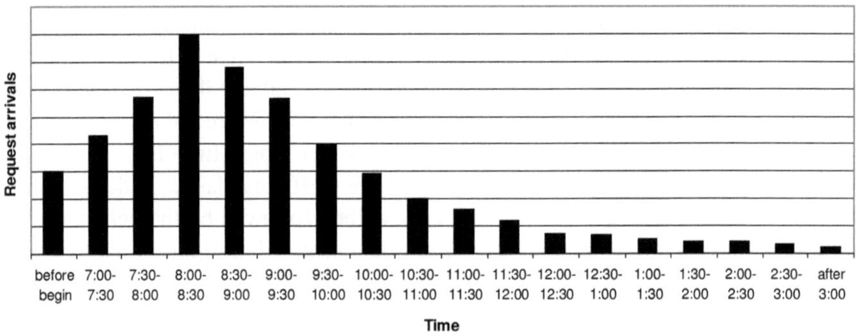

Fig. 8.2 Average request arrival distribution of the request data class S^{REAL}

period for new request arrivals from 6:45 am to 11 am is divided into $\mathcal{Q}$ different disjunct time slices $t_1, \ldots, t_{\mathcal{Q}}$ which do not overlap each other. The parameter $\lambda(a_i, t_j)$ determines the request arrival rate of a time-space Poisson process in region a_i during time slice t_j. Moreover, t_j^{s} defines the start time and t_j^{e} defines the end time of time slice t_j, $j \in \{1, \ldots, \mathcal{Q}\}$. Given the current time slice t_j, the arrival time of the next request is determined using a Poisson process with rate parameter $\lambda(t_j) = \sum_{i=1}^{\mathcal{P}} \lambda(a_i, t_j)$. The region in which this new request will occur is determined according to the region probability values $p(a_i, t_j) = \frac{\lambda(a_i, t_j)}{\lambda(t_j)}$. Finally, within the chosen region a_i, the spatial position of the new request is uniformly drawn. Hence, as suggested by Larsen (2000, p. 45), we utilize time-dependent Poisson processes for generating request arrivals in the request data class S^{GEN}. As in the solution method, parameter values are generated by using an individual stream of random numbers, which are provided by the extended version of the random number generator of Park and Miller (1988) and Park et al. (1993), for each type of parameter.

Based on the rate parameters $\lambda(a_i, t_j)$ used in the applied setting, different levels of spatial and temporal *structural diversity*, i.e., levels of variance in request arrivals, occur. We introduce two *dimensions of structural diversity*: While *RegionDiversity* (*RD*) measures the regional variance of request arrival rates in the service area within one time slice, *TimeDiversity* (*TD*) gives the variance of request arrival rates in each considered region a_i of the service area during the considered planning horizon. Moreover, *TD* also defines variations in $\lambda(t_j)$ from one time slice to the other. In order to evaluate the impact of different levels of structural diversity on the structural quality of request data sets, different request data sets comprising various value settings of both structural diversity dimensions are designed in S^{GEN}. All settings, i.e., request data sets, are iteratively generated by starting from an initial setting with a defined maximum level of structural diversity denoted as $RD = TD = 1.00$. The structural diversity of this initial setting results from individually defined rate parameters $\lambda(a_i, t_j)$. In subsequently generated request data sets with lower levels of structural diversity, the rate parameter values are iteratively reduced by applying a linear averaging that balances the rate parameters $\lambda(a_i, t_j)$ accordingly. In

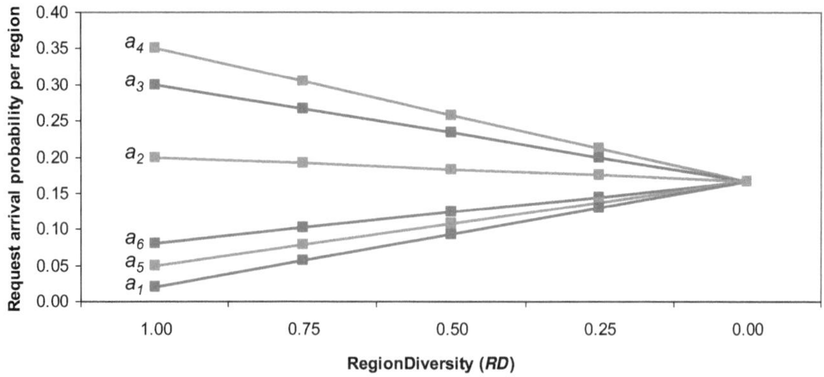

Fig. 8.3 Linear averaging at the example of RegionDiversity (*RD*)

Fig. 8.3, an example of this linear averaging process for resulting request arrival probabilities is illustrated using $\mathcal{P} = 6$ regions $a_1, \ldots, a_6$ and one time slice, i.e., $\mathcal{Q} = 1$. As can be seen, setting $RD = 1.00$ results in a request data set in which the defined probabilities are fully present in each region while $RD = 0.00$ results in the same request arrival probability in each region. With regard to S^{GEN}, by using a step size of 0.25 and completing with a final setting $RD = TD = 0.00$, we obtain 25 request data sets altogether. Apart from negligible distortion effects caused by the real-world character of the road network, setting $RD = TD = 0.00$ produces a complete uniform distribution of request arrivals during the considered planning horizon. Moreover, the setting $RD = 1.00$, $TD = 0.00$ results in a request data set with significantly different request arrival rates in the regions. However, request arrival rates in the regions and the total request arrival rate per time slice remain constant during the planning horizon. In contrast, the setting $RD = 0.00$, $TD = 1.00$ generates a request data set in which request arrivals have the same probability in all regions but the total request arrival rate in a time slice changes between time slices according to the initially defined values for $RD = TD = 1.00$. Algorithm 8.1 illustrates the generation of a request data set of the request data class S^{GEN} with defined levels of structural diversity *RD* and *TD*. In lines 1–13 the mentioned linear averaging of the request arrival rates and request arrival probabilities is performed. Specifically, in line 5, for each time slice t_j the request arrival rate $\lambda'(t_j)$ is calculated according to the chosen *TD* setting. In line 8, the average request arrival probability over all time slices for region a_i denoted as $\bar{p}'(a_i)$ is calculated depending on the initially defined request arrival rates and the selected *RD* setting. Based on this value, the resulting request arrival probability of region a_i in time slice t_j is calculated in line 10 and 11. According to these calculated values, in lines 14–26, the request data set is generated as described above.

In all 25 generated request data sets of S^{GEN}, the number of regions is set to $\mathcal{P} = 18$ and the number of time slices to $\mathcal{Q} = 5$. The request arrival probability parameters utilized for generating the setting $RD = TD = 1.00$ are illustrated in Fig. 8.4. Note that this figure shows a simplified illustration of the road network,

Algorithm 8.1 Generation of a request data set of S^{GEN} with $RD \in [0,1]$ and $TD \in [0,1]$

{Generate request arrival rate parameters based on given structural diversity levels}
$n_{\text{all}} \leftarrow \sum_{j=1}^{\mathcal{Q}} \lambda(t_j)$
for $j \leftarrow 1$ to $\mathcal{Q}$ **do**
  $n \leftarrow n_{\text{all}} \cdot \frac{t_j^{\text{e}} - t_j^{\text{s}}}{t_{\mathcal{Q}}^{\text{e}} - t_1^{\text{s}}}$
  $\lambda'(t_j) \leftarrow n + (\lambda(t_j) - n) \cdot TD$
end for
for $i \leftarrow 1$ to $\mathcal{P}$ **do**
  $\bar{p}'(a_i) \leftarrow \frac{1}{\mathcal{Q}} \cdot \sum_{j=1}^{\mathcal{Q}} (\frac{1}{\mathcal{P}} + (p(a_i, t_j) - \frac{1}{\mathcal{P}}) \cdot RD)$
  for $j \leftarrow 1$ to $\mathcal{Q}$ **do**
    $x \leftarrow \frac{1}{\mathcal{P}} + (p(a_i, t_j) - \frac{1}{\mathcal{P}}) \cdot RD$
    $p'(a_i, t_j) \leftarrow \bar{p}'(a_i) + (x - \bar{p}'(a_i)) \cdot TD$
  end for
end for
{Generate request data set}
for $d \leftarrow 1$ to number of days to be generated **do**
  for $j \leftarrow 1$ to $\mathcal{Q}$ **do**
    $t \leftarrow t_j^{\text{s}}$
    while $t < t_j^{\text{e}}$ **do**
      $t \leftarrow$ Next request arrival time defined by a Poisson process with rate parameter $\lambda'(t_j)$
      if $t < t_j^{\text{e}}$ **then**
        Determine region $k \in \{1, \ldots, \mathcal{P}\}$ in which the request will occur according to $p'(a_i, t_j)$ values, $i \in \{1, \ldots, \mathcal{P}\}$
        Generate new request i at time t with a uniformly distributed location in region k. Determine node $\tilde{n} \in \tilde{N}$ closest to this location in the road network $\tilde{R}$ and set $r_i^{\text{p}} = \tilde{n}$. Furthermore, set request arrival time $r_i^{\text{a}} = t$
      end if
    end while
  end for
end for

i.e., only highways and major roads are included in this illustration. Moreover, in order to generate all request data sets of S^{GEN} with a degree of dynamism that is comparable to the one observed in S^{REAL}, only a few requests arrive before the considered start of the transportation process at 7 am. Specifically, over 90 % of all requests occur after the start of the execution of the transportation process in all generated settings. In contrast to S^{REAL}, in request data sets of S^{GEN} with high levels of structural diversity, request arrival rates increase more rapidly during the planning horizon. Moreover, the peak of request arrivals is moved to the center of the considered planning horizon and the number of request arrivals is maintained

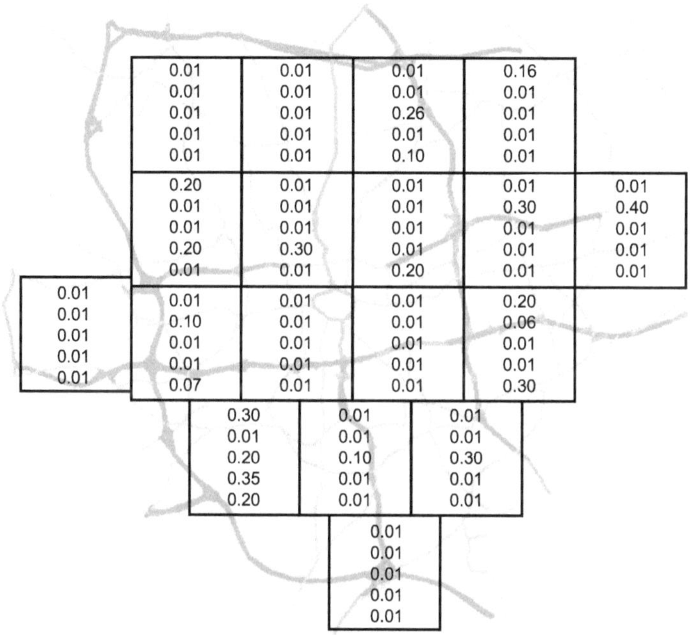

Fig. 8.4 Simplified illustration of the road network with request arrival probabilities $p(a_i, t_j)$ for the structural diversity setting S^{GEN} $RD = TD = 1.00$

on a high level after the peak. This is performed in order to also generate a high system utilization after the request arrival peak. Despite these slight modifications, S^{GEN} possesses characteristics which are similar to S^{REAL}, i.e., a low request arrival rate at the beginning of the planning horizon and a high increase of request arrival rates before the peak. Moreover, the subsequent decrease is significantly lower than its increase before the peak of request arrival rates. The time slices and number of expected request arrivals are illustrated in Table 8.1. Figure 8.5 visualizes the corresponding total request arrival rates of three *TD* settings of S^{GEN}. Note that since regional differences of request arrivals are not illustrated in this diagram, the depicted total request arrival rates are independent of *RD*.

According to the described parameters, Fig. 8.6 visualizes three request data sets of S^{GEN}. As can be seen in this figure, higher levels of structural diversity result in the desired significant differences of temporal and spatial request arrivals. All 25 request data sets of S^{GEN} are illustrated in the Appendix in Fig. B.1. Analogous to S^{REAL}, each of the 25 request data sets of S^{GEN} consists of 30 days. Since request arrivals in the request data sets of S^{GEN} are generated by time-space Poisson processes that are constant from day to day, the proposed forecasting methods are able to generate stochastic knowledge that can reliably forecast future request arrivals. Therefore, we conclude that the request data sets of the request data class S^{GEN} have a low structural distortion.

Table 8.1 Parameter values for the time slices utilized in generating S^{GEN}

Time slice t_j	Start time t_j^{s}	End time t_j^{e}	Request arrival rate $\lambda(t_j)$
0	06:45	08:00	15
1	08:00	09:00	40
2	09:00	09:30	30
3	09:30	10:30	35
4	10:30	11:00	10

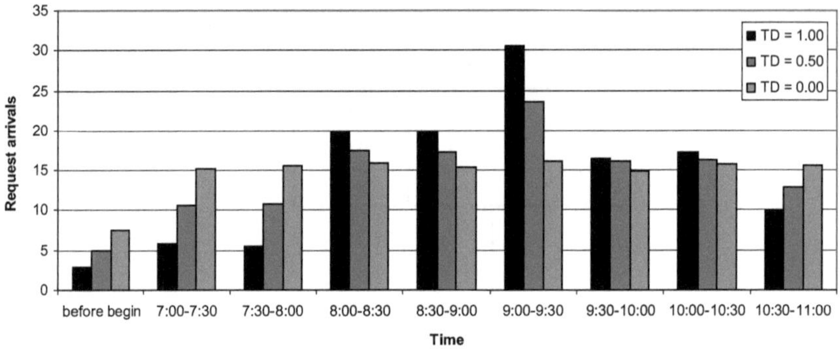

Fig. 8.5 Average request arrival distribution of three *TD* settings of S^{GEN}

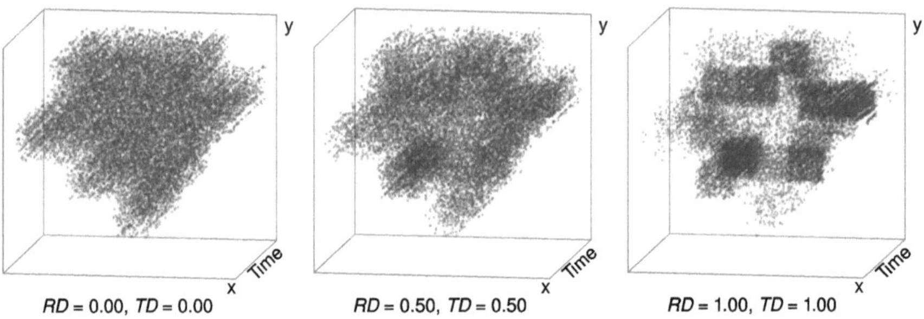

Fig. 8.6 Three request data sets of S^{GEN} with different levels of structural diversity

8.2.5 Number of Conducted Computational Experiments

All computational experiments of both request data classes are separately conducted with both customer inconvenience functions, i.e., linear2X as well as quadratic. As mentioned in Sect. 8.2.1, each available day in both tested request data sets is conducted using 8, 10, and 12 vehicles in order to analyze the impact of the number of available resources on the solution quality. In our experiments, 30 days of each of the 26 request data sets which consist of one request data set for S^{REAL} and 25 request data sets for S^{GEN} are tested. By evaluating the described dummy customer parameter settings as well as the deterministic real-time control approaches, the total number of experiments that are presented in the following computational studies amounts to more than 38,000 test instances. Additionally, preliminary computational experiments and their results are presented in the following sections.

8.2.6 Performance Evaluation Using the Additionally Attainable Improvement

In order to evaluate the performance of different real-time control approaches, we utilize a measure which is based on the value of information presented in Sect. 2.6.6.2. Specifically, instead of evaluating the solution quality of a solution approach on an instance in comparison to its performance on the corresponding offline instance, we compare the results of two solution approaches A_1 and A_2 on the same problem instance denoted by I. In doing so, the additionally attainable improvement of A_2 in comparison to the improvement[2] attained by A_1 on I can be calculated. With $z_A(I)$ denoting the objective function value of the solution on instance I generated by algorithm A, the additionally attainable improvement is defined as:

$$AAI(A_1, A_2, I) = \frac{z_{A_1}(I) - z_{A_2}(I)}{z_{A_1}(I)}$$

As mentioned above, each request data set consists of 30 days which are used in each computational study in order to compare two real-time control approaches A_1 and A_2. Specifically, using a specific request data set combined with a chosen objective function and a defined fleet size results in 30 problem instances denoted as the set $\tilde{I}$. In order to calculate the additionally attainable improvement of A_2 in comparison to A_1 on the instance set $\tilde{I}$, we define the *AAI* as follows:

$$AAI(A_1, A_2, \tilde{I}) = \frac{\frac{\sum_{I \in \tilde{I}} z_{A_1}(I)}{|\tilde{I}|} - \frac{\sum_{I \in \tilde{I}} z_{A_2}(I)}{|\tilde{I}|}}{\frac{\sum_{I \in \tilde{I}} z_{A_1}(I)}{|\tilde{I}|}} = \frac{\sum_{I \in \tilde{I}} z_{A_1}(I) - \sum_{I \in \tilde{I}} z_{A_2}(I)}{\sum_{I \in \tilde{I}} z_{A_1}(I)}$$

For example, in order to determine the performance of the pro-active real-time control approach by comparing its results to the ones obtained by the deterministic

[2] We assume that the application of the real-time control approach A_1 has already led to an improvement in comparison to carrying out the coordination of the transportation process without using any real-time control approaches.

real-time control approach, we calculate $AAI(A_1, A_2, \tilde{I})$ with A_1 as the deterministic real-time control approach and A_2 as the pro-active real-time control approach and with $\tilde{I}$ as the set of conducted problem instances. Whenever the additional improvement of a real-time control approach in comparison to another real-time control approach is described in the computational studies which are presented in what follows, the $AAI(A_1, A_2, \tilde{I})$ is utilized. As mentioned in Sect. 5.3.2.4, the primary objective is to minimize the number of late requests. Due to the prohibitive character of the lateness penalty costs R^{pen} caused by late requests, the number of late requests is separately considered in the computational results in order be able to separately determine the improvements attained in variable customer inconvenience. As shown in the computational results, in almost all scenarios tested with a fleet size of 10 and 12 vehicles, no late requests occur at all so that the number of late requests is mainly of interest in scenarios where a fleet size of 8 vehicles is present.

8.3 Performance of the Tabu Search Solution Method

In this section, we present a performance evaluation of the Tabu Search metaheuristic that has been introduced in Chap. 7. Since this solution method is iteratively applied in the transportation process in order to adapt the current tour plan according to newly incoming requests, its efficiency is of significant importance. Particularly, due to the strong time restrictions of the real-time control, attaining near-optimal to optimal solutions within given limits becomes an essential requirement. Therefore, in order to assess the applicability of the proposed Tabu Search metaheuristic in the real-time control approach, its solution quality is evaluated by comparing the objective function value of attained solutions with the lower bound generated by an exact procedure which is based on the column generation approach of Krumke et al. (2002). In order to allow for comparing the solutions generated by this approach with the ones generated by our approach, we modified this exact approach according to the objective functions of the considered RDOPG applications and replaced the Euclidean-based travel time calculations with the road network based calculations of our proposed real-time control approach. In order to avoid possible distorting effects caused by the real-time environment, we directly compare the solution quality of individual static problem instances that occur during the transportation process of two selected days of the request data class S^{REAL}. Each tested static problem instance is evaluated using a fleet size of 8 and 10 vehicles. The selected days are different with respect to the number of dynamically arriving requests. Consequently, two groups of static problem instances that can be classified into large-sized and moderate-sized problems arise. The first group comprises 137 static problem instances with up to 41 pending requests for a fleet of 10 vehicles and up to 57 pending requests for 8 vehicles. Additionally, there are 96 static problem instances in the second group that consist of up to 26 pending requests for a fleet size of 10 vehicles and 34 pending requests when a fleet of 8 vehicles is used. Note that the use of an increased fleet size results in accelerated request processing which leads to a smaller number of unserviced pending requests in the 10 vehicle case compared to the 8 vehicle scenario. As mentioned in Sect. 4.6, this column generation approach is well suited for solving routing problem instances in which an optimal solution consists of tours of short length. Since this is likely to be the case in the problem instances which are evaluated, as can be seen from the low request to vehicle ratio described above, this exact approach is an adequate method for assessing the performance of the proposed Tabu Search metaheuristic. In the conducted experiments, the computational time available for the Tabu Search procedure is restricted to 10 seconds while a time limit of one hour is given for computing the lower bound. Clearly, if the exact solution approach is able to compute an optimal solution within the given time limit, the value of the lower bound is equal to the objective function value of the found optimal solution.

The evaluation using 10 vehicles indicates that the Tabu Search metaheuristic attains an almost optimal solution for all static problem instances. Specifically, the average solution quality of the Tabu Search procedure is only 0.017 % above the lower bound values which were computed by using the approach of Westphal and Krumke (2008). By restricting the fleet of vehicles to only 8 vehicles, which results in more complex static problem instances, this gap increases to 1.26 %. Note that this increase is solely caused by a few specific static problem instances in which distorting effects for single requests, whose maximum allowed response time is violated, occur.

The provided results indicate that the proposed real-time control approach attains a good solution adaptation performance that is necessary for a real-time application. Consequently, the application of the implemented Tabu Search metaheuristic for solving static problem instances allows a sufficiently fast and efficient tour plan adaptation. Note that, as mentioned in Sect. 2.5.3.2.2, optimally solving individual static problem instances does not guarantee an optimal result for the overall problem considered during the planing horizon. Therefore, by integrating stochastic knowledge into the static problem instances, the proposed Tabu Search metaheuristic supports the generation of solutions that provide an increased flexibility for efficiently integrating expected future requests. If high-quality stochastic knowledge is available, optimally solving individual static problem instances during the transportation process is likely to also result in a high quality solution of the overall problem considered. Note that since the best solution can be found in 10 seconds in most of the cases, the utilization of an anticipation horizon with the aforementioned length of 20 seconds ensures that a time period of 10 seconds is available to the solution method, even if pre-simulation activities such as road network calculations consume a time period t_a^{calc} of several seconds. Since requests arrive minute-wise in our request data sets, no drawback in the attained solution quality occurs in the computational results. However, since the average request response time is between 800 and 1,400 seconds in the evaluated scenarios, an anticipation horizon with a length of 20 seconds represents a small time period by which newly arriving requests would be additionally buffered if second-wise request arrival times were considered. Hence, the length of the anticipation horizon is not evaluated any further in this book. Nevertheless, note that the choice of this parameter generally depends on

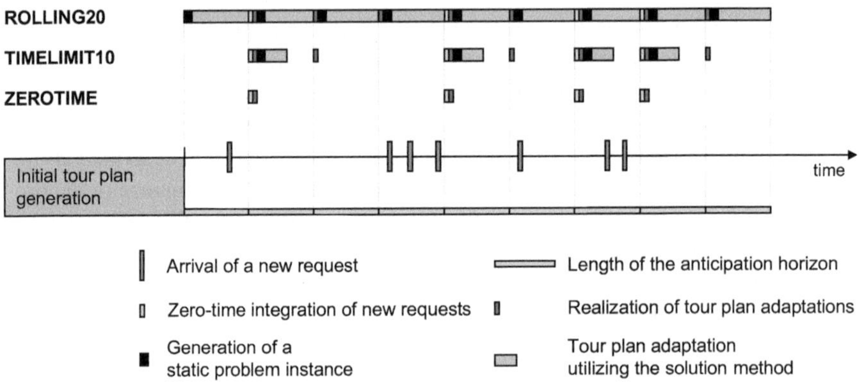

Fig. 8.7 Adaptation handling of the evaluated deterministic real-time control approaches

the considered application. For example, in applications with more pending requests and hence larger static problem instances, more computational time is required in order to attain solutions comprising a high solution quality. In such cases, a trade-off which requires an in-depth evaluation of different lengths of the anticipation horizon may arise. Research work which evaluates different settings for determining the length of the anticipation horizon can be found in, e.g., Ichoua et al. (2000), Bock (2004), and Bock (2010).

8.4 Performance Evaluation of Deterministic Real-Time Control Approaches

In what follows, the deterministic real-time control approach presented in Chap. 5 is designated as ROLLING20 since it applies the solution method in a rolling fashion for the complete length of each anticipation horizon. In the evaluation presented in this section, this approach is tested in direct comparison with two other deterministic real-time control approaches. The first one is called ZEROTIME. It represents a zero time adaptation that integrates newly arrived requests into the tour plan by applying a least cost insertion heuristic, i.e., by using steps 1 and 2 of the update concept presented in Sect. 5.2.2. The second approach is designated as TIMELIMIT10 and extends ZEROTIME by additionally applying the Tabu Search procedure for $\frac{t^{a}}{2} = 10$ seconds whenever new requests have arrived during the previous anticipation horizon. Therefore, TIMELIMIT10 utilizes the same update handling as ROLLING20 (while applying the solution method for half of the time) whenever dynamic events have recently arrived. All three real-time control approaches are illustrated in Fig. 8.7. This preliminary performance evaluation is carried out on the 30 days of S^{REAL} as well as on 30 systematically chosen days from all structural diversity settings of S^{GEN} which are denoted as $S^{\text{GEN}_{\text{all}}}$.

The results that are depicted in Tables 8.2 and 8.3 show that both TIMELIMIT10 and ROLLING20 significantly outperform the ZEROTIME variant for

Table 8.2 Performance of the evaluated deterministic real-time control approaches on S^{REAL}

Vehicles	Percentage improvements attained by	
	TIMELIMIT10 vs. ZEROTIME	ROLLING20 vs. TIMELIMIT10
	linear2X customer inconvenience	
8	32.25 %, pen: 12 vs. 186	0.30 %, pen: 9 vs. 12
10	18.46 %, pen: 0 vs. 8	0.13 %, pen: 0 vs. 0
12	11.18 %, pen: 0 vs. 0	−0.09 %, pen: 0 vs. 0
	quadratic customer inconvenience	
8	54.02 %, pen: 6 vs. 185	0.37 %, pen: 8 vs. 6
10	33.62 %, pen: 0 vs. 4	−0.50 %, pen: 0 vs. 0
12	22.04 %, pen: 0 vs. 0	−0.39 %, pen: 0 vs. 0

Note: Listed is the average improvement, *pen* denotes the total number of late requests

Table 8.3 Performance of different deterministic real-time control approaches on $S^{GEN_{all}}$

Vehicles	Percentage improvements attained by	
	TIMELIMIT10 vs. ZEROTIME	ROLLING20 vs. TIMELIMIT10
	linear2X customer inconvenience	
8	32.45 %, pen: 11 vs. 175	0.22 %, pen: 13 vs. 11
10	14.33 %, pen: 0 vs. 9	−0.09 %, pen: 0 vs. 0
12	9.21 %, pen: 0 vs. 0	−0.15 %, pen: 0 vs. 0
	quadratic customer inconvenience	
8	51.96 %, pen: 14 vs. 158	−0.79 %, pen: 15 vs. 14
10	30.65 %, pen: 0 vs. 5	−0.53 %, pen: 0 vs. 0
12	19.28 %, pen: 0 vs. 1	−0.56 %, pen: 0 vs. 0

Note: Listed is the average improvement, *pen* denotes the total number of late requests

both tested objective functions. Specifically, most improvements are achieved in the 8 vehicle scenarios where resources are strongly limited. This is due to the fact that in this reduced fleet size setting, a very efficient usage of the limited resources which is attained by TIMELIMIT10 and ROLLING20 is of high relevance. This can also be observed in Figs. 8.8 and 8.9 in which the results on S^{REAL} are presented in detail. Here it can be clearly observed that in problem instances with more requests resulting in more complex transportation situations, an efficient usage of vehicle resources yields high improvements. As mentioned above, these attainable improvements are highest when the available fleet size is restricted. Therefore, in accordance with the results achieved in former studies (cf., e.g., Gendreau et al. 1999a and Bock 2010), it can be concluded that the application of elaborated real-time control approaches is advantageous if an efficient execution of complex transporta-

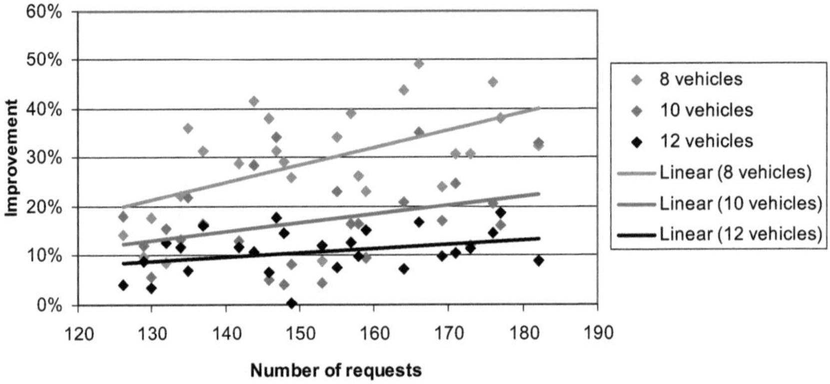

Fig. 8.8 Percentage improvements of TIMELIMIT10 vs. ZEROTIME, linear2X customer inconvenience on S^{REAL}

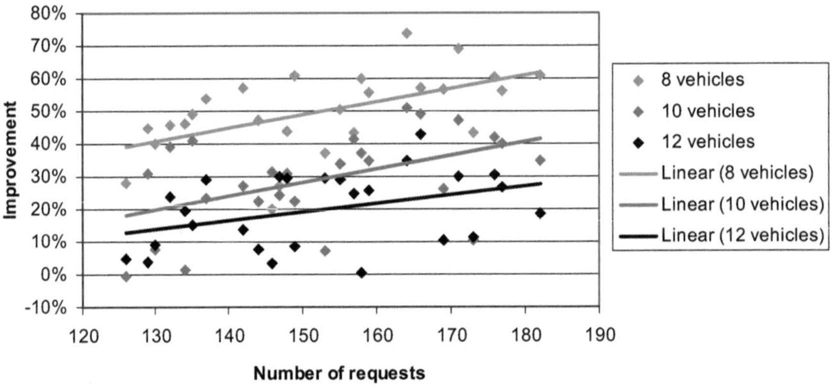

Fig. 8.9 Percentage improvements of TIMELIMIT10 vs. ZEROTIME, quadratic customer inconvenience on S^{REAL}

tion processes is pursued. The corresponding results for $S^{\text{GEN}_{\text{all}}}$ are illustrated in Figs. 8.10 and 8.11.

Furthermore, by directly comparing the results of the real-time control approaches TIMELIMIT10 and ROLLING20 (Tables 8.2 and 8.3), it can be observed that ROLLING20 does not produce any significant improvements in the considered RDOPG applications. Thus, it can be concluded that ROLLING20 does not use the increased computational times efficiently. This result can be explained by analyzing the Tabu Search evaluation in Sect. 8.3 once again. Specifically, the Tabu Search solution approach is utilized in this evaluation in the same way as in TIMELIMIT10. Since solutions in this evaluation that are very close to an optimal solution were attained, no further significant improvements can be achieved by ROLLING20 in this scenario. In what follows, all experiments are carried out using TIMELIMIT10 since this significantly reduces the time required to perform all test instances (see

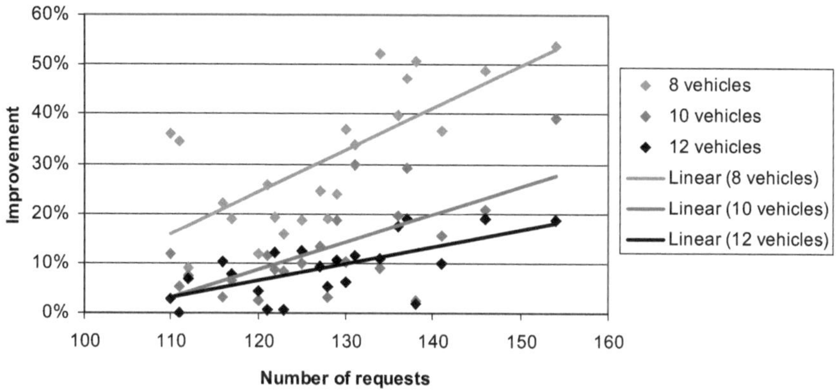

Fig. 8.10 Percentage improvements of TIMELIMIT10 vs. ZEROTIME, linear2X customer inconvenience on $S^{\mathrm{GEN}_{\mathrm{all}}}$

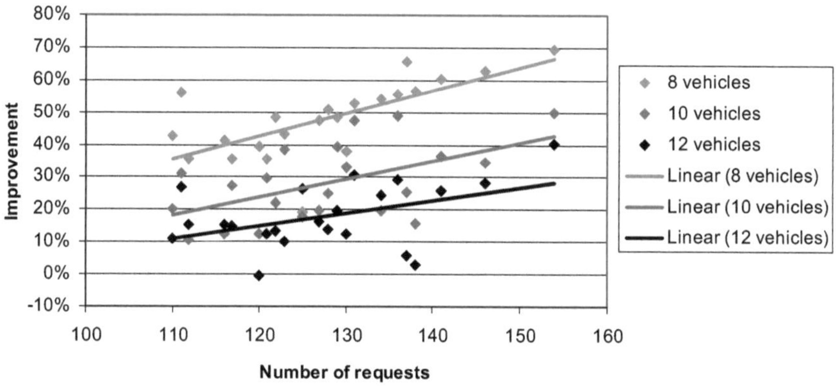

Fig. 8.11 Percentage improvements of TIMELIMIT10 vs. ZEROTIME, quadratic customer inconvenience on $S^{\mathrm{GEN}_{\mathrm{all}}}$

Sect. 8.2.5). Nevertheless, note that other scenarios exist in which ROLLING20 can achieve further improvements compared to TIMELIMIT10 but these are not considered in this book. Specifically, if scenarios with a significantly higher number of vehicles and requests are present, it is unlikely that TIMELIMIT10 will be able to find optimal solutions within the given time limit. Therefore, in such scenarios, the continuous application of the real-time control approach as proposed in ROLLING20 is likely to result in further significant improvements. Note that in the scenarios considered in Sects. 8.8 and 8.9, due to its achieved solution quality TIMELIMIT10 represents a challenging reference setting for attaining further improvements by additionally utilizing stochastic knowledge in the pro-active real-time control approach.

Table 8.4 Performance of TIMELIMIT10 vs. TIMELIMIT10NODIV on S^{REAL}

Vehicles	Objective function value	Travel time	Travel distance	#Reassignments
		linear2X customer inconvenience		
8	17.18 %, pen: 12 vs. 23	6.36 %	8.89 %	45.87 (5.73)
10	17.56 %, pen: 0 vs. 0	7.68 %	9.87 %	54.07 (5.41)
12	14.24 %, pen: 0 vs. 0	8.79 %	11.16 %	46.93 (3.91)
		quadratic customer inconvenience		
8	20.67 %, pen: 6 vs. 20	5.07 %	7.23 %	35.07 (4.38)
10	24.60 %, pen: 0 vs. 0	7.27 %	9.67 %	41.77 (4.18)
12	18.81 %, pen: 0 vs. 0	7.18 %	8.54 %	39.87 (3.32)

Note: Listed is the average improvement of the objective function value, total number of late requests (*pen*), average improvement of travel time and travel distance, and the average number of first request reassignments per test instance and per vehicle *in brackets* observed in TIMELIMIT10

8.5 Performance Evaluation of Allowing Vehicle En-Route Diversion

In this section we evaluate the impact of allowing vehicle en-route diversion on the solution quality in TIMELIMIT10. This is evaluated by disallowing all vehicle en-route diversion activities resulting in a variant of TIMELIMIT10 designated as TIMELIMIT10NODIV. The evaluation of both variants of the deterministic real-time control approach is carried out on all 30 days of S^{REAL}.

The results that are depicted in Table 8.4 show that the application of vehicle en-route diversion significantly improves the solution quality in the tested scenarios. Specifically, depending on the number of available vehicles and the customer inconvenience function, improvements between 14.24 % and 24.60 % can be achieved while the highest improvements can be attained if a quadratic customer inconvenience function is used. Furthermore, in scenarios with a strongly limited fleet size of 8 vehicles, allowing vehicle en-route diversion leads to a significantly lower number of late requests that are serviced after the maximum allowed response time defined by R^{mrt}. These results are in line with the ones obtained by Ichoua et al. (2000) and Richter (2005). Moreover, although the performed travel time as well as travel distance are not considered in the RDOPG objective function, both of these values are also reduced when vehicle en-route diversion is allowed. In Figs. 8.12, 8.13, and 8.14, the attained request response times for all three tested fleet sizes are illustrated. In these figures it can be observed that by allowing vehicle en-route diversion activities, many requests can be serviced earlier in comparison to the setting in which vehicle en-route diversion is forbidden.

Following these results, we conclude that it is advisable to allow the use of the vehicle en-route diversion activities in the considered RDOPG applications. However, allowing vehicle en-route diversion requires more responsiveness from drivers during travel activities since they have to be able to quickly react to changes in

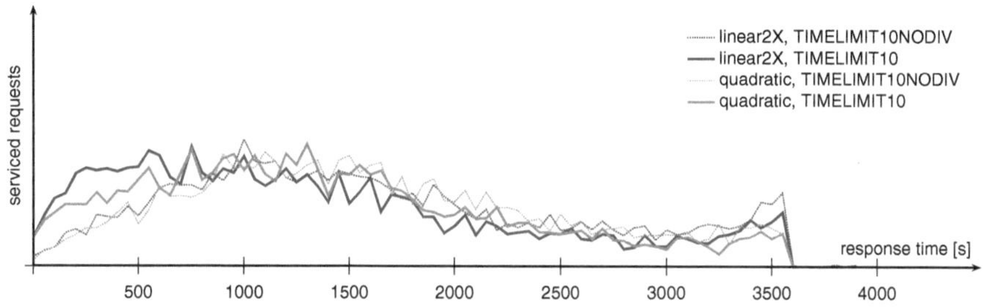

Fig. 8.12 Impacts of allowing vehicle en-route diversion on request response times using S^{REAL} and 8 vehicles grouped by intervals of 50 seconds

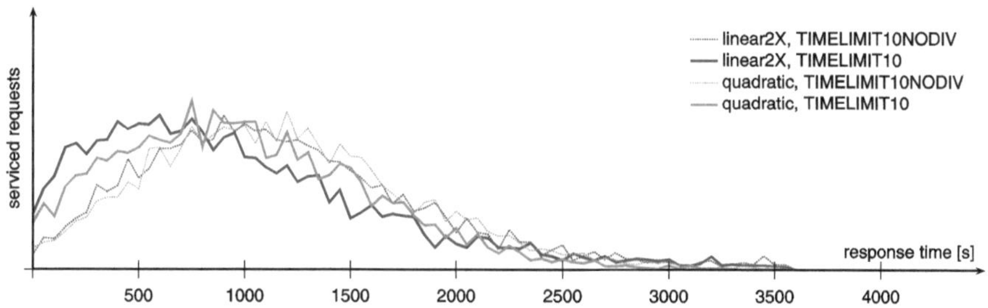

Fig. 8.13 Impacts of allowing vehicle en-route diversion on request response times using S^{REAL} and 10 vehicles grouped by intervals of 50 seconds

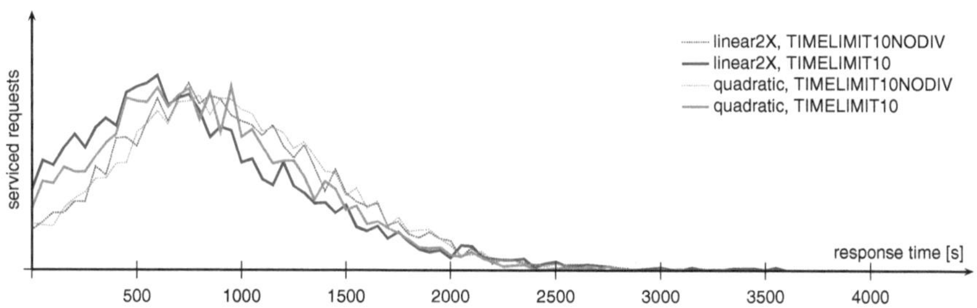

Fig. 8.14 Impacts of allowing vehicle en-route diversion on request response times using S^{REAL} and 12 vehicles grouped by intervals of 50 seconds

their tour that affect the next assigned request. The last column of Table 8.4 shows the number of first request reassignments, i.e., vehicle en-route diversion activities, that have occurred on average per test instance in TIMELIMIT10. Clearly, these values are 0 for TIMELIMIT10NODIV. Note that a first request reassignment is recorded only when the first request of a vehicle is changed during a travel activity. Specifically, a change of the first request is not accounted for if a vehicle is currently servicing a request or waiting idly for new requests. Moreover, note that if the first request of a vehicle is reassigned to another vehicle as first request, this is considered as two reassignment activities since the first request of both vehicles is changed. According to the values depicted in Table 8.4 it can be concluded that vehicle en-route diversion activities occur quite frequently in these computational experiments. When considering the drivers, this large number of en-route diversion activities may lead to a high *driver inconvenience* which is a rarely addressed issue in dynamic routing applications. Hence, in Sect. 8.10, we continue to discuss this issue and introduce a method for reducing the number of reassignment activities and thus accounting for driver inconvenience aspects in the proposed real-time control approach.

Moreover, note that a higher number of available vehicles leads to a slight increase of the performed vehicle en-route diversion activities. This can be explained by considering the fact that when more vehicles are available, they are likely to be closely located to each other. When new requests arrive, reassigning requests to those vehicles that are nearby becomes more attractive with regard to the overall solution quality.

8.6 Measuring Structural Distortion Using the Goodness-of-Fit Test

As mentioned in Chap. 6, our proposed forecasting methods as well as the pro-active real-time control approach assume that request arrivals can be modeled by a set of unknown time-space Poisson processes whose rate parameters are also unknown but remain constant from day to day. Since the expected number of request arrivals per day of generated clusters is determined by calculating the average request arrival rate per cluster on the past n^{f} working days (see Sect. 6.2), the application of the presented pro-active real-time control approach on request data sets which do not fulfill this assumption is likely to result in a poor solution quality. For example, consider a request data set in which requests arrive according to two different patterns which change every day. In this case, forecasting expected future requests by calculating the average of all request arrivals observed on the past n^{f} working days may result in the generation of clusters which predict requests according to a request arrival pattern that is a mixture of both request arrival patterns. However, since this generated average day does not exist, it is likely that the pro-active real-time control approach produces a poor solution quality on both types of days. Hence, it is crucial that the request data fulfills the mentioned assumption to a certain extent. For this reason, our proposed forecasting approach examines the existing request data set. Specifically, the Chi-Square Goodness-of-Fit test described in Sect. 6.2.1 checks the likelihood that request arrivals observed over the past n^{f} working days in potential clusters are generated by Poisson processes and discards them if this is too unlikely. We denote this likelihood as the Poisson quality of a cluster. According to our definition, a high Poisson quality defines a low structural distortion and vice versa.

In order to assess the effectiveness of our approach in identifying structural distortion in request data sets, we generate different request data sets that, according

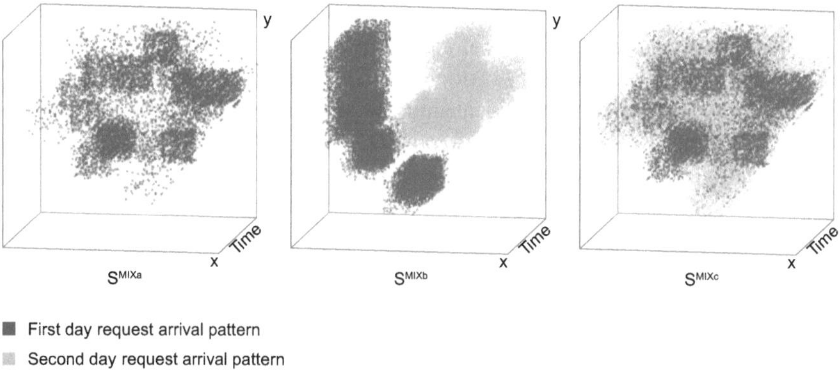

Fig. 8.15 Three request data sets with a high structural distortion

to the assumptions of the proposed forecasting approach, have a high structural distortion. Specifically, we generate four request data sets each having two individual request arrival patterns that alternate each working day. The generated request data sets are:

1. **Request data set S^{MIXa}:** The first pattern is defined as S^{GEN} $RD = TD = 1.00$ whereas no request arrivals are defined in the second pattern. Hence, on each second working day, no requests arrive.
2. **Request data set S^{MIXa2x}:** The first pattern is defined as S^{GEN} $RD = TD = 1.00$ with doubled request arrival rates, i.e., doubled $\lambda(t_j)$ values from Table 8.1 whereas no request arrivals are defined in the second pattern. These doubled request arrival rates are used in order to generate average request arrival rates which are similar to S^{GEN} $RD = TD = 1.00$.
3. **Request data set S^{MIXb}:** In this request data set, the first pattern is defined as a variation of S^{GEN} $RD = TD = 1.00$ in which all requests only arrive in subareas located on the left side of the considered service region. The second pattern is a variation of S^{GEN} $RD = TD = 1.00$ in which all requests arrive in subareas located on the right side of the considered service region.
4. **Request data set S^{MIXc}:** The first pattern is defined as S^{GEN} $RD = TD = 1.00$ whereas the second pattern is defined as S^{GEN} $RD = TD = 0.00$. This request data set is used to assess the performance of the used Chi-Square Goodness-of-Fit test on a request data set in which requests arrive every day at all locations but according to two different request arrival patterns.

The request data sets S^{MIXa}, S^{MIXb}, and S^{MIXc} are illustrated in Fig. 8.15. In addition to the four request data sets above, we evaluate the structural distortion of S^{GEN} $RD = TD = 1.00$, $RD = TD = 0.75$, and of S^{REAL}. For the minimum demanded Poisson quality of a valid cluster denoted by DC^{minPQ}, we evaluate values of $DC^{\text{minPQ}} \in \{0, 0.05, 0.10, 0.20, 0.40\}$. For this purpose, in the applied Chi-Square Goodness-of-Fit test the type I error parameter is set to $\alpha = DC^{\text{minPQ}}$. Moreover, in order to generate clusters which, according to the assumptions of our forecasting approach, have a high probability that real customers will arrive, we apply a minimum demanded dummy customer quality of $DC^{\min\lambda} = 1.8$.

The results of the structural distortion evaluation are illustrated in Table 8.5. As can be seen in this table, a significant number of dummy customers can be generated in all request data sets in the setting $DC^{\text{minPQ}} = 0$ in which no minimum Poisson quality is demanded for the generated dummy customers. Therefore, one could falsely presume a high suitability of the four request data sets (S^{MIXa} to S^{MIXc}) to generate well-usable stochastic knowledge using the proposed forecasting approach. For example, by comparing S^{MIXa2x} and S^{GEN} $RD = TD = 1.00$, the number of generated dummy customers is similar, i.e., 35 to 37. However, as the applied Chi-Square Goodness-of-Fit test reveals, the generated average Poisson quality of these four request data sets is significantly lower than that of the request data sets S^{GEN} and S^{REAL} . Specifically, generated dummy customers of the three request data sets S^{MIXa}, S^{MIXa2x}, and S^{MIXb} have a significantly low Poisson quality of less than 0.01 % which indicates a high structural distortion for these request data sets. Moreover, each of these three request data sets have a *request realization rate* of at most 50 %. Specifically, in each of the corresponding clusters of the generated dummy customers, request arrivals are observed on, at most, 50 % of the n^{f} working days. This can be explained by their request arrival patterns mentioned above. These observed request realization rates which are significantly smaller than the expected request realization rate defined by $PS_{1.8}(X \geq 1) = 83.5$ % also indicate that these three request data sets have a high structural distortion. Consequently, if a minimum Poisson quality of $DC^{\text{minPQ}} = 0.05$ is demanded, no dummy customers can be generated at all in these three request data sets. In contrast to the three previously evaluated request data sets, the average Poisson quality of generated dummy customers in S^{MIXc} using $DC^{\text{minPQ}} = 0$ is higher, i.e., about 5 % compared to an average Poisson quality of less than 0.01 %. Nevertheless, this Poisson quality is still about an order of magnitude smaller than in S^{GEN} and S^{REAL}. Moreover, another indicator that S^{MIXc} has an inferior Poisson quality is that the request realization rate of corresponding clusters is on average 71 %, which is significantly smaller than the expected request realization rate of 83.5 %. Hence, if a minimum Poisson quality of $DC^{\text{minPQ}} = 0.05$ is demanded, almost half of the 13 clusters are discarded. Note that this is a significant difference to S^{GEN} and S^{REAL} in which almost the same number of dummy customers are generated in $DC^{\text{minPQ}} = 0.00$ and $DC^{\text{minPQ}} = 0.05$. By further increasing the minimum demanded Poisson quality to $DC^{\text{minPQ}} = 0.40$, only one dummy customer is generated in S^{MIXc} while in the evaluated request data sets of S^{GEN} and S^{REAL} only a slight decrease of the number of generated dummy customers occurs. Preliminary computational experiments on S^{GEN} and S^{REAL} evaluating the previously mentioned values of DC^{minPQ} showed that the setting $DC^{\text{minPQ}} = 0.40$ yields the best solution quality among the tested values. Consequently, $DC^{\text{minPQ}} = 0.40$ guarantees a low number of erroneously erased clusters while ensuring that inappropriate clusters with a significantly inferior Goodness-of-Fit quality are removed. Note that we refrained from evaluating values for DC^{minPQ} higher than 0.40 for two reasons. First, as mentioned above, this value efficiently removes almost all dummy customers with a low Poisson quality in the

Table 8.5 Evaluating the structural distortion on different generated request data sets using different DC^{minPQ} settings and $DC^{min\lambda} = 1.8$

Request data set	Minimum Poisson quality DC^{minPQ}				
	0	0.05	0.10	0.20	0.40
S^{MIXa}	8	0	0	0	0
	<0.0001	–	–	–	–
	47 % 49 % 50 %	–	–	–	–
S^{MIXa2x}	35	0	0	0	0
	<0.0001	–	–	–	–
	47 % 49 % 50 %	–	–	–	–
S^{MIXb}	19	0	0	0	0
	<0.0001	–	–	–	–
	47 % 49 % 50 %	–	–	–	–
S^{MIXc}	13	7	5	2	1
	0.0495	0.1295	0.167	0.3162	0.5206
	62 % 71 % 78 %	73 % 76 % 78 %	75 % 77 % 78 %	78 % 78 % 78 %	80 % 80 % 80 %
S^{GEN}	15	15	15	15	15
$RD = TD = 0.75$	0.426	0.4314	0.5243	0.5747	0.6386
	77 % 84 % 93 %	77 % 84 % 93 %	77 % 85 % 93 %	80 % 86 % 95 %	80 % 86 % 95 %
S^{GEN}	37	37	37	37	35
$RD = TD = 1.00$	0.5124	0.5139	0.4925	0.581	0.7048
	80 % 86 % 93 %	80 % 86 % 93 %	78 % 86 % 93 %	78 % 86 % 93 %	82 % 86 % 92 %
S^{REAL}	15.3	15	14.9	14.1	13.1
	0.4047	0.4238	0.4537	0.5148	0.6261
	68 % 81 % 92 %	73 % 81 % 92 %	73 % 81 % 93 %	75 % 82 % 93 %	77 % 82 % 92 %

Note: Listed is the average number of generated dummy customers, their average Poisson quality, and the minimum, average, and maximum request realization rate, i.e., the percentage of the n^f working days on which at least one request arrives in one corresponding cluster of the generated dummy customers

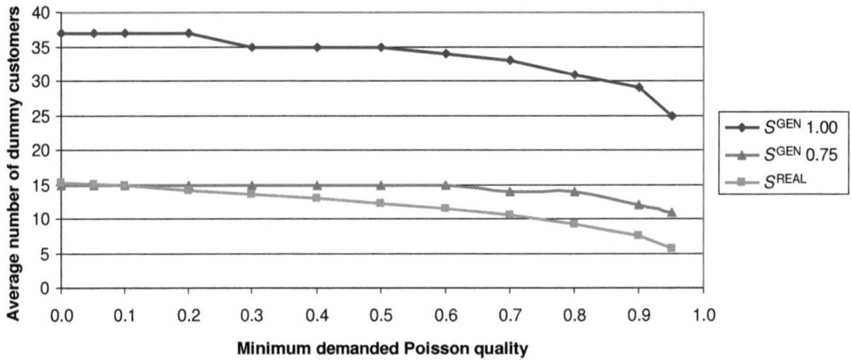

Fig. 8.16 Average number of generated dummy customers on S^{REAL}, S^{GEN} $RD = TD = 0.75$, and S^{GEN} $RD = TD = 1.00$ using $DC^{\min\lambda} = 1.8$

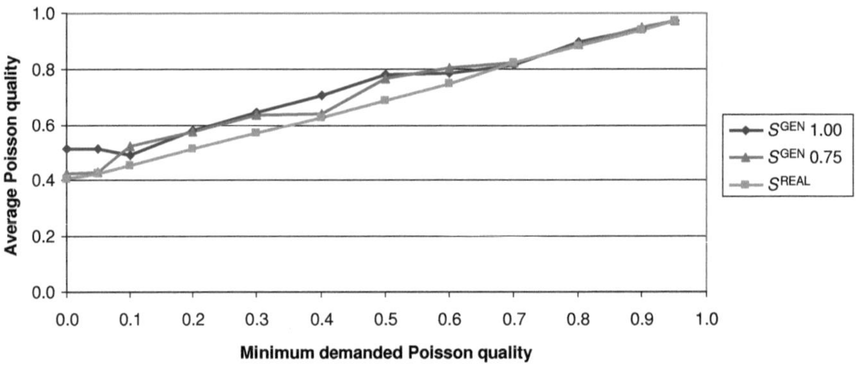

Fig. 8.17 Average Poisson quality on S^{REAL}, S^{GEN} $RD = TD = 0.75$, and S^{GEN} $RD = TD = 1.00$ using $DC^{\min\lambda} = 1.8$

tested request data sets that have a high structural distortion. Second, as illustrated in Fig. 8.16, choosing higher DC^{minPQ} values increasingly reduces the number of dummy customers that can be generated, especially in S^{REAL} so that a trade-off arises between the resulting Poisson quality and the number of generated dummy customers. Since setting DC^{minPQ} to 0.40 results in an average Poisson quality of more than 0.6 (see Fig. 8.17), we conclude that this setting allows for the generation of clusters which sufficiently fulfill the demanded assumptions. Note that we deviate from setting type I error to $\alpha = 0.05$ on purpose (cf. Israel 2008, p. xxii, Sapsford and Jupp 2006, p. 217) in order to only generate stochastic knowledge which has a sufficient Poisson quality.

From the results presented above we conclude that the proposed approach using the Chi-Square Goodness-of-Fit test to determine the Poisson quality of request data sets is an efficient method for determining the structural distortion of an existing request data set. Specifically, it allows for identifying request data sets with a high structural distortion in which a reasonable number of dummy customers can be generated without applying this test. Note that the structural distortion of a request data set depends on the applied forecasting approach. Specifically, by extending the proposed forecasting approach, it becomes possible to efficiently generate and integrate stochastic knowledge out of specific request data sets which have a high structural distortion in our forecasting approach. Nevertheless, for each forecasting approach, there are specific request data sets which have a high structural distortion with regard to the respective forecasting approach. Hence, we recommend determining the structural distortion of given request data sets for all forecasting approaches.

Structural Distortion of S^{REAL} The attained results also provide information on the structural distortion of the real-world request data set S^{REAL}. Specifically, since S^{REAL} possesses a Poisson quality which is similar to both evaluated S^{GEN} request data sets that possess a low structural distortion with certainty, we conclude that S^{REAL} has also a low structural distortion. Note that this makes it reasonable to evaluate S^{REAL} in our main computational experiments in Sect. 8.8.

8.7 Results of Generating Stochastic Knowledge for S^{REAL} and S^{GEN}

In this section, results of generating the dummy customers for the request data classes S^{REAL} and S^{GEN} using the exact clustering approach proposed in Chap. 6 are presented. For generating this stochastic knowledge the previously derived minimum Poisson quality $DC^{\text{minPQ}} = 0.40$ is used.

After all valid clusters are generated, a Mixed Integer Programming (MIP) problem is solved in order to select the largest number of valid clusters which start earliest so that no segments overlap as described in Sect. 6.2.2. All of these formulated MIP problems are solved using ILOG CPLEX 11.0. With regard to the required computational time, MIP problems with higher $DC^{\text{min}\lambda}$ values can be solved within seconds while lower $DC^{\text{min}\lambda}$ values result in significantly longer computational times due to the high number of valid clusters. Specifically, for the lowest evaluated minimum demanded dummy customer quality $DC^{\text{min}\lambda} = 1.0$, up to 2,362 valid clusters were generated for S^{REAL} and 3,272 valid clusters for S^{GEN} $RD = TD = 1.00$. Note that this cluster selection process can often be significantly accelerated. Specifically, by using characteristics of the problem structure (see objective function defined in Eq. (6.1) in Sect. 6.2.2), for the objective function value z of the optimal MIP solution with n dummy customers it holds that $n \cdot M - U \leq z \leq n \cdot M$ with $U = \sum_{i \in C_n^{\text{latest}}} c_i^{\text{startTL}}$ where $C_n^{\text{latest}} \subseteq C$ represents the set of the n clusters of C which start latest. By using this property and consulting the currently known upper bound during the solution process of an MIP problem, a new value of the maximum number of selectable clusters defined by $C^{\max}$ can be dynamically determined. For example, assume $M = 100{,}000$, $U = 18{,}000$, and a currently upper bound of $8{,}532{,}102$, e.g., the objective function value of the currently best known

Table 8.6 Number of generated dummy customers on S^{REAL} depending on the minimum required dummy customer quality $DC^{\min\lambda}$ using $DC^{\min PQ} = 0.40$

Quality of stochastic knowledge ($DC^{\min\lambda}$)				
1.0	1.2	1.5	1.8	2.0
67.7 (13.4 %)	44.6 (12 %)	24.4 (10.4 %)	13.1 (10.1 %)	9.4 (7.7 %)

Note: Listed is the number of dummy customers generated by the proposed exact clustering method. The *values in brackets* denote the additional percentage of dummy customers that were generated compared to a heuristic greedy variant of the exact clustering approach

LP relaxation. Assuming further that the currently best known MIP solution has an objective function value of 8,489,125, the current gap is $\frac{8{,}532{,}102-8{,}489{,}125}{8{,}532{,}102} \approx 5\,\%$. Using the described technique, it can be derived that the number of clusters that can be selected is at most 85 since $8{,}532{,}102 < 86 \cdot 100{,}000 - 18{,}000 = 8{,}582{,}000$. Hence, the objective function value of the known upper bound is at least reduced to 8,500,000 so that the gap is reduced to at most $\frac{8{,}500{,}000-8{,}489{,}125}{8{,}532{,}102} \approx 0.12\,\%$. We utilize this technique in the conducted MIP problems by examining the currently best known upper bound during the solution process and adjusting the value of parameter $C^{\max}$ accordingly. In doing so, all conducted MIP problems can be solved to optimality within two hours of computational time while most of the MIP problems are solved within a few minutes.

In Tables 8.6 and 8.7, the number of dummy customers which were generated using our proposed exact cluster selection approach are presented. Note that in S^{REAL}, for each of the 30 conducted days, the last corresponding $n^{\text{f}} = 60$ working days out of the available real-word request information were used. Hence, for each of the 30 days of S^{REAL}, five MIP problems were generated for the different $DC^{\min\lambda}$ values so that in total 150 MIP problems were solved for S^{REAL}. Consequently, Table 8.6 illustrates the average number of dummy customers generated per scenario and $DC^{\min\lambda}$ value. Since all days of S^{GEN} are generated according to the procedure described in Sect. 8.2.4.2, dummy customers are generated once for all 25 structural diversity settings and $DC^{\min\lambda}$ value. These dummy customers are then used in each of the 30 days. As can be seen in Table 8.7, specific $DC^{\min\lambda}$ values require a minimum level of structural diversity so that in total 87 MIP problems were solved for S^{GEN}. Moreover, the number of generated dummy customers is compared to a heuristic clustering approach which generates clusters in the same manner as the exact approach but greedily selects dummy customers whenever a valid cluster has been found. Although the heuristic clustering approach only requires about one minute for solving each dummy customer selection problem, the results in Tables 8.6 and 8.7 show that it is reasonable to utilize the exact clustering approach since it yields on average 10.7 % more generated dummy customers on S^{REAL} and 6.4 % more on S^{GEN}.

With regard to the base sizes which are used in the selected clusters, in S^{GEN} it can be observed that in request data sets with higher levels of structural diversity the average number of base segments in a selected cluster becomes lower. Specif-

Table 8.7 Number of generated dummy customers on S^{GEN} depending on the level of structural diversity and the minimum required dummy customer quality $DC^{\min\lambda}$ using $DC^{\min PQ} = 0.40$

Structural diversity		Quality of stochastic knowledge ($DC^{\min\lambda}$)				
RD	*TD*	1.0	1.2	1.5	1.8	2.0
0.00	0.00	2 (0 %)				
0.00	0.25	11 (10 %)				
0.00	0.50	21 (23.5 %)	5 (0 %)			
0.00	0.75	36 (9.1 %)	12 (9.1 %)			
0.00	1.00	46 (15 %)	19 (11.8 %)	7 (40 %)	2 (0 %)	
0.25	0.00	5 (0 %)				
0.25	0.25	17 (6.3 %)	3 (0 %)			
0.25	0.50	26 (8.3 %)	8 (14.3 %)			
0.25	0.75	36 (9.1 %)	21 (5 %)	8 (0 %)		
0.25	1.00	45 (9.8 %)	26 (13 %)	10 (11.1 %)	6 (0 %)	3 (0 %)
0.50	0.00	20 (0 %)	5 (0 %)			
0.50	0.25	31 (3.3 %)	10 (0 %)	2 (0 %)		
0.50	0.50	36 (9.1 %)	20 (0 %)	7 (0 %)		
0.50	0.75	44 (7.3 %)	29 (7.4 %)	16 (6.7 %)	8 (0 %)	4 (0 %)
0.50	1.00	54 (17.4 %)	37 (23.3 %)	24 (20 %)	14 (0 %)	11 (0 %)
0.75	0.00	32 (0 %)	15 (0 %)	2 (0 %)		
0.75	0.25	38 (8.6 %)	18 (5.9 %)	6 (0 %)		
0.75	0.50	40 (11.1 %)	28 (7.7 %)	16 (0 %)	6 (0 %)	2 (0 %)
0.75	0.75	54 (10.2 %)	38 (11.8 %)	24 (4.3 %)	15 (0 %)	11 (0 %)
0.75	1.00	70 (9.4 %)	50 (16.3 %)	33 (3.1 %)	27 (8 %)	20 (5.3 %)
1.00	0.00	37 (2.8 %)	22 (4.8 %)	14 (0 %)	1 (0 %)	1 (0 %)
1.00	0.25	45 (4.7 %)	28 (0 %)	12 (0 %)	5 (0 %)	2 (0 %)
1.00	0.50	61 (15.1 %)	42 (10.5 %)	25 (8.7 %)	16 (14.3 %)	9 (0 %)
1.00	0.75	74 (12.1 %)	54 (12.5 %)	34 (3 %)	26 (13 %)	20 (11.1 %)
1.00	1.00	80 (14.3 %)	65 (14 %)	46 (9.5 %)	35 (12.9 %)	29 (3.6 %)

Note: Listed is the number of dummy customers generated by the proposed exact clustering method. The *values in brackets* denote the additional percentage of dummy customers that were generated compared to a heuristic greedy variant of the exact clustering approach. *Empty cells* denote that no dummy customers were generated in this setting

ically, the maximum allowed spatial extension of a cluster of $DC^{mse} = (2 \times 2) = 4$ segments (see Sect. 8.2.3) is always used when scenarios with low levels of structural diversity are present. In contrast, the average number of base segments declines in $RD = TD = 1.00$, depending on the minimum quality of stochastic knowledge, to values from 3.153 segments ($DC^{\min\lambda} = 1.0$) to 3.438 segments ($DC^{\min\lambda} = 2.0$). Hence, higher levels of structural diversity in S^{GEN} do not only allow for generating more dummy clusters but also those which have a higher precision. This increases the value of the generated stochastic knowledge. On S^{REAL}, the average number of base segments in a selected cluster ranges from 3.729 segments ($DC^{\min\lambda} = 1.0$) to 4 segments ($DC^{\min\lambda} = 2.0$).

In Table 8.8, the request realization rates of the generated stochastic knowledge in all 26 request data sets are illustrated. Moreover, for each $DC^{\min\lambda}$ setting, we simulated 100 clusters. For each of these clusters, we generated 60 replications, i.e., 60 days, of request arrivals using a Poisson process with $\lambda = DC^{\min\lambda}$ and measured the request realization rates. The obtained values as well as the arrival probability of at least one request according to the Poisson distribution are also illustrated. The results show that the request realization rates mainly depend on $DC^{\min\lambda}$ but not on the level of structural diversity. The similarity of the results of the 26 request data sets to the simulated as well as estimated values underscores the fact that all request data sets have a low structural distortion. Hence, we conclude that in both evaluated request data classes request arrivals occur according to Poisson processes to an extent which is sufficient to conduct the main computational experiments on these request data classes in what follows.

8.8 Performance Evaluation of the Pro-Active Real-Time Control Approach

In this section, we present the main results of our computational experiments. In these experiments, the performance of the pro-active real-time control approach that utilizes the generated stochastic knowledge is evaluated. As mentioned above, since no benchmark instances for the considered RDOPG applications are available in the literature, this evaluation is carried out by comparing the results obtained by the pro-active real-time control approach with the ones of the deterministic real-time control approach TIMELIMIT10. In order to select an appropriate setting of the pro-active real-time control approach, a preliminary study comparable to the one performed for the deterministic real-time control approach in Sect. 8.4 has been conducted. It reveals that, similar to the results obtained for the deterministic real-time control approach, the continuous application and the time-limit variant of the pro-active real-time control approach produce a comparable solution quality. Hence, in what follows, the computational experiments are carried out by comparing the pro-active real-time control approach applied for 10 seconds when dynamic events have occurred in the previous anticipation horizon (PROACTIVE10) with its deterministic variant TIMELIMIT10. Consequently, both approaches only differ from each other in the stochastic knowledge additionally utilized by PROACTIVE10.

Computational Results on Dummy Customer Removal Decisions Regarding the two tested values of the parameter $DC^{\lambda\text{rem}}$, which defines the time point at which each dummy customer will be individually removed from the tour plan (see Sect. 8.2.3), preliminary tests indicate that an earlier removal of dummy customers

Table 8.8 Request realization rates (in percents) in S^{GEN} and S^{REAL} depending on the parameter $DC^{min\lambda}$ using $DC^{minPQ} = 0.40$

Structural diversity		Quality of stochastic knowledge ($DC^{min\lambda}$) in percents				
RD	*TD*	1.0	1.2	1.5	1.8	2.0
0.00	0.00	60 61 62				
0.00	0.25	58 62 67				
0.00	0.50	60 64 72	65 70 73			
0.00	0.75	58 65 77	67 71 77			
0.00	1.00	60 65 75	67 72 77	72 77 83	83 86 88	
0.25	0.00	60 63 67				
0.25	0.25	60 64 68	68 69 70			
0.25	0.50	58 64 72	68 72 75			
0.25	0.75	58 64 77	67 71 78	75 79 83		
0.25	1.00	60 65 78	65 71 77	75 79 83	82 85 88	87 88 88
0.50	0.00	60 64 70	67 68 72			
0.50	0.25	58 64 72	68 72 78	77 78 80		
0.50	0.50	60 65 75	67 72 78	73 78 83		
0.50	0.75	60 65 78	65 72 78	75 80 85	82 87 92	82 87 92
0.50	1.00	60 66 77	67 73 78	73 80 85	80 85 92	85 89 93
0.75	0.00	58 65 75	67 71 75	78 79 80		
0.75	0.25	60 65 70	65 72 78	72 78 85		
0.75	0.50	60 66 77	63 71 78	75 79 87	78 83 87	85 89 93
0.75	0.75	57 65 73	63 72 78	75 80 85	80 86 95	85 89 92
0.75	1.00	58 65 77	65 71 83	73 80 87	77 85 92	83 89 93
1.00	0.00	57 65 70	63 72 78	72 78 85	87 87 87	87 87 87
1.00	0.25	58 65 70	65 71 80	73 79 83	83 86 88	90 90 90
1.00	0.50	60 66 77	65 72 87	75 80 87	80 85 90	85 89 93
1.00	0.75	58 65 80	65 72 82	72 79 88	80 85 90	82 89 95
1.00	1.00	58 66 80	63 72 80	73 80 90	82 86 92	83 88 95
S^{REAL}		55 63 78	62 70 83	70 77 87	77 82 92	78 85 93
Simulated		48 64 80	58 71 87	62 77 88	72 83 93	75 86 97
$PS_{DC^{min\lambda}}(X \geq 1)$		63.2	69.9	77.7	83.5	86.5

Note: Listed is the minimum, average, and maximum request realization rate, i.e., the percentage of the n^f working days on which at least one request arrives in one corresponding cluster of the generated dummy customers

is more promising in the evaluated RDOPG application. Specifically, setting the parameter $DC^{\lambda rem}$ to 0.50 instead of 0.25 allows a more flexible tour plan adaptation. This results from the fact that when a smaller value for $DC^{\lambda rem}$ is used, the probabil-

Table 8.9 Percentage improvements achieved by PROACTIVE10 vs. TIMELIMIT10 (i.e., improvements attained by utilizing stochastic knowledge) on S^{REAL}

Vehicles	Quality of stochastic knowledge ($DC^{\min\lambda}$)				
	1.0	1.2	1.5	1.8	2.0
			linear2X customer inconvenience		
8	↑ −11.50 % (28)	↑ −7.66 % (21)	↑ −3.12 % (18)	↑ −0.27 % (18)	↑ **1.60** % (**10**)
10	−4.86 % (22)	−1.42 % (20)	−0.68 % (17)	**1.00** % (**16**)	0.01 % (16)
12	−0.67 % (15)	−0.14 % (16)	**1.70** % (**13**)	0.64 % (11)	0.88 % (11)
			quadratic customer inconvenience		
8	↑ −19.91 % (27)	↑ −7.52 % (21)	↑ 0.21 % (14)	↑ 1.17 % (15)	↑ **1.13** % (**13**)
10	−3.99 % (20)	0.70 % (12)	2.76 % (9)	**2.79** % (**10**)	0.71 % (14)
12	6.59 % (9)	**7.93** % (**5**)	6.78 % (9)	7.31 % (8)	6.08 % (8)

Note: Illustrated is the average improvement. The *values in brackets* denote the number of worse test instances. *Bold*: best $DC^{\min\lambda}$. ↑: more late requests than in TIMELIMIT10

ity of a future request arrival in the corresponding cluster becomes too low to justify keeping a vehicle there any longer. Hence, by using higher values for $DC^{\lambda\text{rem}}$, vehicles are available at an earlier time point so that pending requests can be fulfilled faster. In what follows, $DC^{\lambda\text{rem}}$ is set to 0.50 for all experiments carried out using PROACTIVE10.

8.8.1 Results on S^{REAL}

The results that are attained by evaluating the real-world request data class are illustrated in Table 8.9. Specifically, the table shows the additional improvements attained by PROACTIVE10 compared to the results obtained by TIMELIMIT10. Moreover, the number of worse test instances out of the 30 evaluated ones is illustrated. Specifically, according to the objective function described in Sect. 5.3.2.4, solutions generated by PROACTIVE10 and TIMELIMIT10 are compared according to the number of late requests (primary objective) and the total customer inconvenience (secondary objective). Therefore, a solution generated by PROACTIVE10 is denoted as worse if either (a) it comprises more late requests or (b) it comprises an equal number of late requests but a larger total variable customer inconvenience than the one generated by TIMELIMIT10.

As indicated by the boldly printed results, under specific conditions, significant improvements can be achieved by the additional use of the stochastic knowledge although, as mentioned above, the underlying request distributions in S^{REAL} are completely unknown. Particularly, in scenarios with a larger fleet size (i.e., 10 and 12 vehicles), the application of the quadratic customer inconvenience function enables a successful integration of the generated stochastic knowledge. This is due to the fact that the over-linear increase of the quadratic inconvenience function leads to a lower number of requests that are significantly postponed in favor of other requests as can be observed in the linear2X inconvenience function. Consequently, generated tour plans are more balanced according to the resulting request response times so that violations of the maximum allowed request response time can be less observed in case the quadratic customer inconvenience is used. In contrast to this, the application of the linear2X inconvenience function more frequently leads to distorting effects, where some requests suddenly become extremely urgent due to an imminent violation of the maximum allowed response time. Since in these scenarios additional delays by further arriving requests may cause considerable deteriorations, tour plans generated if the quadratic inconvenience function is used are more robust when delays caused by incorrectly forecasted requests or unexpected additional requests occur. Moreover, reductions of resulting response times due to correctly forecasted requests have more significant effects in the quadratic customer inconvenience case.

In situations where more vehicles are available, the negative effect of incorrectly forecasted request arrivals that becomes more likely with lower $DC^{\min\lambda}$ values has less impact on both tested inconvenience functions. Hence, also more uncertain information can be gainfully utilized so that the best results are achieved by using smaller values of parameter $DC^{\min\lambda}$. In contrast, in scenarios where resource bottlenecks can be observed (8 vehicles), the mentioned robustness of generated tour plans when using a quadratic inconvenience function diminishes. Specifically, since the limited resources are very busy with servicing the pending requests, the resulting request response times increase towards the maximum allowed response time for a large number of requests. Therefore, as depicted above for the linear2X customer inconvenience case, an additional delay caused by incorrectly forecasted requests often results in a substantial increase of customer inconvenience caused by the high penalty costs (see Sect. 5.3.2.4). Consequently, no improvement can be attained for all tested settings with the smallest fleet size. Hence, in case of a small fleet size, the integration of stochastic knowledge is reasonable only if the stochastic knowledge about future requests possesses a certain value which we consider later in this section. Otherwise, the attempt to integrate stochastic knowledge often causes deteriorating effects which lead to a significant postponement of pending requests and thus to a higher number of late requests and a reduced solution quality.

The described circumstances can also be observed by analyzing the distribution of the attained request response times in the experiments. Specifically, when using a small fleet size, Fig. 8.18 illustrates that the utilization of uncertain stochastic knowledge (i.e., setting $DC^{\min\lambda} = 1.0$) further increases the workload of the already very busy vehicles. Since the uncertainty of this knowledge does not allow for a faster service of many requests, a high number of late serviced customers occurs which can be observed by the high peak that occurs shortly before the maximum allowed request response time of $R^{\mathrm{mrt}} = 3{,}600$ seconds. This is especially true when the linear2X customer inconvenience function is used which, as mentioned above, tends to postpone some requests in order to achieve a better average request response time which becomes problematic when future requests unexpectedly arrive. As can be seen in Fig. 8.19, the use of more certain stochastic knowledge generated by

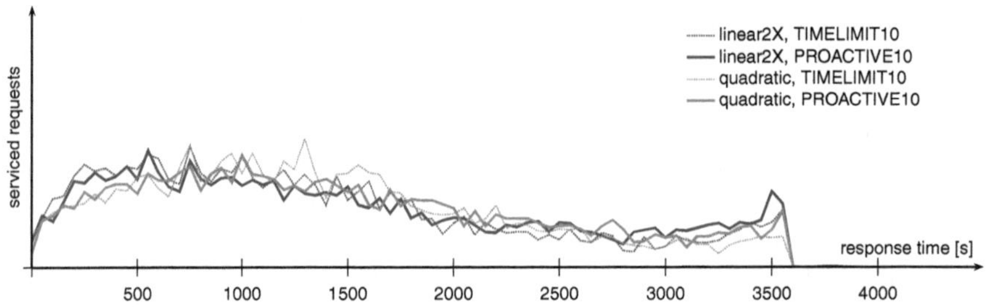

Fig. 8.18 Request response times on S^{REAL} using 8 vehicles and $DC^{\min\lambda} = 1.0$ grouped by intervals of 50 seconds

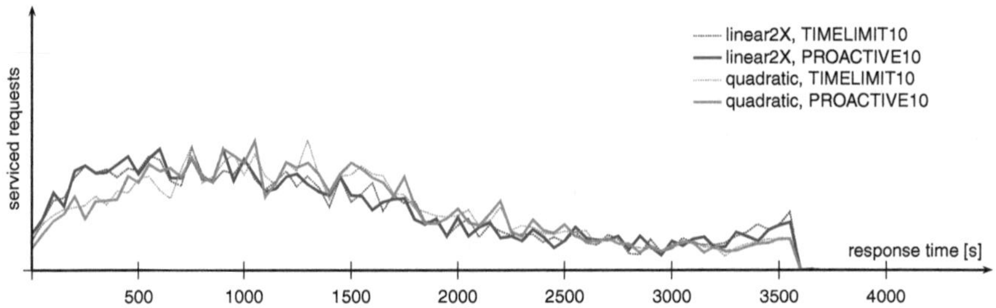

Fig. 8.19 Request response times on S^{REAL} using 8 vehicles and $DC^{\min\lambda} = 2.0$ grouped by intervals of 50 seconds

higher $DC^{\min\lambda}$ values may reduce these negative effects. Nevertheless, due to the restricted fleet size, the busy vehicle situation remains so that no improvements can be attained in this situation either.

In contrast, in scenarios where 10 vehicles are available, a high number of requests that are almost late can still be observed when uncertain stochastic knowledge is utilized, but to a significantly lower extent compared to the 8 vehicle case (see Figs. 8.20 and 8.21). Hence, if the linear inconvenience function is used, improvements of up to 1.0 % can be achieved whereas improvements of up to 2.79 % can be achieved if the quadratic inconvenience function is used. When 12 vehicles are used, the use of more uncertain stochastic knowledge can gain significant improvements. Especially if the quadratic inconvenience function is used, a meaningful number of requests can be serviced earlier (see Figs. 8.22 and 8.23). As can be seen in Table 8.9, significant improvements in the solution quality can be attained for both evaluated inconvenience functions. Specifically, if the linear inconvenience function is used, improvements of up to 1.7 % can be achieved whereas improvements of up to 7.93 % can be achieved if the quadratic inconvenience function is used. The complete figures depicting the results of the computational experiments on S^{REAL} are illustrated in the Appendix in Sect. C.1.

Structural Quality of the Real-World Request Data A visual analysis of the utilized request data set reveals that S^{REAL} comprises a request data structure which

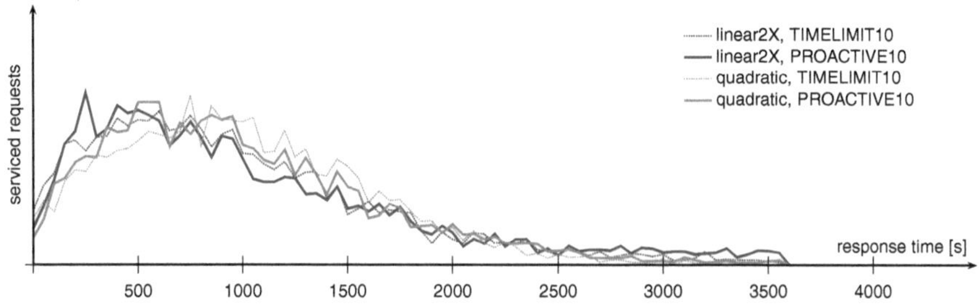

Fig. 8.20 Request response times on S^{REAL} using 10 vehicles and $DC^{\min\lambda} = 1.0$ grouped by intervals of 50 seconds

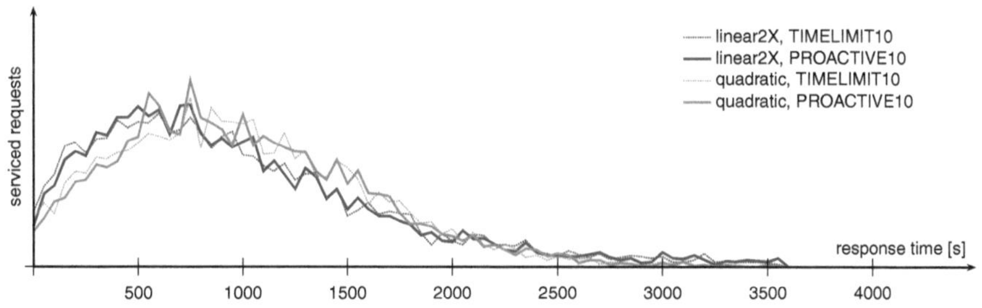

Fig. 8.21 Request response times on S^{REAL} using 10 vehicles and $DC^{\min\lambda} = 2.0$ grouped by intervals of 50 seconds

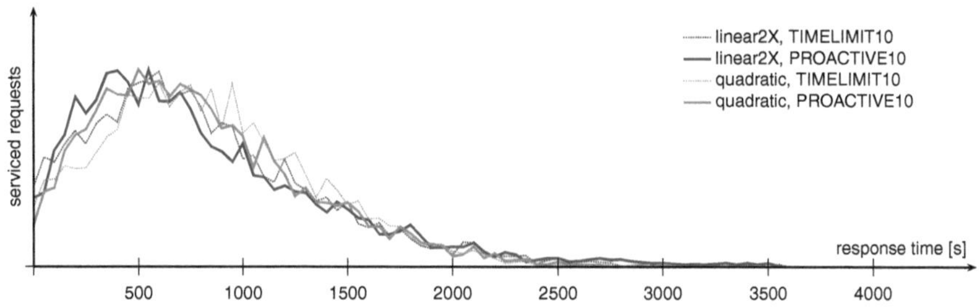

Fig. 8.22 Request response times on S^{REAL} using 12 vehicles and $DC^{\min\lambda} = 1.0$ grouped by intervals of 50 seconds

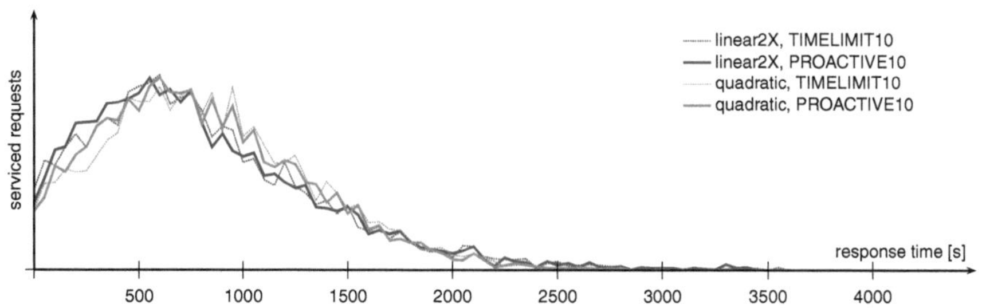

Fig. 8.23 Request response times on S^{REAL} using 12 vehicles and $DC^{\min\lambda} = 2.0$ grouped by intervals of 50 seconds

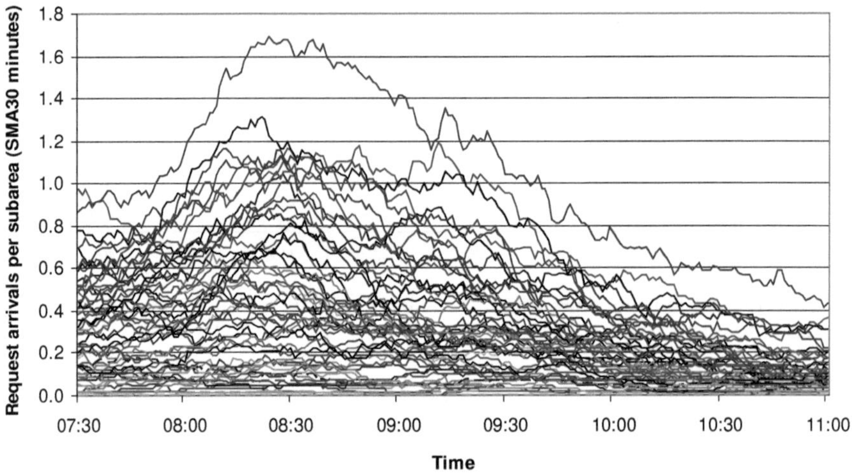

Fig. 8.24 Request arrival rates in the 72 subareas during the considered planning horizon in S^{REAL}. All values are calculated using a simple moving average of 30 minutes

limits the performance attainable by applying PROACTIVE10. Specifically, although the arrival rate of requests varies during the day (see again Fig. 8.2), the occurrence probability of request arrivals in each subarea of the considered service area remains almost constant. This can be derived from Fig. 8.24 in which the request arrival rates in 72 subareas during the considered planning horizon are illustrated using a simple moving average with a length of 30 minutes. Note that these 72 subareas correspond to the ones which were also used for generating stochastic knowledge (see also Fig. 6.1). The most important aspect which can be derived from this figure is that the relative ordering of request arrival rates in the subareas in which most of the requests arrive remains almost constant during the day. As a consequence, vehicles are suitably located to service future arriving requests once they have spread over the service area. In such cases, by applying the idle vehicle waiting strategy (see Sect. 5.2.4), TIMELIMIT10 already makes efficient use of this positioning so that attainable improvements of utilizing stochastic knowledge are strongly limited. Therefore, we state that the request data set of S^{REAL} has a *low structural diversity* that makes an advantageous application of PROACTIVE10 unlikely. Moreover, the subareas in which the highest request arrivals can be observed are adjacently located in the middle of the considered service area. This can be explained by the fact that most of the subsequent delivery requests occur in the downtown city area where the population density in the considered service area is highest. Consequently, the generation of stochastic knowledge with high quality, i.e., using high $DC^{\min\lambda}$ values, results in dummy customers which are only placed in these subareas. Since the depot is also located there, vehicles are by default well positioned for servicing future requests so that the additional value of stochastic knowledge is strongly limited. Further computational experiments on S^{REAL} using stochastic knowledge solely generated from past request information of the same weekday did not result in any significant improvements either.

Besides the low structural diversity, individual days of S^{REAL} may possess structural differences due to its real-world character. Specifically, request arrivals in this request data set may contain additional request arrival structures so that the request arrivals cannot entirely be represented by a set of time-space Poisson processes that are constant from day to day. Therefore, stochastic knowledge generated by using the last $n^{\mathrm{f}} = 60$ weekdays may represent a rather unreliable prediction of future requests. However, as shown by our analysis in Sect. 8.6, it could not be determined that S^{REAL} possesses a high structural distortion. This is underscored by another test study in which a request data set denoted as S^{REAL^*} is generated based on the request arrival characteristics observed in S^{REAL} as shown in Fig. 8.24. Therefore, possible existing day-to-day differences of S^{REAL} are eliminated so that S^{REAL^*} has a low level of structural distortion with certainty. Computational results conducted on S^{REAL^*} show a comparable solution quality to the one obtained on S^{REAL}. Hence, we conclude that if there were any day-to-day differences in S^{REAL}, they would only represent a minor reason for the limited improvements attained by PROACTIVE10 in comparison to those attained by TIMELIMIT10.

With regard to our definition of structural quality depending on structural distortion and structural diversity, we conclude that a low structural distortion is a prerequisite in order to efficiently apply the proposed forecasting method and the pro-active real-time control approach. As shown at the example of S^{REAL}, we further state that a low structural diversity generally impedes an efficient utilization of stochastic knowledge in pro-active real-time control approaches. Accordingly, we state that the structural quality of request data mainly depends on the structural diversity. Hence, S^{REAL} also has a *low structural quality* which limits the value of attainable stochastic knowledge thereof. Crucial characteristics of S^{REAL} which lead to the low structural diversity and hence the limited improvements that were attained are further analyzed in Sect. 8.9.

8.8.2 Results on S^{GEN}

In this section, we present the computational results obtained by applying the deterministic and the pro-active real-time control approach on the 25 generated request sets of the request data class S^{GEN}.

In Table 8.10, the additional improvements attained by PROACTIVE10 in comparison to TIMELIMIT10 for all structural diversity settings of S^{GEN} as well as the number of worse test instances out of the corresponding 30 test instances are illustrated. From the results in this table, it can be derived that the level of structural diversity significantly influences the improvement rates attainable by PROACTIVE10. This can be explained by the fact that if higher levels of structural diversity are present in a request data set, stochastic knowledge has a higher value. Specifically, since in this case request arrival rates in individual regions strongly vary during the

Table 8.10 Percentage improvements achieved by PROACTIVE10 vs. TIMELIMIT10 (i.e., improvements attained by utilizing stochastic knowledge) on S^{GEN}

Vehicles	RegionDiversity (*RD*)	TimeDiversity (*TD*) 0.00	0.25	0.50	0.75	1.00
		linear2X customer inconvenience				
8	0.00	0.08 % (1.0, ∗)	↓ 1.23 % (1.0, ∗)	↑ −0.18 % (1.2, ∗)	↓ 1.95 % (1.2, ∗)	↓ 0.26 % (1.5, ∗)
	0.25	↑ −0.21 % (1.0, ∗)	1.77 % (1.0, ∗)	↑ 0.35 % (1.0, ∗)	2.51 % (1.2, ∗)	↓ 4.48 % (1.5, ∗)
	0.50	1.46 % (1.0, 8)	↓ 1.85 % (1.2, ∗)	↓ 3.92 % (1.2, ∗)	↓ 2.92 % (1.5, 8)	↓ 3.85 % (1.5, ∗)
	0.75	−0.55 % (1.5, ∗)	1.65 % (1.5, ∗)	↓ 5.06 % (1.5, 7)	↓ 7.94 % (1.5, 6)	↓ 10.83 % (1.5, 1)
	1.00	↑ −0.33 % (1.0, ∗)	↓ 2.98 % (1.0, 7)	6.30 % (1.5, 5)	↓ 8.92 % (1.5, 5)	↓ 19.51 % (2.0, 1)
10	0.00	0.51 % (1.0, ∗)	1.86 % (1.0, 7)	2.30 % (1.2, 8)	3.22 % (1.2, 8)	↓ 2.28 % (1.2, ∗)
	0.25	0.79 % (1.0, ∗)	2.27 % (1.0, 7)	4.07 % (1.0, 6)	↓ 5.21 % (1.2, 4)	↑ 4.84 % (1.2, 6)
	0.50	4.72 % (1.0, 3)	5.37 % (1.0, 3)	5.19 % (1.0, 4)	6.44 % (1.5, 5)	↓ 10.02 % (1.5, 0)
	0.75	4.44 % (1.0, 3)	6.12 % (1.0, 2)	7.84 % (1.2, 4)	11.14 % (1.0, 2)	14.61 % (1.5, 1)
	1.00	3.87 % (1.0, 3)	6.75 % (1.0, 2)	9.65 % (1.2, 0)	14.39 % (1.5, 1)	22.95 % (2.0, 0)
12	0.00	1.22 % (1.0, 9)	3.18 % (1.0, 5)	2.04 % (1.0, ∗)	4.52 % (1.0, 3)	5.95 % (1.0, 6)
	0.25	0.87 % (1.0, ∗)	2.37 % (1.0, ∗)	4.50 % (1.0, 2)	6.11 % (1.0, 4)	6.38 % (1.2, 3)
	0.50	5.48 % (1.0, 2)	6.21 % (1.0, 1)	8.05 % (1.0, 0)	9.04 % (1.2, 0)	11.34 % (1.5, 0)
	0.75	6.09 % (1.0, 1)	8.46 % (1.0, 0)	9.97 % (1.0, 1)	15.85 % (1.0, 0)	20.27 % (1.2, 0)
	1.00	5.81 % (1.0, 3)	9.19 % (1.0, 1)	11.86 % (1.2, 0)	17.99 % (1.2, 0)	26.72 % (1.5, 0)

Table 8.10 (Continued)

Vehicles	RegionDiversity (*RD*)	TimeDiversity (*TD*) 0.00	0.25	0.50	0.75	1.00
		quadratic customer inconvenience				
8	0.00	1.48 % (1.0, ∗)	3.02 % (1.0, ∗)	↓ 3.43 % (1.0, 9)	↓ 4.21 % (1.2, 7)	↑ 4.57 % (1.2, ∗)
	0.25	↑ 2.46 % (1.0, ∗)	↓ 5.44 % (1.0, 7)	↓ 6.72 % (1.0, ∗)	↓ 7.17 % (1.2, 9)	↓ 3.00 % (1.5, ∗)
	0.50	↑ 4.14 % (1.2, 8)	↑ 5.66 % (1.0, 9)	↓ 10.45 % (1.0, 6)	↓ 9.20 % (1.2, 3)	↓ 7.06 % (1.5, ∗)
	0.75	5.51 % (1.0, ∗)	↓ 5.26 % (1.0, 8)	↑ 9.99 % (1.2, 6)	↓ 14.63 % (1.5, 3)	↓ 16.75 % (1.5, 2)
	1.00	8.26 % (1.0, 7)	9.98 % (1.2, 2)	↓ 15.16 % (1.5, 3)	↓ 22.64 % (1.5, 3)	↓ 29.86 % (2.0, 0)
10	0.00	1.37 % (1.0, ∗)	6.40 % (1.0, 6)	7.79 % (1.0, 6)	9.16 % (1.0, 5)	5.08 % (1.2, ∗)
	0.25	1.80 % (1.0, ∗)	8.19 % (1.0, 5)	10.13 % (1.0, 2)	↓ 14.49 % (1.2, 0)	↓ 10.05 % (1.5, 3)
	0.50	9.08 % (1.0, 3)	11.32 % (1.0, 2)	15.31 % (1.0, 1)	14.05 % (1.0, 4)	↓ 17.89 % (1.5, 1)
	0.75	9.59 % (1.0, 0)	13.60 % (1.0, 1)	19.88 % (1.0, 1)	20.08 % (1.0, 1)	27.30 % (1.5, 0)
	1.00	12.20 % (1.0, 4)	14.38 % (1.0, 2)	19.56 % (1.2, 2)	25.78 % (1.5, 1)	39.33 % (2.0, 0)
12	0.00	2.34 % (1.0, 8)	5.02 % (1.0, 8)	7.74 % (1.0, 3)	13.16 % (1.0, 1)	15.36 % (1.0, 1)
	0.25	0.42 % (1.0, ∗)	8.08 % (1.0, 2)	14.04 % (1.0, 2)	17.05 % (1.0, 0)	15.05 % (1.0, 2)
	0.50	8.72 % (1.0, 2)	13.60 % (1.0, 0)	16.70 % (1.0, 1)	18.24 % (1.0, 1)	21.64 % (1.2, 1)
	0.75	10.17 % (1.0, 2)	15.68 % (1.0, 1)	21.01 % (1.0, 0)	29.25 % (1.0, 0)	37.15 % (1.2, 0)
	1.00	11.48 % (1.0, 2)	15.40 % (1.0, 0)	24.00 % (1.0, 0)	33.72 % (1.0, 0)	47.24 % (1.2, 0)

Note: Illustrated is the average improvement. The *values in brackets* denote the best $DC^{\min\lambda}$ value and the number of worse test instances (∗: more than 9 worse).
↑: more total late requests than in TIMELIMIT10, ↓: less total late requests than in TIMELIMIT10

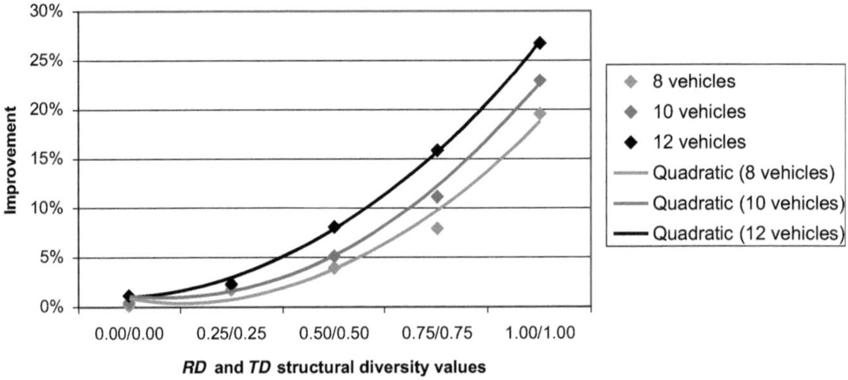

Fig. 8.25 Attainable improvements on S^{GEN} at different levels of structural diversity using linear2X customer inconvenience

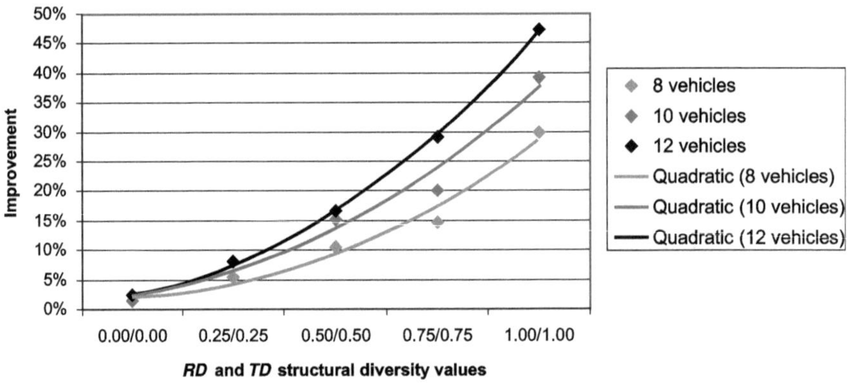

Fig. 8.26 Attainable improvements on S^{GEN} at different levels of structural diversity using quadratic customer inconvenience

day, vehicle positions resulting from last serviced requests do not represent a good positioning for future request arrivals. Hence, future arriving requests are likely to be serviced with a high response time and thus a high customer inconvenience. It is worth mentioning that increasing both levels of structural diversity conjectures a quadratic dependency between the level of structural diversity and attainable improvements, independent of the utilized customer inconvenience function and fleet size (see Figs. 8.25 and 8.26).

In Figs. 8.27 and 8.28, a visual analysis of two request data sets of S^{GEN} is performed. In these tables, the impacts of high levels of structural diversity on temporal and spatial changes in request arrival rates can be clearly observed. Specifically, according to the values defined in S^{GEN} (see again Fig. 8.4), the highest request arrival rates occur in different subareas during the day which are also located far away from each other. Therefore, the subarea with the highest request ar-

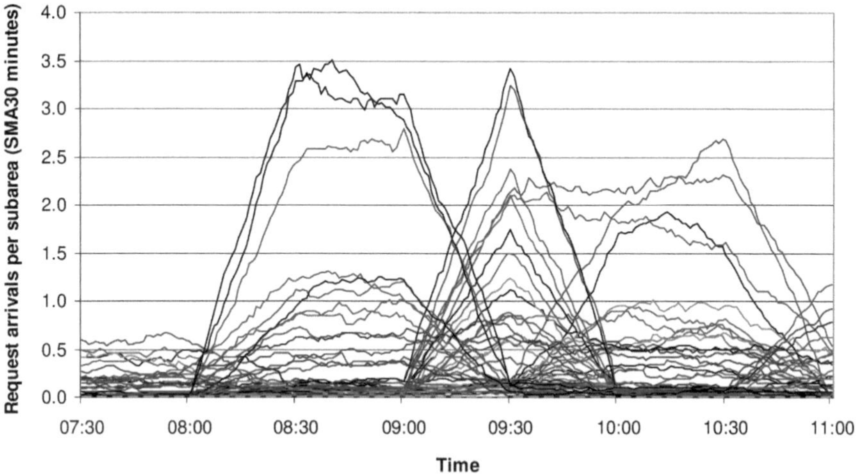

Fig. 8.27 Request arrival rates in the 72 subareas during the considered planning horizon in S^{GEN} $RD = TD = 1.00$. All values are calculated using a simple moving average of 30 minutes

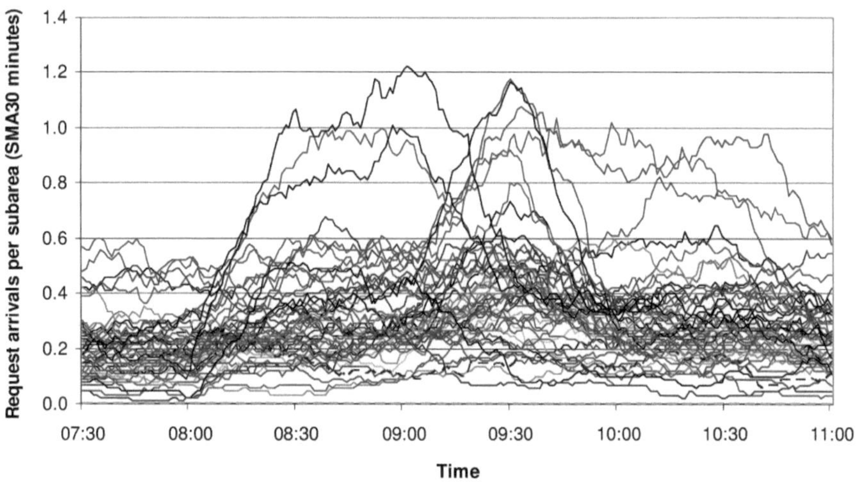

Fig. 8.28 Request arrival rates in the 72 subareas during the considered planning horizon in S^{GEN} $RD = TD = 0.50$. All values are calculated using a simple moving average of 30 minutes

rival rates significantly changes in the considered service area during the day. Note that this significantly distinguishes S^{GEN} from the request data class S^{REAL} (see again Fig. 8.24). Since the idle vehicle waiting strategy that is applied in TIMELIMIT10 relies on the value of these positions, TIMELIMIT10 is significantly outperformed by PROACTIVE10 which efficiently repositions vehicles to future request-likely areas in advance if high levels of structural diversity are present. This has an even higher impact since the highest request arrival rates are about twice as high in S^{GEN} $RD = TD = 1.00$ as in S^{REAL} (i.e., 3.5 compared to 1.7). However, also in the request data set S^{GEN} $RD = TD = 0.50$, in which the highest request arrival rates are about 1.4 and therefore lower than in S^{REAL}, improvements between 3.75 % and 18.32 % could be attained (see Table 8.10). These improvements are significantly higher than the ones achieved for S^{REAL}. Hence, it can be concluded that the value of generatable stochastic knowledge and thus the structural quality of a request data set is significantly determined by variations of request arrivals in the subareas during the planning horizon. As can be seen in Fig. 8.28, the mentioned changes in request arrival rates in the individual subregions also exist in the request data set S^{GEN} $RD = TD = 0.50$. As a result, since S^{GEN} has a low structural distortion, the level of structural diversity present in a request data set of S^{GEN} directly influences its structural quality. This interrelation is further discussed in Sect. 8.9. Moreover, in comparison to S^{REAL}, substantial improvements can also be achieved even with the smallest fleet size of 8 vehicles in S^{GEN}. Hence, the application of the pro-active real-time control approach is also advantageous in case of a small fleet size if a specific minimum level of structural diversity is present.

Furthermore, different impacts on the achieved solution quality according to the chosen $DC^{\min\lambda}$ setting can be observed. Specifically, if low levels of structural diversity is present, low $DC^{\min\lambda}$ settings achieve the best results. This can be explained by considering Table 8.7 which illustrates the number of generated dummy customers depending on the levels of structural diversity and $DC^{\min\lambda}$. It becomes obvious that a minimum level of structural diversity is necessary to generate a sufficient amount of stochastic knowledge of the minimum demanded reliability. For example, using the setting $DC^{\min\lambda} = 2.0$, minimum levels of structural diversity of, e.g., $RD = 0.50,\ TD = 1.00$ or $RD = TD = 0.75$ is necessary. Therefore, low levels of structural diversity settings do not allow for the generation of stochastic knowledge which comprises requests expected with a high reliability.

Two further aspects become relevant when a small fleet size of 8 vehicles is used. First, as shown in Table 8.10, applying PROACTIVE10 reduces the number of late requests in almost all structural diversity settings. This implies that also in TIMELIMIT10 the vehicles are very busy in the evaluated test instances of S^{GEN}. Second, similar aspects as observed in S^{REAL} may become important. Specifically, if a request data set with low levels of structural diversity exists, the utilization of stochastic knowledge should be carefully considered. For example, considering the setting $RD = 0.00,\ TD = 1.00$, only stochastic knowledge of lower quality is available due to the low level of RD in this setting (see Table 8.7). This leads to a reduced solution quality due to the limited resources as depicted in Fig. 8.29. Note that in this figure, the reduced number of requests serviced just before the maximum allowed response time in PROACTIVE10 is caused by additional requests that are serviced after the maximum allowed response time. Nevertheless, it is worth mentioning that when a larger fleet size is used (e.g., 10 vehicles), the generated stochastic knowledge can be advantageously utilized even in this request data set (see Fig. 8.30).

In contrast to the discussed effects in scenarios where a low structural diversity is present, in scenarios with a higher structural diversity, a significant number

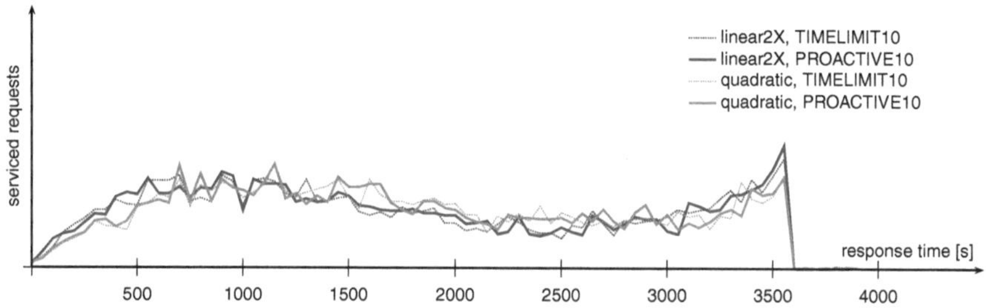

Fig. 8.29 Request response times on S^{GEN} using 8 vehicles and $DC^{\min\lambda} = 1.5$ for $RD = 0.00$ and $TD = 1.00$ grouped by intervals of 50 seconds

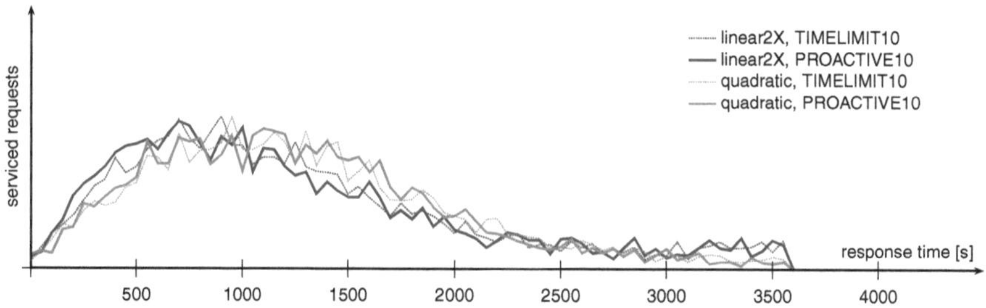

Fig. 8.30 Request response times on S^{GEN} using 10 vehicles and $DC^{\min\lambda} = 1.2$ for $RD = 0.00$ and $TD = 1.00$ grouped by intervals of 50 seconds

of future requests can be reliably forecasted. By efficiently integrating this high-quality stochastic knowledge into the pro-active real-time control approach, significant reductions of customer inconvenience can be attained. Specifically, larger $DC^{\min\lambda}$ values perform best for fleet sizes of 8 and 10 vehicles. If the fleet size is enlarged to 12 vehicles and vehicles are therefore less utilized, further improvements are possible by also integrating less reliable stochastic knowledge into the pro-active real-time control approach. This does not apply for smaller fleet sizes since in these settings vehicles are busier and a misplacement of vehicles caused by an incorrect forecast has substantial negative consequences. For the structural diversity setting $RD = TD = 1.00$, the request response time distributions for the $DC^{\min\lambda}$ settings of 1.0, 1.5, and 2.0 are illustrated in Figs. 8.31, 8.32, and 8.33 for all three evaluated fleet sizes. As can be seen in Fig. 8.31, the integration of stochastic knowledge significantly reduces the number of almost late serviced requests when a restricted fleet size is present. Moreover, in all three fleet sizes, a considerably higher number of requests is quickly serviced when stochastic knowledge is utilized. The request response times attained on additional request data sets of S^{GEN} are illustrated in the Appendix in Sect. C.2. The request response times of all request data sets of S^{GEN} can be found online.[3]

[3] Available at http://extras.springer.com.

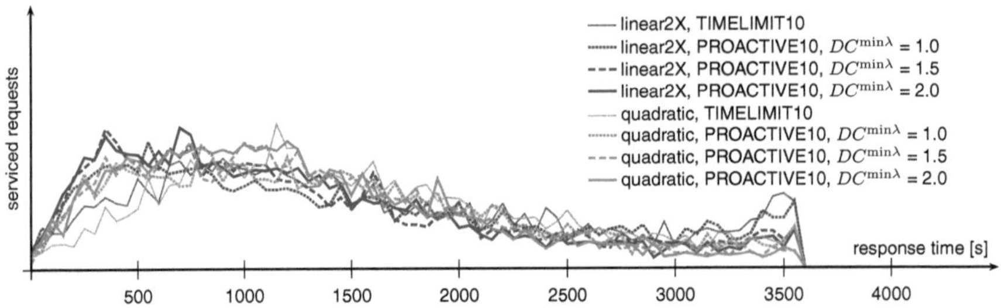

Fig. 8.31 Request response times on S^{GEN} using 8 vehicles and $DC^{\min\lambda} = 1.0$, 1.5, and 2.0 for $RD = 1.00$ and $TD = 1.00$ grouped by intervals of 50 seconds

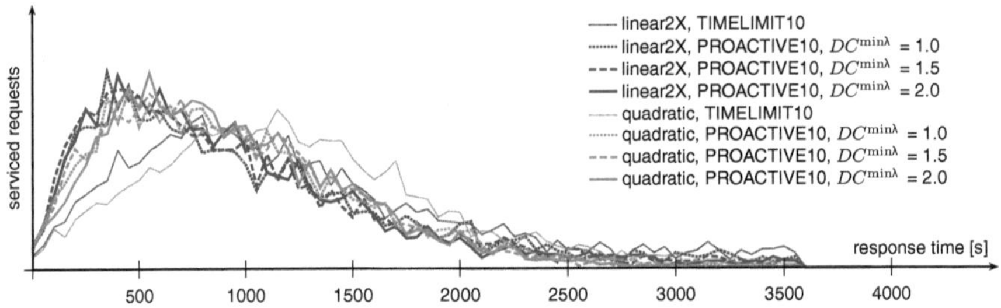

Fig. 8.32 Request response times on S^{GEN} using 10 vehicles and $DC^{\min\lambda} = 1.0$, 1.5, and 2.0 for $RD = 1.00$ and $TD = 1.00$ grouped by intervals of 50 seconds

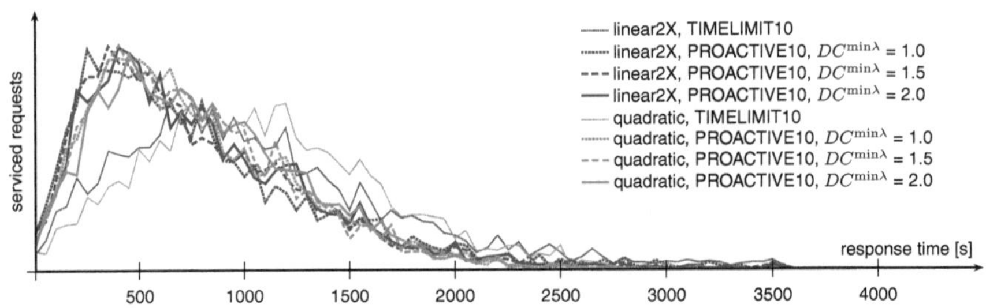

Fig. 8.33 Request response times on S^{GEN} using 12 vehicles and $DC^{\min\lambda} = 1.0$, 1.5, and 2.0 for $RD = 1.00$ and $TD = 1.00$ grouped by intervals of 50 seconds

To sum up the computational results of the pro-active real-time control approach, two main criteria can be highlighted that are necessary for an efficient application of our proposed pro-active real-time control approach. First, a sufficient structural quality is of significant importance. It presupposes a minimum level of structural diversity as well as a low level of structural distortion. According to the computational results, we conclude that the level of structural diversity mainly defines the structural quality of the individual request data sets of S^{GEN}, i.e., request data sets with high *RD* and *TD* structural diversity values possess a high structural quality, and vice versa. Second, the *number of available resources* (i.e., the fleet size) directly influences the minimum quality of the stochastic knowledge required for attaining promising results. Specifically, if an increased fleet size is available, the minimum required structural quality is decreased since the negative effects of incorrectly forecasted requests diminish due to lower utilization of the vehicles.

8.9 The Degree of Structural Diversity

In the previous section, a two-dimensional structural diversity was introduced in order to generate request data sets with different structural quality. In our computational results, the impact of the structural diversity present in different request data sets of S^{GEN} on the attainable stochastic knowledge and on the attainable solution quality improvements has been evaluated. Particularly, it was shown that an efficient generation and integration of stochastic knowledge requires a minimum level of RegionDiversity (*RD*) as well as TimeDiversity (*TD*). However, although this provides new significant insights into the importance of structural quality of request data with regard to the efficiency of pro-active real-time control approaches, the proposed two-dimensional structural diversity does not provide a general definition of structural quality. Specifically, by applying it as presented in Sect. 8.2.4.2, different scenarios with individual levels of structural quality are *newly generated* by linearly transforming an initial setting of a defined highest structural diversity which is given by $RD = TD = 1.00$. Therefore, the generated request data sets neither allow for a classification of *given* request data sets, e.g., real-world data sets such as S^{REAL} nor do they cover all relevant characteristics. For instance, the initial setting $RD = TD = 1.00$ is not the maximum structural diversity (and hence structural quality) that can exist in a request data set of the considered RDOPG application. Instead, this setting represents a request data set in which the freely chosen maximum diversity values defined in Fig. 8.4 are fully present. Such a general measure for structural quality may increase the practicability of the proposed pro-active real-time control approach since it is likely to significantly support the identification of scenarios where the integration of stochastic knowledge is reasonable. As mentioned in Sect. 4.8, this is also motivated by Bent and van Hentenryck (2005) and Ghiani et al. (2009) who suggest analyzing the impact of the quality of the available past request information on the performance of pro-active real-time control approaches. Furthermore, such a measure may help to choose appropriate parameters for the pro-active real-time control approach according to the considered real-world application. Consequently, we propose a first approach for determining the structural quality of arbitrary given request data sets in what follows. We evaluate the applicability of this approach on S^{REAL} and S^{GEN}.

In order to determine the structural quality of a given request data set, we propose a measure denoted as the *degree of structural diversity (dosd)*. In contrast to the previously used two-dimensional structural diversity, the *dosd* is a one-dimensional value. It ranges from 0 to 1. While 0 corresponds to a request data set with no structural diversity and hence does not allow for efficiently integrating stochastic knowledge due to the existing low structural quality, the value 1 defines a theoretical upper bound on the maximum structural diversity which a request data set can have.

Based on the insights of the previous section, the main idea of the *dosd* is that a high degree of structural diversity is present if there exists significant spatial variations in request arrivals in the subareas during the considered planning horizon. Specifically, if the subareas in which high request arrival rates subsequently occur are located far away from each other, the value of using stochastic knowledge and hence the *dosd* becomes high. Moreover, also the number of requests that occur in the individual subareas is important. Specifically, the more requests occur in a subarea, the higher the impact on the solution quality since these requests are serviced with a high response time leading to a high customer inconvenience. Furthermore, changes in the request arrival rates in the subareas change the vehicles' positions within the considered service area accordingly. Therefore, in order for a request data set to obtain a high *dosd*, changes in the request arrival rates should occur only after a certain time period so that vehicles will have traveled to another position in the meantime.

For a given request data set, the *dosd* is determined as follows. The rate parameters $\lambda(p_s)$ of the time-space Poisson processes attached to each segment $s \in S$ are calculated as described in Sect. 6.2.1 for a specified time frame of the given request data set, e.g., the last n^{f} working days. The considered planning horizon of the corresponding real-world application is divided into T time periods with a length of m minutes each. For each time period $t \in \{1, \ldots, T\}$, the tuple (b_t^x, b_t^y) defines the geographical position of the barycenter of requests that have arrived in time period t according to the subarea separation depicted in Fig. 6.1. Moreover, n_t gives the number of requests which occurred on average in time period t in the considered time frame of the request data set. Furthermore, we assume that the considered service area is divided into $x \cdot y$ subareas. In order to calculate the *dosd*, we track how the barycenter of request arrivals changes between consecutive time periods. Specifically, the Euclidean distance between the barycenters of two consecutive time periods $t_1, t_2 \in T$ is computed and weighted with the number of request arrivals of t_2 given by n_{t_2}. By calculating the weighted distances for all time period pairs and normalizing the sum of weighted distances with an upper bound on the distance defined by $maxDist = \sqrt{x^2 + y^2}$, we obtain the following definition of the *dosd*:

$$dosd = \frac{\sum_{t=1}^{T-1} n_{t+1} \cdot \sqrt{(b_t^x - b_{t+1}^x)^2 + (b_t^y - b_{t+1}^y)^2}}{maxDist \cdot \sum_{t=1}^{T-1} n_{t+1}}$$

Note that since in our application all vehicles start their tour at an identical depot, the first time period is used to determine individual starting positions of the vehicles. However, since in many request data sets considered in our RDOPG applications only a small portion of all requests arrive in the first time period, n_1 is comparably small and its omission does not cause significant changes to the calculated *dosd*. According to this definition, variations in request arrivals of a given request data set are quantified by the weighted length of an approximate path that the majority of the vehicle fleet travels in order to service the requests. Consequently, the larger the resulting distances between consecutive barycenters and the higher the number of requests, the more valuable the utilization of stochastic knowledge becomes. This is due to the fact that in such situations a deterministic real-time control approach is increasingly forced to relocate a larger number of vehicles in each time period which is likely to result in higher response times and hence increased customer inconvenience. Therefore, parameter m is approximately set to the average duration during which the vehicles have completed their travel activity between two consecutive barycenters.

In order to assess the practicability of the proposed *dosd*, we evaluate it on the request data classes S^{REAL} and S^{GEN}. Note that all request data sets of these request data classes have a comparable number of requests. Consequently, if a small number of vehicles is available, they are highly utilized while a large number of available vehicles results into the vehicles becoming less utilized in all of these request data sets. Hence, the values of the *dosd* calculated for the request data sets of both request data classes are comparable, i.e., the stochastic knowledge which can be generated from them has a similar value. Specifically, different request data sets with a similar *dosd* value are likely to generate a similar improvement in the solution quality when the pro-active real-time control approach is used compared to using the deterministic real-time control approach. The advantage that the *dosd* has over computational experiments is that it makes an instant estimation of the efficiency of pro-active real-time control approaches on given request data sets possible. In order to statistically prove the above statement, we directly compare the respective *dosd* value calculated for each request data set of both request data classes with the improvement rates which have been attained by using PROACTIVE10 in the previously described computational experiments. In this comparison, we use the best improvement rates attained among the evaluated parameter settings for $DC^{\min\lambda}$ which are illustrated in Tables 8.9 and 8.10. As presented in the computational experiments, these best improvement rates were attained by conducting the five $DC^{\min\lambda}$ settings on each of the 25 request data sets of S^{GEN} as well as for the request data set of S^{REAL}. In order to analyze the correlation between both variables, we apply six linear regression models in total—one for each of both customer inconvenience functions combined with each of the three evaluated fleet sizes on the 25 request data sets of S^{GEN}. For calculating the *dosd*, we set $m = 30$ resulting in eight time periods of the considered planning horizon which has a duration of four hours as mentioned above.

The results of this evaluation are illustrated in Figs. 8.34 and 8.35. As can be seen from these figures, the proposed *dosd* is an effective measure for determining the structural quality of request data sets and can therefore identify scenarios in which the application of stochastic knowledge is advantageous. Specifically, for S^{GEN}, a detailed correlation analysis using the Pearson product-moment correlation coefficient r (cf. Rodgers and Nicewander 1988) provides values between 0.9469 and 0.9647 for the linear2X customer inconvenience function and values between 0.9161 and 0.9491 for the quadratic customer inconvenience function,

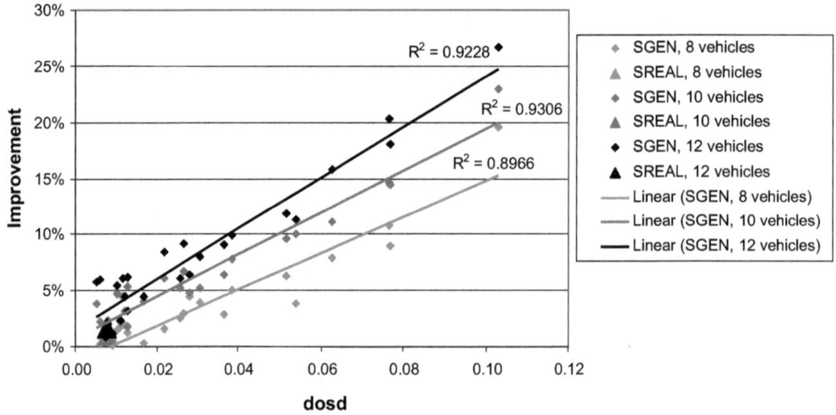

Fig. 8.34 Results of the degree of structural diversity using the linear2X customer inconvenience function

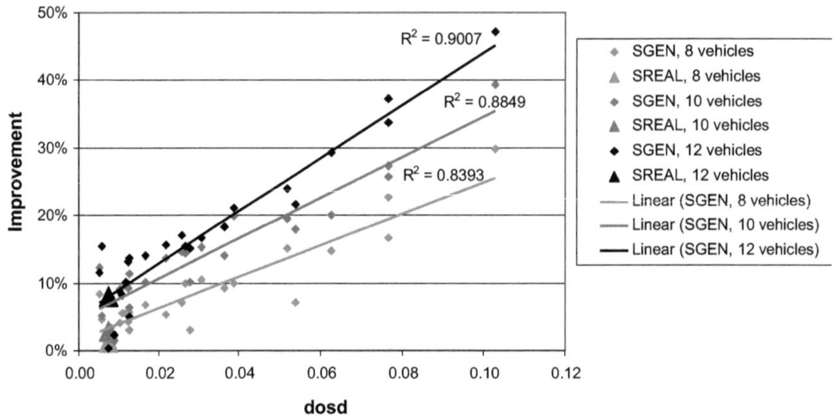

Fig. 8.35 Results of the degree of structural diversity using the quadratic customer inconvenience function

depending on the utilized fleet size. Accordingly, the corresponding coefficient of determination R^2 (defined by r^2) is illustrated in both figures. Consequently, a strong linear relationship between the *dosd* and the improvement that is attainable by the pro-active real-time control approach is statistically proven. This indicates a good quality for using the *dosd* on S^{GEN}. The results also reproduce the previous observation in Sect. 8.8 that the attainable improvement rates increase with the available fleet size. Moreover, the figures show that the *dosd* represents a good measure for estimating attainable improvements on the real-world request data set S^{REAL}. Specifically, the determined low *dosd* value for S^{REAL} reflects the observations in Sect. 8.8 with regard to the low structural data quality of S^{REAL}. Since the *dosd* provided good results on S^{REAL} and S^{GEN}, it is likely that the *dosd* is

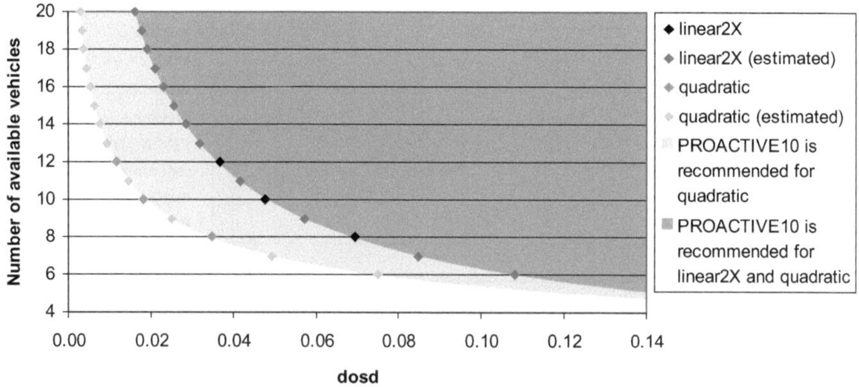

Fig. 8.36 Managerial recommendations based on the *dosd* and S^{GEN} demanding a minimum improvement of 10 %

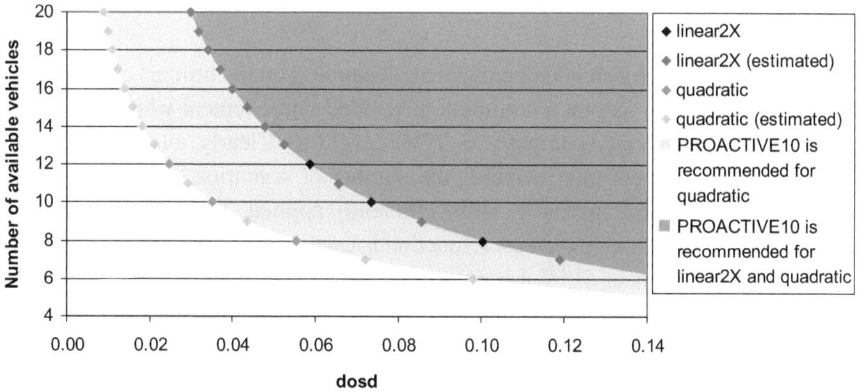

Fig. 8.37 Managerial recommendations based on the *dosd* and S^{GEN} demanding a minimum improvement of 15 %

also a suitable measure for determining the structural quality of other request data sets. According to the attained results, it can be concluded that a *dosd* ≥ 0.06 of a request data set strongly indicates that the existing structural quality enables an advantageous application of the proposed pro-active real-time control approach on this request data set. In accordance with the computational results provided in Sect. 8.8, this is particularly true if a larger fleet size is available. Moreover, since the attainable improvements utilized in this evaluation represent the best parameter settings of the tested ones, appropriate parameter values for generating stochastic knowledge and for utilizing the proposed pro-active real-time control approach can be chosen according to the *dosd* of a given request data set.

Based on the results of the *dosd* on S^{GEN}, a two-dimensional matrix providing managerial recommendations can be derived. Specifically, Figs. 8.36, 8.37, and 8.38 illustrate the scenario constellations in which the utilization of the pro-active

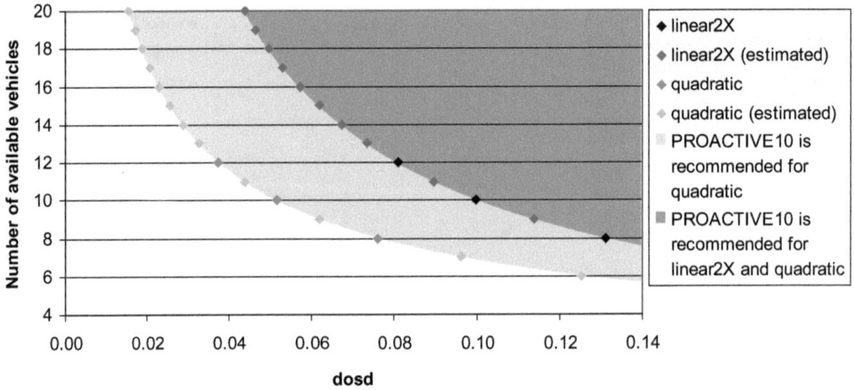

Fig. 8.38 Managerial recommendations based on the *dosd* and S^{GEN} demanding a minimum improvement of 20 %

real-time control approach is recommended, depending on the utilized customer inconvenience function and on a minimum demanded improvement which can be attained by PROACTIVE10 compared to TIMELIMIT10. Clearly, a lower minimum demanded improvement rate increases the number of scenarios in which the pro-active real-time control approach can be efficiently applied. This is also true when the quadratic customer inconvenience function is used. Moreover, based on the evaluated request data sets of S^{GEN} it is conjectured that differences in the applicability of PROACTIVE10 with regard to the linear2X and the quadratic objective function decrease with higher degrees of structural diversity. Specifically, in request data sets with a low structural quality, using the quadratic customer inconvenience function has a beneficial impact since, as mentioned above, it avoids situations in which many requests have a high response time close to the R^{mrt} due to incorrectly forecasted requests. In contrast, if the request data set comprises a high structural quality, the number of incorrectly forecasted requests is significantly reduced. Hence, since a high structural diversity reduces the number of incorrectly forecasted requests and increased the value of the generatable stochastic knowledge, this issue is expected to become less significant.

To summarize the results obtained from this evaluation, the proposed *dosd* is an effective general measure for assessing the structural quality of arbitrary given request data sets. Utilizing it for a given request data set makes it possible to estimate the solution quality improvements which are additionally attainable by utilizing PROACTIVE10 compared to TIMELIMIT10. Moreover, it supports the identification of parameter settings that can be recommended for efficiently applying stochastic knowledge in real-time control approaches. These insights significantly improve the practicability of the proposed pro-active real-time control approach.

8.10 Driver Inconvenience Aspects in the Real-Time Control Approaches

In this section, we continue to analyze the impact on the solution quality when allowing vehicle en-route diversion activities. Specifically, in the previously presented computational experiments, vehicle en-route diversion is allowed without any penalty costs. Hence, vehicle en-route diversion activities are considered by the solution method as additional decisions that are available without any drawbacks. However, in practical applications, allowing an arbitrary number of vehicle en-route diversion activities may expose a huge amount of inconvenience to the vehicle drivers. This may result in a lack of motivation for work so that the positive effects that can be attained by utilizing the pro-active real-time control approach cannot be completely realized. On the other hand, as presented in Sect. 8.5, allowing vehicle en-route diversion provides an effective way of significantly increasing the solution quality. Therefore, although a complete prohibition of vehicle en-route diversion activities does not seem to be reasonable, it is advisable to include mechanisms that limit the number of performed vehicle en-route diversion activities. In order to do so, it is recommended to evaluate the trade-off of limiting vehicle en-route diversion activities and the attained solution quality.

In this section, we propose a method for limiting the number of performed vehicle en-route diversion activities. This limitation operates by only permitting vehicle en-route diversion activities whose improvement with regard to the objective function value is above a certain *threshold*. Specifically, penalty costs for performing vehicle en-route diversion activities are temporally included in the objective function during the solution process. Hence, only en-route diversion activities, whose improvement is larger than these penalty costs, are performed. Therefore, in this computational study, we do not only take into account inconvenience on the customer's side, but also inconvenience aspects of the deployed drivers.

Note that due to the different evolutions of both tested objective functions as described in Sect. 5.3.2.4, these penalty costs have to be individually defined for each customer inconvenience function. Specifically, the previously mentioned threshold is defined by a total decrease in request response times that is required to be achieved in order to justify a vehicle en-route diversion activity. This response time-based approach has been chosen since it is assumed that the effort to save a certain amount of request response time is equal across scenarios with a different customer inconvenience function. For this threshold, several values which are illustrated in Table 8.11 are evaluated. For example, a defined response time threshold of 36 seconds means that the total improvement of the objective function value by performing the vehicle en-route diversion activity is required to be larger than 0.01 in the linear2X case and 0.0001 in the quadratic case, respectively.

The computational experiments in this section are carried out using the real-time control approaches TIMELIMIT10 and PROACTIVE10 on the request data set S^{GEN} $RD = TD = 1.00$ with a fleet size of 10 vehicles while evaluating both customer inconvenience functions. For PROACTIVE10, according to the computational results depicted in Table 8.10, the best dummy customer quality settings

Table 8.11 Evaluated response time thresholds and penalty costs included in the objective function for vehicle en-route diversion activities depending on the applied customer inconvenience function

Response time threshold [seconds]	Objective function penalty costs	
	linear2X	quadratic
(unrestricted) 0	0.0	0.0
36	0.01	0.0001
72	0.02	0.0004
180	0.05	0.0025
360	0.1	0.01
720	0.2	0.04
1800	0.5	0.25
3600	1.0	1.0

represented by $DC^{\min\lambda} = 2.0$ for the linear2X as well as for the quadratic customer inconvenience function are used. The upper graph in Fig. 8.39 shows the performed number of vehicle en-route diversion activities if different threshold parameter settings are applied to linear2X. In the same figure, the respective changes in the objective function value are presented in the lower graph. Analogously, Fig. 8.40 shows the results that have been achieved using the quadratic customer inconvenience function. Note that for the approach TIMELIMIT10, the number of vehicle en-route diversion activities is recorded according to the rules described in Sect. 8.5. For the approach PROACTIVE10, a change of a dummy customer to a real request is not counted as a vehicle en-route diversion activity when a dummy customer is currently assigned as the first request to a vehicle. Specifically, due to our defined utilization of dummy customer requests, the purpose of dummy customers is to act as place-holders for an expected number of real requests arriving in the corresponding request-likely area. When a vehicle is currently traveling to the location of a dummy customer request and its next request is changed to a real request, it is likely that this change happens due to the fact that a real customer request which is located near the dummy customer has recently arrived. Therefore, since it is the purpose of dummy customers to guide vehicles into request-likely areas without knowing the exact position of arriving requests, such a change is not considered as a deterioration of the transportation process. It is assumed that drivers are notified by the on-board units when they are currently traveling to a dummy customer so that they are aware of a likely change of this dummy customer request. Similarly, a change of the first assigned request, due to the removal of a dummy customer from the tour plan (see Sect. 6.3.1), is not counted as a vehicle en-route diversion activity either. Nevertheless, a reassignment of a still active dummy customer to another dummy customer is considered as a vehicle en-route diversion activity that is penalized in the objective function and also recorded in the computational results (see Table 8.12).

In Figs. 8.41 and 8.42, the request response times resulting from the application of particular penalty costs are illustrated. As can be seen, higher penalty costs lead to

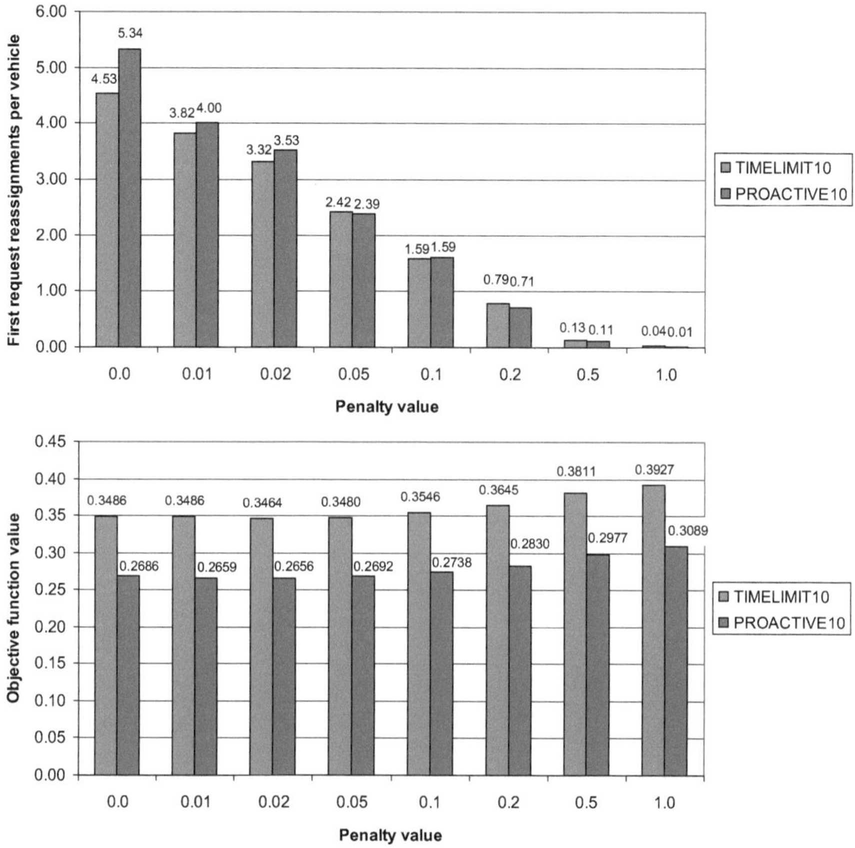

Fig. 8.39 Results of limiting vehicle en-route diversion activities on S^{GEN} $RD = TD = 1.00$ with $DC^{\min\lambda} = 2.0$ using the linear2X customer inconvenience function and a fleet size of 10 vehicles

longer response times which results in a later service of requests. Note that this is in accordance with the results obtained in Sect. 8.5. Following the results presented in both figures and Table 8.12, it can be observed that both proposed real-time control approaches often make use of vehicle en-route diversion activities when vehicle en-route diversion activities can be performed at no additional costs. Note that this is similar to the results observed on S^{REAL} presented in Sect. 8.5.

Furthermore, the results show that our proposed threshold-based approach can be successfully applied in order to limit the number of performed vehicle en-route activities. Interestingly, forbidding vehicle en-route diversion activities that would gain only low objective function improvements significantly reduces the number of vehicle en-route diversion activities but leads to only slight increases of the attained average customer inconvenience. Moreover, in scenarios where the linear2X customer inconvenience function is used, applying small threshold parameter values of up to 0.02 does not increase the attained customer inconvenience at all but rather

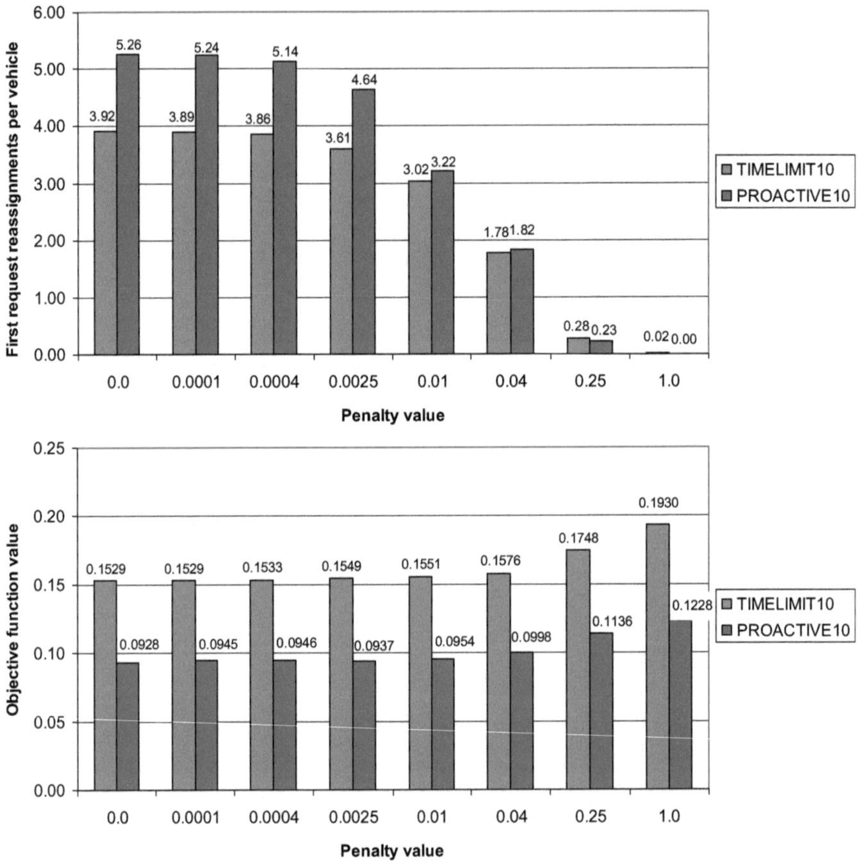

Fig. 8.40 Results of limiting vehicle en-route diversion activities on S^{GEN} $RD = TD = 1.00$ with $DC^{\min\lambda} = 2.0$ using the quadratic customer inconvenience function and a fleet size of 10 vehicles

even reduces it while significantly reducing the number of vehicle en-route diversion activities. This can be explained by the fact that disallowing low-improving vehicle en-route diversion activities does not have a significant impact on the solution quality. In contrast, by prohibiting these vehicle en-route diversion activities vehicles are less disturbed while traveling to their next request so that, in the end, this may even lead to a slight improvement in the achieved customer inconvenience. As a positive effect, the number of vehicle en-route diversion activities can be significantly reduced at the cost of negligible drawbacks while significantly reducing driver inconvenience.

When vehicle en-route diversion activities are allowed without additional penalty costs, the number of performed vehicle en-route diversion activities is significantly higher in PROACTIVE10 compared to TIMELIMIT10. This is especially true for the quadratic customer inconvenience function. This can be explained by the fact that due to the quadratic increase of customer inconvenience, even many small ve-

Table 8.12 Results for TIMELIMIT10 and PROACTIVE10 under different penalty costs on S^{GEN} $RD = TD = 1.00$ using 10 vehicles

Penalty	TIMELIMIT10		PROACTIVE10	
	#ra per vehicle	Objective function value	#ra per vehicle	Objective function value
		linear2X customer inconvenience		
0.0	4.53	0.3486	5.34 (−17.9 %)	0.2686 (22.9 %)
0.01	3.82 (15.5 %)	0.3486 (0.0 %)	4.00 (11.6 %)	0.2659 (23.7 %)
0.02	3.32 (26.7 %)	0.3464 (0.6 %)	3.53 (22.0 %)	0.2656 (23.8 %)
0.05	2.42 (46.5 %)	0.3480 (0.2 %)	2.39 (47.2 %)	0.2692 (22.8 %)
0.1	1.59 (64.9 %)	0.3546 (−1.7 %)	1.59 (64.8 %)	0.2738 (21.5 %)
0.2	0.79 (82.6 %)	0.3645 (−4.6 %)	0.71 (84.3 %)	0.2830 (18.8 %)
0.5	0.13 (97.1 %)	0.3811 (−9.3 %)	0.11 (97.5 %)	0.2977 (14.6 %)
1.0	0.04 (99.1 %)	0.3927 (−12.6 %)	0.01 (99.7 %)	0.3089 (11.4 %)
		quadratic customer inconvenience		
0.0	3.92	0.1529	5.26 (−34.3 %)	0.0928 (39.3 %)
0.0001	3.89 (0.8 %)	0.1529 (0.0 %)	5.24 (−33.8 %)	0.0945 (38.2 %)
0.0004	3.86 (1.4 %)	0.1533 (−0.3 %)	5.14 (−31.2 %)	0.0946 (38.1 %)
0.0025	3.61 (7.9 %)	0.1549 (−1.3 %)	4.64 (−18.5 %)	0.0937 (38.7 %)
0.01	3.02 (22.9 %)	0.1551 (−1.4 %)	3.22 (17.8 %)	0.0954 (37.6 %)
0.04	1.78 (54.6 %)	0.1576 (−3.1 %)	1.82 (53.5 %)	0.0998 (34.7 %)
0.25	0.28 (92.9 %)	0.1748 (−14.3 %)	0.23 (94.1 %)	0.1136 (25.7 %)
1.0	0.02 (99.4 %)	0.1930 (−26.2 %)	0.00 (100.0 %)	0.1228 (19.7 %)

Note: Listed is the average number of first request reassignments (#ra) per vehicle and the average objective function value based on the applied penalty costs for vehicle en-route diversion activities. For both values, the percentage improvement of the respective penalty threshold parameter setting, *presented in brackets*, is compared to TIMELIMIT10 in which all vehicle en-route diversion activities are allowed

hicle en-route diversion activities lead to an improvement of individual system situations during the execution of the transportation process. However, when applying small threshold parameter values of up to 0.02 in the linear2X and 0.0025 in the quadratic case, this significantly reduces the number of performed vehicle en-route diversion activities thereby reducing driver inconvenience. At the same time, these small threshold parameter values provide a quality of customer inconvenience that is similar to the quality attainable when all vehicle en-route diversion activities are allowed. In contrast, when higher threshold parameter values are used, this results in a significant increase in customer inconvenience since vehicle en-route diversion activities by which high improvements of the solution quality could be achieved are also increasingly forbidden. Since almost all vehicle en-route diversion activities are removed if the highest penalty costs of 1.0 are applied, this setting produces an increase in the customer inconvenience that is similar to the already considered case in which vehicle en-route diversion is completely forbidden (see Sect. 8.5). Note that

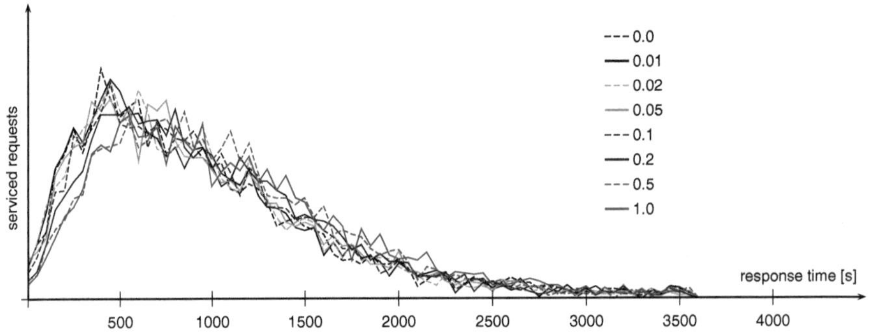

Fig. 8.41 Impacts of assigning penalty costs to vehicle en-route diversion activities on request response times attained by PROACTIVE10 on S^{GEN} $RD = TD = 1.00$ using 10 vehicles and the linear2X customer inconvenience function grouped by intervals of 50 seconds

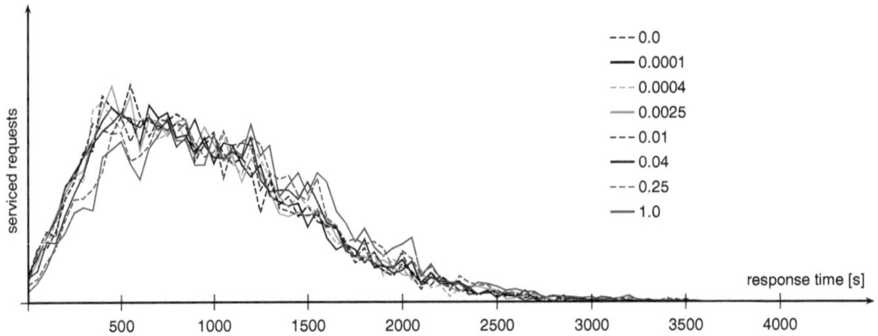

Fig. 8.42 Impacts of assigning penalty costs to vehicle en-route diversion activities on request response times attained by PROACTIVE10 on S^{GEN} $RD = TD = 1.00$ using 10 vehicles and the quadratic customer inconvenience function grouped by intervals of 50 seconds

in this case, in which the highest reduction of driver inconvenience generated by vehicle en-route diversion activities can be attained, PROACTIVE10 still achieves a significant improvement of the customer inconvenience of 11.4 % or 19.7 % if the linear2X or the quadratic customer inconvenience function is applied respectively, compared to TIMELIMIT10 when all vehicle en-route diversion activities are permitted.

From another perspective, when considering TIMELIMIT10 with a threshold parameter value of 0 as the reference, PROACTIVE10 achieves a lower number of vehicle en-route diversion activities for the linear2X customer inconvenience starting from values of 0.01 and for the quadratic inconvenience function also starting from values of 0.01. When this threshold parameter value is used, PROACTIVE10 achieves a better solution quality with regard to customer inconvenience of 23.7 % in the linear2X and 37.6 % in the quadratic customer inconvenience case. Moreover, when applying the same threshold parameter value to TIMELIMIT10 and PROACTIVE10, the number of vehicle en-route diversion activities becomes comparable at 0.05 in the linear2X case and 0.04 in the quadratic case. From these threshold parameter values up to the highest tested threshold parameter values, the attained improvements of customer inconvenience by PROACTIVE10 compared to TIMELIMIT10 remain constant at about 21–22 % in the linear2X case and 35–38 % in the quadratic case. Hence, for these threshold parameter values, PROACTIVE10 provides significant improvements with regard to the achievable customer inconvenience while attaining a level of driver inconvenience reduction which is at least as high as is achieved by TIMELIMIT10.

To sum up, the application of the pro-active real-time control approach PROACTIVE10 is able to significantly reduce customer inconvenience and driver inconvenience at the same time compared to TIMELIMIT10. With regard to the effects of setting the introduced threshold parameter on driver inconvenience, which is the focus of this section, different recommendations of setting this parameter in accordance to the pursued aim can be derived from the results of this computational study:

- If the minimization of customer inconvenience is the only aim and driver inconvenience aspects can be completely neglected, allowing all vehicle en-route diversion activities is generally advisable. Furthermore, as mentioned above, if the linear2X customer inconvenience function is used, small applied threshold parameter values can slightly improve the solution quality of TIMELIMIT10 as well as of PROACTIVE10. Furthermore, the use of these small threshold parameter values also significantly reduces driver inconvenience. In this case, using PROACTIVE10 achieves improvements of up to 23.8 % or 39.3 % when the linear2X or the quadratic customer inconvenience function is used, respectively.
- If the minimization of customer inconvenience and driver inconvenience are both important, moderate values for the threshold parameter are recommended. By utilizing a threshold parameter value of 0.1 if the linear2X customer inconvenience function is applied and 0.04 if the quadratic customer inconvenience function is used, vehicle en-route diversion activities can be reduced by more than 50 % while customer inconvenience is only increased by 1.9 % (linear2X) or 7.1 % (quadratic). This still results in an attainable improvement by PROACTIVE10 of 21.5 % using the linear2X customer inconvenience function and 34.7 % if the quadratic customer inconvenience function is applied compared to TIMELIMIT10 where all vehicle en-route diversion activities are allowed.
- If the minimization of driver inconvenience is highly prioritized, forbidding all vehicle en-route diversion activities is attainable by setting the threshold parameter to 1.0 or higher. In this case, the customer inconvenience is significantly increased in TIMELIMIT10 and PROACTIVE10. Nevertheless, also in this case PROACTIVE10 provides a significantly reduced customer inconvenience compared to TIMELIMIT10. In addition to this, the results achieved by PROACTIVE10 are still better than the ones attained by TIMELIMIT10 in which all vehicle en-route diversion activities are permitted. In particular, PROACTIVE10 still achieves a significant improvement of customer inconvenience of 11.4 % or 19.7 % if the linear2X or if the quadratic customer inconvenience function is applied, respectively.

8.11 Summary and Answers to the Three Central Research Questions

In this chapter, extensive computational experiments were conducted in order to evaluate the deterministic and the pro-active real-time control approach which were proposed for coordinating the transportation process of RDOPG applications. After describing the discrete event-based simulator utilized to carry out the experiments and after presenting the tested environment including evaluated parameters, the two request data classes on which the computational experiments are conducted were introduced. Specifically, while the first request data class S^{REAL} consists of a real request data set derived from a real-world application dealing with the subsequent delivery of newspapers, the second request data class S^{GEN} consists of 25 different generated request data sets. These generated request data sets possess individually designed levels of structural diversity which allows a differentiated analysis of improvements attainable by the proposed pro-active real-time control approach when different levels of structural quality exist.

Three preliminary computational studies were performed. In the first of these studies, the performance of the proposed Tabu Search solution method was evaluated. Specifically, its attained results on various static problem instances were compared with the ones obtained by a lower bound of an appropriate exact solution approach known in the literature. It was shown that applying the proposed Tabu Search approach for 10 seconds yields near-optimal results for individual static problem instances which arise during the considered RDOPG transportation processes. In the second study, the performance of the deterministic real-time control approach proposed in Chap. 5 was evaluated against two other deterministic real-time control approaches. This study revealed that the deterministic real-time control approach denoted as TIMELIMIT10, which invokes the Tabu Search solution method for 10 seconds whenever new requests have arrived in the previous anticipation horizon, achieves a performance that is comparable to the original deterministic real-time control approach denoted as ROLLING20. This result can be explained by the fact that, as shown in the first evaluation, TIMELIMIT10 already achieves a near-optimal solution quality within the given time limit so that no further significant improvements can be attained by applying the proposed solution method for a longer time period as performed in ROLLING20. Hence, TIMELIMIT10 was used as the reference setting in the further computational studies. Due to its high solution quality, TIMELIMIT10 represents a challenging reference setting for attaining additional improvements by utilizing the pro-active real-time control approach. In the third preliminary study, the impacts of allowing vehicle en-route diversion on the solution quality were evaluated. A comparison of TIMELIMIT10 to its variant which prohibits vehicle en-route diversion activities showed that, depending on the fleet size and the utilized customer inconvenience function, it is possible to attain improvements between 14 % and 25 % while simultaneously reducing travel time and travel distance when vehicle en-route diversion is allowed. After this study, an evaluation of the Poisson quality approach in our forecasting methods, which uses a Goodness-of-Fit test to determine the structural distortion of request data sets, was conducted. This evaluation revealed that this approach allows for efficiently identifying request data sets which have a high structural distortion. Moreover, information on the stochastic knowledge generated using S^{REAL} and S^{GEN} were provided.

In the main computational experiments, the results from the proposed exact clustering method with regard to the generated stochastic knowledge were presented. This generated knowledge was then used to analyze the improvements which can be additionally attained by utilizing the pro-active real-time control approach PROACTIVE10 compared to TIMELIMIT10 on S^{REAL} and S^{GEN}. This analysis was carried out in order to answer the first two questions posed in Sect. 1.3.1:

Question 1: Does the integration of information about expected future request arrivals into a deterministic real-time control approach, thus forming a pro-active real-time control approach, allow for a significant improvement of the solution quality by reducing customer inconvenience?

Answer: As shown in the computational experiments in Sect. 8.8, utilizing the generated stochastic knowledge in the proposed pro-active real-time control approach makes it possible to significantly reduce customer inconvenience and thus significantly improve the solution quality. Specifically, on the request data class S^{REAL}, depending on the available fleet size and utilized customer inconvenience function, improvements of up to 7.93 % were achieved by PROACTIVE10 compared to TIMELIMIT10. However, in problem instances with a limited fleet size of 8 vehicles, no improvements could be attained at all. On the request data class S^{GEN}, in almost all tested request data sets comprising different levels of structural diversity, significant improvements were achieved. In the setting with the highest tested levels of structural diversity represented by $RD = TD = 1.00$, 19.51 % to 26.72 % of additional improvement could be attained under the linear2X customer inconvenience function depending on the fleet size. In scenarios in which the quadratic customer inconvenience function was utilized, improvements of between 29.86 % and 47.24 % were achieved. According to the observed differences regarding attainable improvements on S^{REAL} and S^{GEN} by PROACTIVE10 we can conclude that the existence of specific characteristics in the request data sets fosters improvements that can be attained by utilizing pro-active real-time control approaches. In this book, we denoted these characteristics, which are discussed in what follows, as the structural quality of a request data set.

Question 2: If question 1 yields a positive result, under which criteria does this apply? For example, what impact does the structural quality of existing request data and the resulting stochastic knowledge which is generated on the basis of this request data have on the solution quality? Furthermore, which characteristics are suitable for measuring the structural quality of available request data?

Answer: In Sects. 8.8 and 8.9, it was shown that certain prerequisites need to be fulfilled in a request data set in order to enable PROACTIVE10 to attain significant improvements. Hence, when dealing with the coordination of the operative transportation processes of RDOPG applications, it is advantageous to know under which conditions an integration of stochastic knowledge is useful. According to the computational results, we identified characteristics of real-world scenarios under which the application of the proposed pro-active real-time control approach is recommended. Specifically, two aspects that mainly influence an efficient application of the proposed pro-active real-time control approach are illustrated:

- **Structural quality of the available request data:** This criterion describes the quality of the stochastic knowledge that can be generated out of the available request data set. Based on our results, it can be subdivided into two criteria:
  - Structural distortion: As mentioned above, our proposed forecasting methods assume that request arrivals can be modeled by a set of unknown time-space Poisson processes whose rate parameters are also unknown but remain constant from day to day. Hence, the application of the pro-active real-time control approach on request data sets which do not fulfill this assumption is likely to result in a poor solution quality. Consequently, the existence of a low structural distortion in the request data set is a crucial pre-requisite for deriving stochastic knowledge of high value and hence for efficiently applying the proposed pro-active real-time control approach.
  - Structural diversity: As illustrated by the computational results on S^{REAL} and S^{GEN}, it has been shown that the value of the stochastic knowledge utilized in the pro-active real-time control approach increases with higher structural diversity. Specifically, when higher levels of structural diversity are present in the available request data set, an integration of stochastic knowledge into the deterministic real-time control approach becomes increasingly beneficial. This can be explained by the fact that higher variations in request arrival rates in different regions of the considered service area during the day increase the value of information about expected future requests. In these cases, more substantial improvements can be attained due to a higher value which the stochastic knowledge possesses. In contrast, request data sets with a low structural diversity limit the improvements that can be additionally attained by utilizing stochastic knowledge since, in this case, the deterministic real-time control approach can take advantage of keeping idle vehicles at their last serviced request locations using the idle vehicle waiting strategy. This is because in request data sets with a low structural diversity future requests are likely to occur close to requests that have already arrived.
- **Number of available resources:** In addition to the structural quality of available request data sets, the degree of utilization of the available vehicles has also been proven as a decisive aspect for an efficient application of the proposed pro-active real-time control approach. Specifically, the available fleet size directly affects the minimum required structural quality of existing request data which leads to promising results. In case of a limited fleet size, vehicles are very busy with servicing the pending requests. Hence, it is recommended to only integrate reliably forecasted stochastic knowledge since the integration of uncertain stochastic knowledge is likely to lead to postponement of pending requests and therefore to a reduced solution quality. In contrast, vehicles are less busy with servicing

the pending requests in scenarios where larger fleet sizes are available. Hence, the minimum required structural quality of a request data set decreases since an incorrectly forecasted request does not cause significant deteriorations in the execution of the transportation process.

We conclude that the structural distortion of a request data set depends on the applied forecasting approach while the structural diversity represents a general criterion which significantly determines attainable improvements by using pro-active real-time control approaches. Therefore, with regard to structural distortion, it is possible to also efficiently derive and utilize stochastic knowledge for additional request data sets which are classified as having a high structural distortion in our forecasting approach by extending the proposed forecasting approach. Moreover, the structural diversity of the request data set mainly determines its structural quality if a low structural distortion is present. Hence, structural diversity was further evaluated in this book. Besides the request data class S^{GEN} for which the level of structural diversity can be defined during the generation of the individual request data sets, two methods were proposed for analyzing the structural diversity of given request data sets. First, according to the qualitative assessment of request data sets performed by the visual analysis of S^{REAL} in Fig. 8.24 and S^{GEN} in Figs. 8.27 and 8.28, we concluded that request data sets with a high structural quality have high variations in request arrival rates in the subareas during the day which allows for an efficient application of the proposed pro-active real-time control approach. In order to further define structural quality, we introduced the degree of structural diversity (*dosd*) for quantitatively measuring the structural diversity of given request data sets (see Sect. 8.9). The effectiveness of this measure was evaluated by analyzing the *dosd* values for all 25 request data sets of the request data class S^{GEN}. Specifically, linear regression models analyzing the correlation between the *dosd* and the improvement attained in the computational experiments were applied to both customer inconvenience functions and the three tested fleet sizes. The resulting correlation coefficients were between 0.9161 and 0.9606 indicating that the *dosd* is an effective measure for determining the structural quality of given request data sets. Moreover, an analysis of the *dosd* values determined for the real-world request data class S^{REAL} also yielded good results. Therefore, it was concluded that for the first time the proposed *dosd* provides a reasonable starting point for analyzing the structural quality of given request data sets. As mentioned above, this is an essential factor for estimating the improvements that can be attained by utilizing pro-active real-time control approaches. Moreover, it provides decision support for efficiently customizing the proposed pro-active real-time control approach by choosing appropriate $DC^{\min\lambda}$ values for generating stochastic knowledge depending on the available request data set, customer inconvenience function, and fleet size.

In a further computational study, we continued to investigate the impacts of vehicle en-route diversion activities on the solution quality and the resulting driver inconvenience:

Question 3: What impact does vehicle en-route diversion have on the solution quality? Since this increases driver inconvenience, is there a conflict between customer inconvenience and driver inconvenience and does a useful trade-off between both objectives exist?

Answer: As shown in Sect. 8.5, allowing vehicle en-route diversion significantly improves the solution quality since vehicles can quickly react to newly arriving requests which are close to the current vehicles' positions. However, as illustrated in Table 8.5, this imposes a high number of request reassignment activities while drivers are currently traveling to their next assigned customer request. This increases driver inconvenience. For this reason, a threshold-based approach for limiting vehicle en-route diversion activities was proposed in order to reduce driver inconvenience. By using this approach, vehicle en-route diversion activities are limited to the ones which achieve objective function value improvements which are higher than the penalty incurred by the specified threshold parameter value. It was shown that the application of small threshold parameter values significantly reduces the number of vehicle en-route diversion activities while only marginally increasing the resulting customer inconvenience. When using higher threshold parameter values, a further decrease of performed vehicle en-route diversion activities could be achieved but with a significant increase in the attained customer inconvenience. Furthermore, the largest threshold parameter value of 1.0 that we evaluated effectively prohibited almost all vehicle en-route diversion activities which resulted in the lowest achievable driver inconvenience with regard to the considered vehicle en-route diversion activities.

Furthermore, this computational study showed that customer inconvenience and driver inconvenience are conflicting aims. Nevertheless, parameter settings for the proposed threshold-based approach which achieve useful trade-offs between both conflicting aims were determined. Specifically, compared to scenarios in which only customer inconvenience is considered, driver inconvenience can be significantly reduced while only marginally increasing customer inconvenience by using small threshold parameter values whose actual value depends on the utilized customer inconvenience function. Moreover, in order to prioritize customer inconvenience, the threshold parameter value should be set to 0.0 while a prioritization of driver inconvenience can be achieved by setting the threshold parameter value to 1.0.

The computational results also showed that for request data sets with a high structural quality, using the threshold-based approach with PROACTIVE10 can reduce customer inconvenience and driver inconvenience at the same time compared to the results achieved by TIMELIMIT10. It was concluded that the proposed threshold-based approach is an effective method in limiting the number of vehicle en-route diversion activities to a desired level.

Chapter 9
Summary and Outlook on Future Work

In this chapter, the evaluated RDOPG applications and the achieved results of this book are presented. First, a *summary* about the work in this book, attained results, and contributions is given. Finally, an *outlook on future work* that provides avenues for future research is given.

9.1 Summary

In many business relationships, the unexpected occurrence of events may lead to unfavorable situations which cause a high inconvenience or high costs for affected customers. Two examples of such situations are the unsuccessful delivery of newspapers to subscribers and the breakdown of important machines during production processes. Hence, in these situations, customers demand quick assistance from a service company assumed to be determined in advance in order to resolve the problematic situation. What is crucial is that since the occurred situation has already caused a high amount of inconvenience to the customer, a delayed delivery of these goods causes additional significant inconvenience. Therefore, distribution processes of such goods are defined as real-time distribution of perishable goods (RDOPG) applications in this book. The considered RDOPG applications belong to the class of dynamic routing problems in which uncertainty arises from newly arriving requests during the execution of the transportation process. Specifically, in the RDOPG applications, only a few customer requests are known in advance so that a high degree of dynamism is present.

Since marketing research has revealed that customer satisfaction relates to how a company deals with a complaint, it is the topmost priority of the considered service company to fulfill or exceed the expected service quality in order to generate high customer satisfaction and to establish long-term customer connectivity. Moreover, since in the considered RDOPG applications customer inconvenience directly starts at the point in time when a customer places a request at the service company and since it depends on the perceived request response time, attained service quality can

F. Ferrucci, *Pro-active Dynamic Vehicle Routing*, Contributions to Management Science, 267
DOI 10.1007/978-3-642-33472-6_9, © Springer-Verlag Berlin Heidelberg 2013

be increased by providing a fast service of pending requests. Hence, RDOPG applications are characterized by a high level of customer orientation so that service quality is the topmost aim in these applications. Specifically, a late supply of required goods to the customers causes inconvenience and penalty costs that are significantly higher than costs arising from the operative distribution processes. Therefore, the objective solely consists of minimizing the mentioned customer inconvenience.

Motivated by these challenges, two real-time control approaches for efficiently coordinating the transportation process of RDOPG applications are proposed in this book. The deterministic real-time control approach which utilizes a newly developed solution method for solving individual problem instances arising during the execution of the transportation process is proposed. Based on this deterministic real-time control approach, a new pro-active real-time control approach is proposed whose purpose is to pro-actively position vehicles according to expected future request arrivals so that further arriving requests can be efficiently serviced with reduced customer inconvenience. In order to do so, information about expected future requests referred to as stochastic knowledge is utilized. In the literature it was already shown that the integration of stochastic knowledge can significantly increase the solution quality attained in different dynamic routing problems. However, in these approaches it is often assumed that stochastic knowledge is directly available which is not the case in many real-world applications. Therefore, the generation of stochastic knowledge is an important aspect since in many real-world applications no stochastic knowledge is given a priori but can be derived from available past request information. Hence, one of the main contributions of this book is to propose new methods for generating stochastic knowledge by analyzing past request information and integrating this stochastic knowledge into the real-time control approach. Since the proposed approach generates the stochastic knowledge from past request information and only assumes the existence of time-space Poisson processes that are constant from day to day, it can be applied to many practical RDOPG applications.

In order to evaluate the attainable improvements of the solution quality by using the pro-active real-time control approach, its results are compared in detail in an extensive study to the ones generated by the deterministic real-time control approach. The impacts on the results are evaluated using different parameter settings for the real-time control approaches as well under different request data conditions arising on real-world and on different generated request data sets. Attained results are utilized in order to derive suitable measures for determining the structural quality of given request data sets. In what follows, we provide a summary of the individual chapters of this book.

In Chap. 1, a motivation was given by describing challenges found in real-world processes that are represented by the considered RDOPG applications. Moreover, the contributions of this book were described. In Chap. 2, a comprehensive introduction to routing problems was presented. Besides defining the general tasks found in routing problems, general differences between arc- and node-based problems were described. Since the considered RDOPG applications in the book belong to the node-based routing problems, several classic node-based routing problems such as the Traveling Salesman Problem (TSP), Vehicle Routing Problem (VRP), and Pickup and Delivery Problem (PDP) were introduced. Furthermore, since the considered RDOPG applications can be represented as variants of the VRP, in the remainder of Chap. 2, characteristics and variants of VRPs were discussed. These characteristics include solution complexity issues and possible solution methods. In the main part of this chapter, different variants of the VRP were described. Specifically, variants of the static VRP that differ from each other in terms of problem attributes such as transportation requests, number of depots, available vehicles, as well as different pursued objective functions were discussed. As mentioned above, RDOPG applications belong to the class of dynamic routing problems due to newly arriving requests during the execution of the transportation process. Therefore, after describing static routing problems in which the relevant information is constant, dynamic routing problems in which the relevant information changes were presented. For this purpose, different classifications in the literature that deal with differences between static and dynamic routing problems were described. Since dynamic problems can be distinguished according to the type of relevant information revelation, appropriate existing classifications were also presented. In order to aggregate the results that are relevant for the considered RDOPG applications from these classifications, a unified approach for classifying routing problems was developed. In this unified approach, the type of changes that occur during the execution of the transportation process in the relevant information as well as their availability and processing and the type of available stochastic knowledge are considered. According to this classification, different possible variants of VRPs which are known in the literature were introduced. Afterwards, in accordance with the considered RDOPG applications, Dynamic Vehicle Routing Problems (DVRPs) in which a centralized coordination of the tour plan utilizing real-time control approaches is applied were considered next. Specifically, characteristics of such routing problems were presented. Besides typical objectives pursued in dynamic routing problems and a description of dynamic events considered in dynamic routing problems, a measure proposed by Lund et al. (1996) and Larsen (2000) for determining the difficulty level of solving individual problem instances in real-time routing problems was introduced. Based on this measure denoted as the degree of dynamism, a three-echelon classification proposed by Larsen (2000) was introduced and various real-world dynamic routing applications were described according to their degree of dynamism. In order to evaluate the performance of dynamic routing approaches, two different measures known in the literature were presented. Finally, a general classification scheme for characterizing the routing problems was described. This scheme was adopted from the literature and extended in order to cover additional relevant problem characteristics which are important for RDOPG applications.

In Chap. 3 the considered problem represented by the RDOPG applications was described in detail. One of the main characteristics that distinguishes RDOPG applications from many other dynamic routing applications is the objective function that consists either of a linear or a quadratic customer inconvenience function. This customer inconvenience is different from penalty costs which only occur when a request is serviced late since customer inconvenience directly starts with the request arrival in the system. If a maximum allowed request response time is not regarded, high lateness penalty costs occur. Furthermore, vehicle en-route diversion is allowed. Specifically, allowing to re-route a vehicle while it is traveling to a request is likely to result in an improved solution quality since this enables quick service of a newly incoming request which is closely located to the vehicle. Moreover, past request information is available and can be used to generate stochastic knowledge about expected future requests which is important in order to apply the pro-active real-time control approach. After this problem description, various other characteristics of the RDOPG applications, such as its problem complexity, were described and the RDOPG applications were classified according to the general classification scheme which was presented in Chap. 2.

Based on this problem description, an extensive literature review was conducted in Chap. 4 considering various known approaches for dynamic routing problems which cover important aspects of RDOPG applications. After describing the first approaches for dynamic routing problems that have been published in the literature, selected research work dealing with dynamic routing problems was presented. Furthermore, general techniques for providing an increased flexibility in dynamic routing problems were introduced. These techniques were subdivided according to whether they require stochastic knowledge or not. This distinction was later on used when presenting the approaches in the literature that provide an increased flexibility. In the literature review, different approaches including strategic approaches and reoptimization-based approaches were considered. From this literature review it can be observed, as mentioned above, that the majority of the approaches presented in the literature which utilize stochastic knowledge assume that it is directly available by prescribed probability distributions. In addition to methods for increasing flexibility in real-time control approaches, the literature review also presents approaches in the literature which cover other aspects that play an important role in the considered RDOPG applications. These aspects include objective functions that are comparable to the one utilized. Also described were real-time control concepts, road networks for mapping vehicle movements and travel time as well as travel distance calculations, and appropriate technical equipment for implementing real-time control approaches.

After this literature review, the real-time control concept utilized in the RDOPG applications was introduced in Chap. 5. First, a general categorization of different control concepts that are often applied in logistic and production systems was presented. After this categorization, a general challenge of real-time control approaches, i.e., the concurrency between tour plan execution and tour plan adaptation, was described. This is important for many real-time applications since the controlled process is often executed simultaneous to its adaptation. Since transportation processes of RDOPG applications also belong to these types of real-time control processes, the anticipation horizon concept (cf. Bock 2010) which is used for dealing with this challenge was described. Moreover, the update handling used for integrating dynamically arriving requests and generating individual problem instances as well as the idle vehicle waiting strategy were introduced. By making use of this strategy, idle vehicles wait for further arriving requests at their last serviced request location instead of directly traveling back to the depot. In this context, the applied vehicle scheduling strategy was presented. Using this strategy, vehicles directly drive to their next assigned request location since pending requests can always be directly serviced when they arrive in the system. Furthermore, the mathematical model used to represent individual problem instances was presented including all constraints that need to be fulfilled for the considered RDOPG applications. Moreover, both considered objective functions denoted as linear2X and quadratic were defined. Finally, the introduced RDOPG real-time control concept was categorized according to the classification at the beginning of the chapter.

The previously described real-time control approach is deterministic since it only considers requests that have recently arrived during tour plan adaptation but does not utilize stochastic knowledge about requests expected in the future. In order to utilize the available stochastic knowledge, new methods were presented in Chap. 6 for extending the deterministic real-time control approach to the new proposed pro-active real-time control approach. This pro-active real-time control approach integrates the generated stochastic knowledge into the tour plan coordination process by using dummy customers. In contrast to real requests, dummy customers are not serviced but are used to guide vehicles into request-likely areas where future requests are expected with a high probability according to the stochastic knowledge. For this purpose, new methods for generating stochastic knowledge from real past request information were described. This generation is performed by partitioning the considered service area into subareas and segments. For each of these segments, request arrivals are modeled by a time-space Poisson process whose rate parameter is calculated by analyzing the number of request arrivals in the respective segment on a specific number of considered past working days. Afterwards, dummy customers are created by aggregating segments to clusters that fulfill certain quality criteria. Among others, one criterion to be fulfilled consists of requiring that one or more requests actually arrives in place of a dummy customer at least with a specified probability. Moreover, as mentioned above, in the utilized methods for generating stochastic knowledge it is assumed that requests in the individual segments arrive according to time-space Poisson processes. In order to check this assumption, for each dummy customer the likelihood that past request arrivals observed in the corresponding cluster are generated by a time-space Poisson processes is evaluated. If this is too unlikely, the dummy customer is discarded. In order to generate as many valid dummy customers as possible, an exact two-step solution method was presented. In the first step, all valid dummy customers are generated while in the second step the maximum number of non-overlapping dummy customers is selected by utilizing an MIP solver. Moreover, modifications of the real-time control approach required for integrating and handling dummy customers during the execution of the transportation process were presented. This also included the description of an extended vehicle scheduling strategy. Using this strategy, a vehicle waits at its current position if its next assigned request is a dummy customer and if its departure at the current time in the transportation process would otherwise make the vehicle arrive earlier than desired at the dummy customer's location.

In order to efficiently solve the individual static problems generated by the real-time control approach under the tight computational time restrictions imposed by the real-time nature of the considered RDOPG applications, a sophisticated heuristic solution method was proposed in Chap. 7. This solution method is based on a Tabu Search metaheuristic that utilizes different neighborhood operators which are chosen by a deterministic stage-based selection scheme. After explaining the general procedure of the Tabu Search metaheuristic and describing differences between intensifying and diversifying neighborhood operators, the utilized neighborhood operators were described in detail. After this description, the applied tabu-active attribute in the tabu list that is used to identify tabu-active solutions was explained. Specifically, in order to allow an almost unique identification of individual solutions, a fingerprint consisting of two quickly computable checksums was used as the tabu-active attribute. This also allows the identification of solutions across consecutive problem instances. After a description of the initial tour plan generation process, the stage-based selection scheme was presented. This scheme selects the neighborhood operator which is to be applied in each iteration of the solution process. After a predefined number of iterations without finding a new best global solution, the current stage is switched to the next higher one. Therefore, neighborhood operators providing a higher level of diversification are applied in the subsequent iterations. When a new best global solution is found, the stage is reset to the first one so that an intensifying neighborhood operator is applied in the next iterations.

In Chap. 8, the extensive computational experiments that are performed to evaluate the proposed variants of the real-time control approach and to answer the central research questions of this book were presented. This includes a description of the discrete event-based simulator and the tested environment consisting of the utilized road network as well as certain parameters, i.e., the length of the anticipating horizon, tested fleet sizes, applied Tabu Search parameters, and evaluated dummy customer parameters for generating and integrating stochastic knowledge into the pro-active real-time control approach. Furthermore, the two request data classes on which the evaluations are conducted were introduced. The first request data class consists of a real request data set derived from a real-world application dealing with the subsequent delivery of newspapers. The second request data class consists of different generated request data sets that possess specific designed structural quality aspects. Therefore, the second request data class allows a differentiated analysis of the attainable improvements of the proposed pro-active real-time control approach under different levels of structural diversity present in the designed request data sets of this request data class.

Before the main computational study, several preliminary studies were performed. First, the performance of the Tabu Search solution method was evaluated by comparing its attained results with the ones obtained by a lower bound of an exact solution approach for a related dynamic vehicle routing problem. In this study, it was shown that the proposed Tabu Search metaheuristic yields near-optimal results within 10 seconds for individual problem instances that arise during the transportation process of the considered RDOPG applications. Hence, we concluded that the proposed Tabu Search metaheuristic allows a sufficiently fast and efficient adaptation of the individual problem instances. In a second preliminary study, the performance of the deterministic real-time control approach was evaluated against two related deterministic real-time control approaches. The results revealed that the variant denoted as TIMELIMIT10, which invokes the solution method for 10 seconds whenever new requests have arrived in the previous anticipation horizon, achieves a performance that is comparable to the original approach which continuously applies the solution method for the complete length of the anticipation horizon. An analysis showed that this can be explained by the fact that TIMELIMIT10 already achieves a near-optimal solution quality within the given time limit so that no significant further improvements can be attained by applying the proposed solution method for a longer time period. Hence, TIMELIMIT10 was used as the reference setting which is challenging due to the utilization of the described solution method. The next preliminary study dealt with the impacts of allowing vehicle en-route diversion on the attained solution quality in the deterministic real-time control approach. The results showed that, depending on the fleet size and utilized customer inconvenience function, improvements in the solution quality between 14 % and 25 % can be attained compared to the solution quality when vehicle en-route diversion activities are prohibited while reducing travel time and travel distance at the same time. In a further study, the performance of the Poisson quality approach in our forecasting methods, which uses a Goodness-of-Fit test to determine the structural distortion of request data sets, was evaluated. This test was used to check the assumption that request arrivals can be modeled by a set of unknown time-space Poisson processes whose rate parameters are also unknown but remain constant from day to day. Hence, a low structural distortion of the existing request data set is important in order to allow for an efficient application of the proposed pro-active real-time control approach. Evaluations conducted on different request data sets that have different levels of structural distortion showed that the proposed Poisson quality approach is an efficient method for determining the structural distortion of request data sets. The evaluations also indicated that S^{REAL} has, similar to S^{GEN}, a low structural distortion. Moreover, a suitable value for the minimum demanded Poisson quality parameter for generating stochastic knowledge was derived. Using this parameter value, information on the stochastic knowledge generated for the request data sets evaluated our main computational experiments was given. This also included information about the process of generating this stochastic knowledge.

In the main computational experiments, first the results of the generated stochastic knowledge by utilizing the proposed forecasting method were presented. Thereafter, a solution quality comparison of the pro-active real-time control approach and its deterministic variant was presented. In these experiments, the achieved results of evaluating the real-world request data class S^{REAL} were provided. The results revealed that depending on the used customer inconvenience function as well as on the fleet size, different dummy customer settings with regard to the minimum demanded quality $DC^{\min\lambda}$ yielded the best results. Specifically, when few vehicles are present, the utilization of stochastic knowledge should be reconsidered in general. This is especially important if the linear2X customer inconvenience is applied since in this case, the deterministic real-time control approach tends to postpone the service of requests in order to service other requests earlier. This often causes problems when further unexpected requests arrive so that some requests cannot be serviced within the maximum allowed response time. This situation is even worsened when uncertain stochastic knowledge is used in the pro-active real-time control approach. Therefore, the utilization of stochastic knowledge only comprising dummy customers, for which at least one request will actually arrive in place of the dummy customer with a high probability, is recommendable in this situation. When more vehicles are present, the utilization of generated stochastic knowledge can attain significant improvements. Moreover, the use of forecasted stochastic knowledge that also comprises less reliable information can lead to additional improvements of the solution quality. This can be explained by the fact that due to the larger fleet size, available vehicles are less utilized so that uncertain information can be beneficially used without disturbing the service of pending requests. With the largest tested fleet size, improvements of up to 1.7 % in the linear customer inconvenience case and 7.93 % in the quadratic customer inconvenience case could be attained.

Furthermore, a detailed visual analysis of S^{REAL} revealed that the relative request arrival rates in all subareas of the service area do not significantly vary during the day. Therefore, since newly arriving requests keep occurring at similar locations as before, the deterministic real-time control approach already advantageously positions the vehicles in advance for servicing future requests by making use of the presented idle vehicle waiting strategy. Therefore, this results in a low structural diversity for S^{REAL} so that additional improvements attainable by utilizing stochastic knowledge in the pro-active real-time control approach are limited in this request data set. A further potentially problematic issue given by the fact that it may not be possible to reasonably represent request arrivals in S^{REAL} by unknown time-space Poisson processes, which are constant from day to day, could not be identified as a significant problem. Note that this is in line with the previously attained results of analyzing the structural distortion of S^{REAL} that indicated a low structural distortion for this request data set. Hence, the low structural diversity of S^{REAL} seems to be the major reason for the prohibition of further improvements attainable on S^{REAL} by using the pro-active real-time control approach.

The second evaluated request data class S^{GEN} was generated using time-space Poisson processes with individual request arrival rates in different subregions. In this data, these arrival probabilities as well as the total request arrival rate significantly change during the day. Note that higher arrival rate variations reduce the probability that vehicles are closely located to future request arrivals by just waiting at the location of the last serviced request. Therefore, in this case, an a priori repositioning of vehicles according to expected future request arrivals based on stochastic knowledge increasingly becomes advantageous. The mentioned variation in request arrival rates was defined according to two dimensions that are called RegionDiversity (*RD*) and TimeDiversity (*TD*). Utilizing these two dimensions, the structural diversity of this request data class was defined. Overall, 25 request data sets with different levels of structural diversity were generated. The computational results showed that in request data sets with higher levels of structural diversity, significantly higher solution quality improvements can be attained by using the pro-active real-time control approach compared to the results obtained by the deterministic real-time control approach. Specifically, while in almost all tested structural diversity settings improvements could be achieved, in the highest structural diversity setting, 19.51 % to 26.72 % of additional improvement could be attained under the linear2X customer inconvenience function, depending on the fleet size. When the quadratic customer inconvenience function was utilized, improvements of between 29.86 % and 47.24 % were achieved. As performed on S^{REAL}, a visual analysis of this request data class was provided. In this analysis, significant structural differences between the evaluated real-world and a generated request data set comprising a high level of structural diversity were illustrated. Specifically, while S^{REAL} possesses a low structural diversity, the structural diversity of the designed request data sets depends on the *RD* and *TD* parameter settings used for generating the corresponding request data set. By applying this structural diversity classification on the evaluated request data sets combined with the above described Poisson quality approach for determining structural distortion, differences in the attained improvement of the solution quality by using the pro-active real-time control approach in comparison to its deterministic variant became explainable.

Based on the insights of the main computational study with regard to the level of structural diversity and the improvement achievable by applying generated stochastic knowledge in the pro-active real-time control approach, we extended the previously described visual analysis. Specifically, a measure called the degree of structural diversity (*dosd*) was introduced for determining the structural quality of given request data sets. This measure is based on the fact that stochastic knowledge possesses a higher value if the relative request arrival rates in the individual subareas of the service area significantly change during the considered planning horizon. The *dosd* ranges from 0 to 1 representing the lowest or highest structural diversity which a given request data set can possess according to this measure. The effectiveness of this measure was evaluated by analyzing the *dosd* values generated for all 25 designed request data sets of S^{GEN} with regard to the additional improvements that have been obtained in the main computational experiments by the pro-active real-time control approach. In total, six different data sets were analyzed resulting from the two customer inconvenience functions evaluated under the three different fleet sizes. A linear regression model was applied to each of the six data sets and the correlation coefficient was calculated. The resulting correlation coefficients were between 0.9161 and 0.9647 which statistically proved the significance of the *dosd*. Moreover, good results were also obtained for S^{REAL}. Therefore, it was concluded that the proposed *dosd* provides a reasonable starting point for generally measuring the structural quality of a given request data set. As shown above, this is an essential factor for estimating the improvements attainable by utilizing pro-active real-time control approaches.

Besides the evaluations dealing solely with the minimization of customer inconvenience, we investigated another computational study which additionally considers the number of vehicle en-route diversion activities performed during the transportation process. It was concluded that this number is an important factor in real-world applications since drivers need to react to assigned vehicle en-route diversion activities very quickly. Therefore, a high number of vehicle en-route diversion activities imposes a high inconvenience on the drivers which may result in reduced work motivation. In order to reduce driver inconvenience by lowering the number of vehicle en-route diversion activities, a threshold-based approach for limiting these vehicle en-route diversion activities was proposed. By using this approach, vehicle en-route diversion activities are limited to the ones which achieve objective function value improvements which are higher than the penalty incurred by the threshold parameter value. In computational experiments performed by using the deterministic real-time control approach, the impact of different values of the proposed threshold parameter on the solution quality was analyzed. It was shown that the application of small threshold parameter values significantly reduces the number of vehicle en-route diversion activities while only marginally increasing the resulting customer inconvenience. When using higher threshold parameter values, a further decrease of performed vehicle en-route diversion activities could be observed but with a significant increase in the attained customer inconvenience. Furthermore, the highest investigated threshold parameter value effectively prohibited all vehicle en-route diversion activities which resulted in the lowest achievable driver inconvenience with regard to the number of performed vehicle en-route diversion activities.

Moreover, the utilization of the proposed pro-active real-time control approach was also recommended when considering driver inconvenience aspects. Specifically, the computational results showed that the pro-active real-time control approach is able to reduce customer inconvenience and driver inconvenience at the same time when appropriate threshold parameter values are used. It was concluded that the proposed threshold-based approach is an effective method in limiting the number of vehicle en-route diversion activities to a desired level. For the different pursued aims, i.e., prioritizing customer inconvenience or driver inconvenience or considering a balance of both contradicting aims, adequate values for the threshold parameter approach were provided.

The work presented in this book extends the literature of stochastic and dynamic vehicle routing problems by a new pro-active real-time control approach and an approach for generating stochastic knowledge out of past request information. All approaches have been extensively tested under different conditions. Furthermore, several evaluations which consider the impacts of the structural request data quality on the solution quality as well as a method for limiting vehicle en-route diversion activities have been proposed and impacts on the solution quality have been assessed. The achieved results are likely also to provide additional insights into other variants of dynamic routing problems which are related to the considered RDOPG applications.

9.2 Outlook on Future Work

The attained results provide promising starting points for avenues of further research which are presented in what follows. First, extensions of the proposed methods for generating stochastic knowledge seem to be useful. For instance, the integration of geographical information that is available in the road network could have a significant impact on the performance of the pro-active approach. Specifically, the performed generation of compact clusters on predefined segments might restrict the generation of stochastic knowledge and its applicability if natural barriers such as rivers or channels exist. Therefore, an extension of the segment-based clustering methods used for generating stochastic knowledge should be examined. Another extension may improve the performance of the pro-active real-time control approach if the existing request data does not consist of request arrival characteristics which are constant from day to day but instead consists of different recurring types of days in which request arrivals significantly vary and the type of the upcoming day cannot be determined in advance. For example, this is the case when the requests arrive on different days according to individual sets of time-space Poisson processes. In such cases, further improvements are likely to be attained by integrating sophisticated pattern generation and recognition methods into the cluster generation and the pro-active real-time control approach. Specifically, as an extension to the proposed forecasting approach which assumes one pattern that is constant from day to day, different day patterns can be learned based on past request information. During the execution of the transportation process, an appropriate day pattern is identified according to requests that have already arrived during the current day. After this identification, adequate stochastic knowledge is dynamically integrated into the pro-active real-time control approach during the execution of the transportation process. As mentioned in Sect. 8.6, an extension of the proposed forecasting methods requires further developing methods for the purpose of identifying structural distortion. Furthermore, using methods for an ex-post comparison of predicted requests and those that have occurred during the transportation process can be useful in fine-tuning the stochastic knowledge.

Moreover, as mentioned in the literature review, other pro-active approaches exist in the literature for utilizing stochastic knowledge. For example, sampling-based approaches and approximate dynamic programming approaches have also been successfully applied to dynamic routing problems. In this context, ideas inspired from these types of approaches for generating and utilizing stochastic knowledge can be integrated into the proposed pro-active real-time control approach. In doing so, an additional increase in solution quality improvements is likely to be attained. Furthermore, as can be seen in the computational experiments, the impact of different parameters used in the generation of the stochastic knowledge, for example the impacts of a minimum demanded rate parameter of each cluster on the solution quality have been evaluated in detail. In addition, there exist other parameters whose impact on the solution quality could be analyzed in order to improve the methods used for generating stochastic knowledge. For example, a more detailed analysis of chosen parameter values such as the minimum demanded Poisson quality or the number of working days n^{f} used could be investigated. Moreover, the development of methods which are suitable for assessing whether practical request data sets fulfill requirement (c) of Poisson processes (see Sect. 6.2.1) could be researched.

Furthermore, the integration of other sources of dynamism into the pro-active real-time control approach can increase its practicability. For example, this includes the consideration of traffic jams, time-dependent travel times, or vehicle breakdowns. Moreover, as described in Sect. 8.1, in the computational experiments all vehicle activities are performed in line with the anticipated activities. Hence, the real-time control approach as well as the simulator could be extended in order to deal with uncertainties in anticipated and actually performed vehicle activities. For this purpose, the development of methods which synchronize the anticipated tour plan with the current system situation at the end of each anticipation horizon is advantageous. Furthermore, as mentioned in the considered RDOPG application dealing with the subsequent delivery of newspapers, the service time of requests is assumed to be known and constant for all requests (see Sect. 3.1). By analyzing past request information according to service times required for servicing requests and by including this additional stochastic knowledge into the real-time control approach, the proposed approaches can be also be applied to other types of RDOPG applications. This is especially useful if the request data has specific characteristics, e.g., if requests located in the same subregion of the service area have a homogeneous service time which significantly varies between subregions or within a subregion over different time periods during the considered planning horizon. Other scenarios that can take advantage of this extension are those in which different types of requests for which estimations about expected service times can be individually derived exist.

The previously mentioned integration of time-dependent travel times into the real-time control approach is also an important aspect in the considered urban transportation system environment. Adequate methods for generating and integrating suitable stochastic knowledge for these time-dependent travel times could be developed, e.g., for predicting required travel times according to weekday-dependent speed profiles observed in the past. This information could be directly considered in the real-time control approach for determining the fastest routes that vehicles should travel between requests depending on the current time of the day. The relevance of such a system could be evaluated by comparing its attainable results to the results of the current version of the real-time control approach that does not make use of time-dependent travel time information.

Moreover, the proposed pro-active real-time control approach can be extended to other types of dynamic routing problems, such as Dynamic Pickup and Delivery Problems (DPDP). In doing so, other real-world applications which also consider customer inconvenience and service quality aspects, such as urban express courier services, could be considered. Due to differences in the problem structure, each request consists of two request locations instead of one as is the case in the currently considered RDOPG applications. Therefore, the methods used for generating stochastic knowledge are required to be extended according to different aspects. For example, the expected pickup-delivery pairs need to be identified in order to anticipate future requests. This information is also required in order to calculate the corresponding expected total request service time which results from the sum of service times at both request locations and the expected travel time required for traveling between both locations. Since in other problem variants such as the DPDP the neighborhoods of the applied neighborhood operators may be more complex and hence more time-consuming to evaluate, the integration of further techniques for time-efficiently calculating neighborhood move values as proposed by Badeau et al. (1997) can be advantageous in order to still be able to quickly generate solutions of high quality. Moreover, other variants of VRPs mentioned in Sect. 2.4 provide additional ideas for extending the considered problem.

Furthermore, the considered objective function can be extended. Specifically, besides only considering customer inconvenience, one extension has been proposed that evaluates the possibility of reducing driver inconvenience represented by the number of performed vehicle en-route diversion activities and the impact on customer inconvenience. Other objectives, for example the minimization of travel time or travel distance, could also be integrated. Computational experiments which evaluate the impact of also considering other objectives on the customer inconvenience could lead to new insights. Additionally, apart from the evaluated linear2X and quadratic objective function for modeling the evolution of customer inconvenience, other functions could also be investigated.

In the computational experiments, a first approach for calculating the structural diversity for given request data sets referred to as the degree of structural diversity (*dosd*) was presented. Since the existence of a high structural quality has been identified as a generally important characteristic for pro-active real-time control approaches which pursue the aim of improving the system performance by utilizing stochastic knowledge, it could be interesting to develop this measure further. Specifically, the integration and extension of the Poisson quality approach proposed for determining the structural distortion into the *dosd* could be investigated. Furthermore, the impact of different other performance characteristics on the attainable solution quality, such as the request to vehicle ratio or road network characteristics present in a scenario, or the ratio of generated dummy customers to the total number of customer requests could be evaluated and also integrated into this measure. In doing so, the *dosd* can be extended in ways which make it possible to enable a comparison even of request data sets or scenarios with significantly different characteristics.

As mentioned by Stenfors et al. (2007), "Tools which support creativity are needed". As a consequence, a further enhancement of the proposed approaches towards more interactive real-time control approaches could be advantageous in increasing the acceptance and practicability of a decision support system which utilizes the developed forecasting methods and the pro-active real-time control approach in practice. This especially includes interactive control possibilities in the decision support system for the dispatcher. By enabling such control possibilities, transportation process situations which require individual decisions that cannot be easily included in the real-time control approach in advance could be manually handled by the dispatcher. In doing so, for example, the possibility that a customer may want to be serviced by a specific service worker could be considered. Furthermore, decision support of the approach could be improved by directly integrating the dispatcher's knowledge into the real-time control of the transportation process. Specifically, the proposed approach could be extended so that the dispatcher can additionally provide stochastic knowledge to the system or review stochastic knowledge which has been automatically generated by the proposed approach. Moreover, instead of choosing a tour plan directly, the real-time control approach could provide several alternative tour plans to the dispatcher who selects a desired tour plan and may also apply further modifications to it. Another possibility of increasing the interactivity can be realized by integrating additional actions into the on-board units of the vehicles so that service workers are able to inform the central dispatching center about the progress of request activities. For instance, a service worker can inform the dispatching center when the current customer request requires more time for service than planned. Furthermore, a new estimated completion time of the current service activity could be provided.

Moreover, the validation of attainable solution quality improvements by applying the proposed pro-active real-time control in real-world applications is another field of research which might be very interesting for theory and practice.

Appendix A
Additional Facts Related to Chap. 1

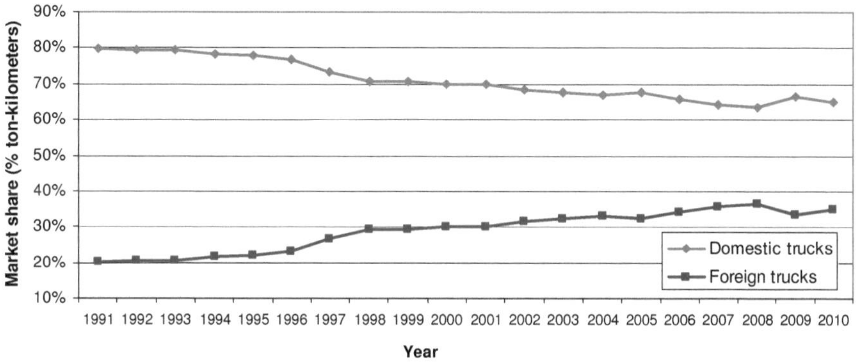

Fig. A.1 Market share of domestic and foreign trucks in Germany for domestic and cross-border transportation from 1991 to 2010 (cf. BGL e.V. 2011)

Table A.1 Development of transportation performance in billion ton-kilometers in the EU-27 from 1995 to 2008 (cf. European Commission 2010, p. 108)

Year	Road	Sea	Rail	Inland waterways	Pipelines	Air	Total
1995	1,289	1,146	386	122	115	2.0	3,060
1996	1,303	1,160	392	120	119	2.1	3,095
1997	1,352	1,193	410	128	118	2.2	3,202
1998	1,414	1,232	393	131	125	2.3	3,297
1999	1,470	1,268	384	129	124	2.3	3,376
2000	1,519	1,314	404	134	127	2.5	3,499
2001	1,556	1,334	386	133	133	2.5	3,544
2002	1,606	1,355	384	133	128	2.4	3,608
2003	1,625	1,378	392	124	130	2.4	3,651
2004	1,747	1,427	416	137	132	2.5	3,861
2005	1,800	1,461	414	139	136	2.6	3,952
2006	1,854	1,505	440	138	135	2.7	4,076
2007	1,915	1,532	453	147	127	2.8	4,176
2008	1,878	1,498	443	145	124	2.7	4,090
Increase 1995–2008	45.7 %	30.7 %	14.7 %	19.0 %	8.0 %	35.0 %	33.7 %

Note: *Air* and *Sea*: only domestic and intra-EU-27 transport; provisional estimates. *Road*: national and international haulage by vehicles registered in the EU-27

DOI 10.1007/978-3-642-33472-6, © Springer-Verlag Berlin Heidelberg 2013

Table A.2 Newspaper advertising expenditures in the United States of America from 1950 to 2010 (cf. Newspaper Association of America 2011)

Year	Newspaper print total		Newspaper online total		Combined newspaper print and online total	
	$Millions	%Change	$Millions	%Change	$Millions	%Change
1950	$2,070	8.3 %				
1951	$2,251	8.7 %				
1952	$2,464	9.5 %				
1953	$2,632	6.8 %				
1954	$2,685	2.0 %				
1955	$3,077	14.6 %				
1956	$3,223	4.7 %				
1957	$3,268	1.4 %				
1958	$3,176	−2.8 %				
1959	$3,526	11.0 %				
1960	$3,681	4.4 %				
1961	$3,601	−2.2 %				
1962	$3,659	1.6 %				
1963	$3,780	3.3 %				
1964	$4,120	9.0 %				
1965	$4,426	7.4 %				
1966	$4,865	9.9 %				
1967	$4,910	0.9 %				
1968	$5,232	6.6 %				
1969	$5,714	9.2 %				
1970	$5,704	−0.2 %				
1971	$6,167	8.1 %				
1972	$6,939	12.5 %				
1973	$7,481	7.8 %				
1974	$7,842	4.8 %				
1975	$8,234	5.0 %				
1976	$9,618	16.8 %				
1977	$10,751	11.8 %				
1978	$12,213	13.6 %				
1979	$13,863	13.5 %				
1980	$14,794	6.7 %				
1981	$16,527	11.7 %				
1982	$17,694	7.1 %				
1983	$20,581	16.3 %				
1984	$23,522	14.3 %				
1985	$25,170	7.0 %				

Table A.2 (Continued)

Year	Newspaper print total		Newspaper online total		Combined newspaper print and online total	
	$Millions	%Change	$Millions	%Change	$Millions	%Change
1986	$26,990	7.2 %				
1987	$29,412	9.0 %				
1988	$31,197	6.1 %				
1989	$32,368	3.8 %				
1990	$32,280	−0.3 %				
1991	$30,349	−6.0 %				
1992	$30,639	1.0 %				
1993	$31,869	4.0 %				
1994	$34,109	7.0 %				
1995	$36,092	5.8 %				
1996	$38,075	5.5 %				
1997	$41,330	8.5 %				
1998	$43,925	6.3 %				
1999	$46,289	5.4 %				
2000	$48,670	5.1 %				
2001	$44,305	−9.0 %				
2002	$44,102	−0.5 %				
2003	$44,939	1.9 %	$1,216		$46,156	
2004	$46,703	3.9 %	$1,541	26.7 %	$48,244	4.5 %
2005	$47,408	1.5 %	$2,027	31.5 %	$49,435	2.5 %
2006	$46,611	−1.7 %	$2,664	31.4 %	$49,275	−0.3 %
2007	$42,209	−9.4 %	$3,166	18.8 %	$45,375	−7.9 %
2008	$34,740	−17.7 %	$3,109	−1.8 %	$37,848	−16.6 %
2009	$24,821	−28.6 %	$2,743	−11.8 %	$27,564	−27.2 %
2010	$22,795	−8.2 %	$3,042	10.9 %	$25,838	−6.3 %

Note: Numbers are rounded

Appendix B
Additional Information on Request Data Sets of Request Data Class S^{GEN}

B.1 Parameters for Generating Selected Request Data Sets of S^{GEN}

F. Ferrucci, *Pro-active Dynamic Vehicle Routing*, Contributions to Management Science,
DOI 10.1007/978-3-642-33472-6,

Table B.1 Resulting request arrival probabilities of selected request data sets in S^{GEN} with different levels of structural diversity

j	Request arrival probabilities $p'(a_i, t_j)$ per time slice in each segment a_i, $i \in \{1, \dots, 18\}$																	$\lambda'(t_j)$
									$RD = TD = 1.00$									
1	0.01	0.01	0.01	0.16	0.20	0.01	0.01	0.01	0.01	0.01	0.01	0.01	0.20	0.01	0.30	0.01	0.01	0.013	15.00
2	0.01	0.01	0.01	0.01	0.01	0.01	0.01	0.30	0.40	0.10	0.01	0.01	0.06	0.01	0.01	0.01	0.01	0.01	40.00
3	0.01	0.01	0.26	0.01	0.01	0.01	0.01	0.01	0.01	0.01	0.01	0.01	0.01	0.01	0.20	0.10	0.30	0.01	30.00
4	0.01	0.01	0.01	0.01	0.20	0.30	0.01	0.01	0.01	0.01	0.01	0.01	0.01	0.01	0.35	0.01	0.01	0.01	35.00
5	0.01	0.01	0.10	0.01	0.01	0.01	0.20	0.01	0.01	0.07	0.01	0.01	0.30	0.01	0.20	0.01	0.01	0.01	10.00
									$RD = TD = 0.75$									
1	0.0214	0.0214	0.0341	0.1114	0.1425	0.0323	0.0285	0.0323	0.0360	0.0270	0.0214	0.0214	0.1481	0.0214	0.2224	0.0248	0.0323	0.0214	20.8088
2	0.0214	0.0214	0.0341	0.0270	0.0356	0.0323	0.0285	0.1954	0.2554	0.0776	0.0214	0.0214	0.0694	0.0214	0.0593	0.0248	0.0323	0.0214	37.6471
3	0.0214	0.0214	0.1748	0.0270	0.0356	0.0323	0.0285	0.0323	0.0360	0.0270	0.0214	0.0214	0.0413	0.0214	0.1661	0.0754	0.1954	0.0214	26.3235
4	0.0214	0.0214	0.0341	0.0270	0.1425	0.1954	0.0285	0.0323	0.0360	0.0270	0.0214	0.0214	0.0413	0.0214	0.2505	0.0248	0.0323	0.0214	33.8971
5	0.0214	0.0214	0.0848	0.0270	0.0356	0.0323	0.1354	0.0323	0.0360	0.0608	0.0214	0.0214	0.2044	0.0214	0.1661	0.0248	0.0323	0.0214	11.3235
									$RD = TD = 0.50$									
1	0.0328	0.0328	0.0498	0.0778	0.0993	0.0473	0.0423	0.0473	0.0523	0.0403	0.0328	0.0328	0.1068	0.0328	0.1558	0.0373	0.0473	0.0328	26.6176
2	0.0328	0.0328	0.0498	0.0403	0.0518	0.0473	0.0423	0.1198	0.1498	0.0628	0.0328	0.0328	0.0718	0.0328	0.0833	0.0373	0.0473	0.0328	35.2941
3	0.0328	0.0328	0.1123	0.0403	0.0518	0.0473	0.0423	0.0473	0.0523	0.0403	0.0328	0.0328	0.0593	0.0328	0.1308	0.0598	0.1198	0.0328	22.6471
4	0.0328	0.0328	0.0498	0.0403	0.0993	0.1198	0.0423	0.0473	0.0523	0.0403	0.0328	0.0328	0.0593	0.0328	0.1683	0.0373	0.0473	0.0328	32.7941
5	0.0328	0.0328	0.0723	0.0403	0.0518	0.0473	0.0898	0.0473	0.0523	0.0553	0.0328	0.0328	0.1318	0.0328	0.1308	0.0373	0.0473	0.0328	12.6471
									$RD = TD = 0.25$									
1	0.0442	0.0442	0.0569	0.0592	0.0703	0.0550	0.0513	0.0550	0.0588	0.0498	0.0442	0.0442	0.0759	0.0442	0.1002	0.0475	0.0550	0.0442	32.4265
2	0.0442	0.0442	0.0569	0.0498	0.0584	0.0550	0.0513	0.0732	0.0832	0.0554	0.0442	0.0442	0.0672	0.0442	0.0820	0.0475	0.0550	0.0442	32.9412
3	0.0442	0.0442	0.0725	0.0498	0.0584	0.0550	0.0513	0.0550	0.0588	0.0498	0.0442	0.0442	0.0640	0.0442	0.0939	0.0532	0.0732	0.0442	18.9706

Table B.1 (Continued)

j	Request arrival probabilities $p'(a_i, t_j)$ per time slice in each segment a_i, $i \in \{1, \ldots, 18\}$																		$\lambda'(t_j)$
4	0.0442	0.0442	0.0569	0.0498	0.0703	0.0732	0.0513	0.0550	0.0588	0.0498	0.0442	0.0442	0.0640	0.0442	0.1033	0.0475	0.0550	0.0442	31.6912
5	0.0442	0.0442	0.0625	0.0498	0.0584	0.0550	0.0632	0.0550	0.0588	0.0535	0.0442	0.0442	0.0822	0.0442	0.0939	0.0475	0.0550	0.0442	13.9706
	$RD = TD = 0.00$																		
1	0.0556	0.0556	0.0556	0.0556	0.0556	0.0556	0.0556	0.0556	0.0556	0.0556	0.0556	0.0556	0.0556	0.0556	0.0556	0.0556	0.0556	0.0556	38.2353
2	0.0556	0.0556	0.0556	0.0556	0.0556	0.0556	0.0556	0.0556	0.0556	0.0556	0.0556	0.0556	0.0556	0.0556	0.0556	0.0556	0.0556	0.0556	30.5882
3	0.0556	0.0556	0.0556	0.0556	0.0556	0.0556	0.0556	0.0556	0.0556	0.0556	0.0556	0.0556	0.0556	0.0556	0.0556	0.0556	0.0556	0.0556	15.2941
4	0.0556	0.0556	0.0556	0.0556	0.0556	0.0556	0.0556	0.0556	0.0556	0.0556	0.0556	0.0556	0.0556	0.0556	0.0556	0.0556	0.0556	0.0556	30.5882
5	0.0556	0.0556	0.0556	0.0556	0.0556	0.0556	0.0556	0.0556	0.0556	0.0556	0.0556	0.0556	0.0556	0.0556	0.0556	0.0556	0.0556	0.0556	15.2941
	$RD = 1.00$, $TD = 0.00$																		
1	0.01	0.01	0.078	0.04	0.086	0.068	0.048	0.068	0.088	0.04	0.01	0.01	0.116	0.01	0.212	0.028	0.068	0.01	38.2353
2	0.01	0.01	0.078	0.04	0.086	0.068	0.048	0.068	0.088	0.04	0.01	0.01	0.116	0.01	0.212	0.028	0.068	0.01	30.5882
3	0.01	0.01	0.078	0.04	0.086	0.068	0.048	0.068	0.088	0.04	0.01	0.01	0.116	0.01	0.212	0.028	0.068	0.01	15.2941
4	0.01	0.01	0.078	0.04	0.086	0.068	0.048	0.068	0.088	0.04	0.01	0.01	0.116	0.01	0.212	0.028	0.068	0.01	30.5882
5	0.01	0.01	0.078	0.04	0.086	0.068	0.048	0.068	0.088	0.04	0.01	0.01	0.116	0.01	0.212	0.028	0.068	0.01	15.2941
	$RD = 0.00$, $TD = 1.00$																		
1	0.0556	0.0556	0.0556	0.0556	0.0556	0.0556	0.0556	0.0556	0.0556	0.0556	0.0556	0.0556	0.0556	0.0556	0.0556	0.0556	0.0556	0.0556	15.00
2	0.0556	0.0556	0.0556	0.0556	0.0556	0.0556	0.0556	0.0556	0.0556	0.0556	0.0556	0.0556	0.0556	0.0556	0.0556	0.0556	0.0556	0.0556	40.00
3	0.0556	0.0556	0.0556	0.0556	0.0556	0.0556	0.0556	0.0556	0.0556	0.0556	0.0556	0.0556	0.0556	0.0556	0.0556	0.0556	0.0556	0.0556	30.00
4	0.0556	0.0556	0.0556	0.0556	0.0556	0.0556	0.0556	0.0556	0.0556	0.0556	0.0556	0.0556	0.0556	0.0556	0.0556	0.0556	0.0556	0.0556	35.00
5	0.0556	0.0556	0.0556	0.0556	0.0556	0.0556	0.0556	0.0556	0.0556	0.0556	0.0556	0.0556	0.0556	0.0556	0.0556	0.0556	0.0556	0.0556	10.00

Note: Numbers are rounded

B.2 Visualization of All Request Data Sets of S^{GEN}

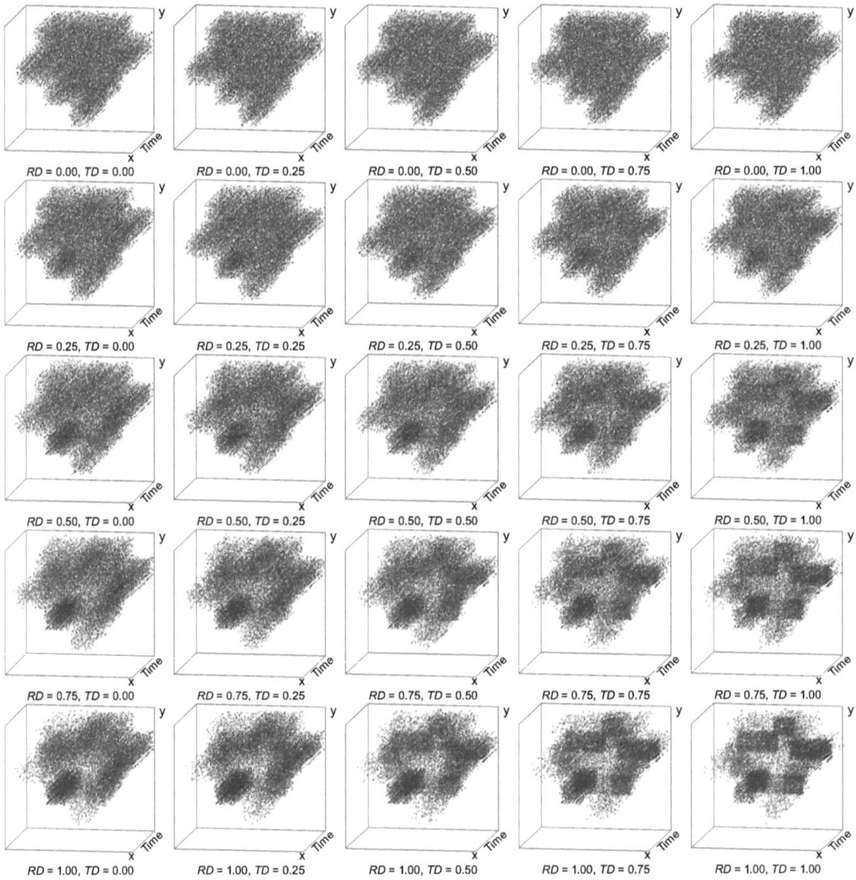

Fig. B.1 All request data sets of S^{GEN} with different levels of structural diversity

Appendix C
Request Response Times of the Examined Request Data Classes

C.1 Request Response Times of S^{REAL}

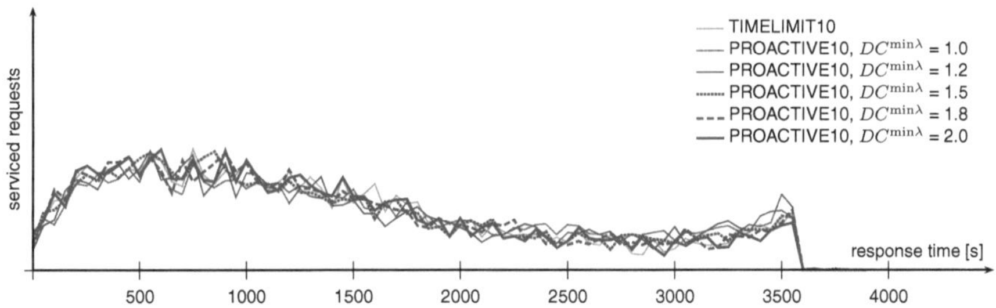

Fig. C.1 Request response times on S^{REAL} using 8 vehicles and $DC^{\min\lambda} = 1.0$, 1.2, 1.5, 1.8, and 2.0, linear2X customer inconvenience grouped by intervals of 50 seconds

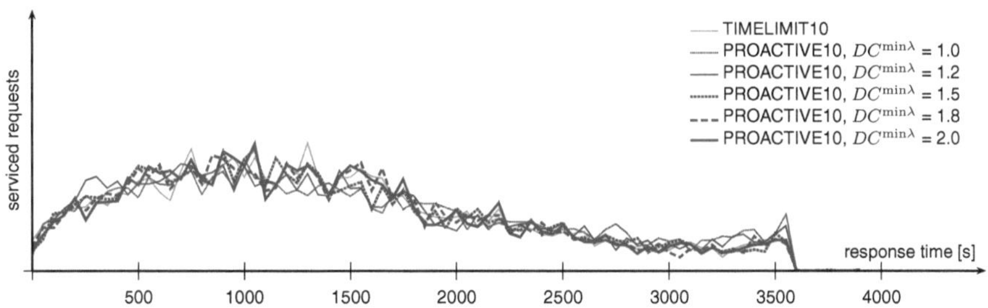

Fig. C.2 Request response times on S^{REAL} using 8 vehicles and $DC^{\min\lambda} = 1.0$, 1.2, 1.5, 1.8, and 2.0, quadratic customer inconvenience grouped by intervals of 50 seconds

F. Ferrucci, *Pro-active Dynamic Vehicle Routing*, Contributions to Management Science, DOI 10.1007/978-3-642-33472-6, © Springer-Verlag Berlin Heidelberg 2013

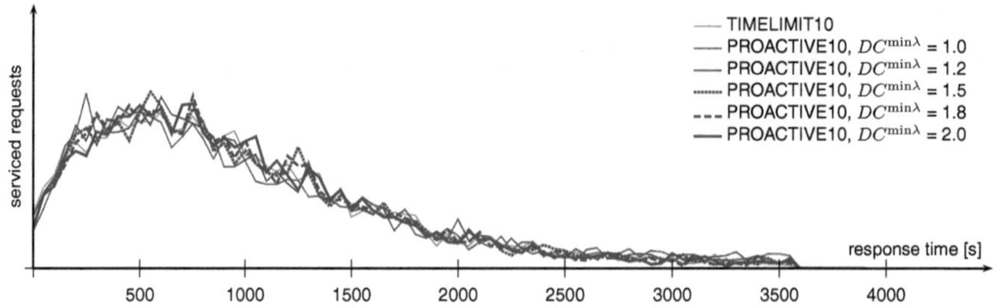

Fig. C.3 Request response times on S^{REAL} using 10 vehicles and $DC^{\min\lambda} = 1.0, 1.2, 1.5, 1.8$, and 2.0, linear2X customer inconvenience grouped by intervals of 50 seconds

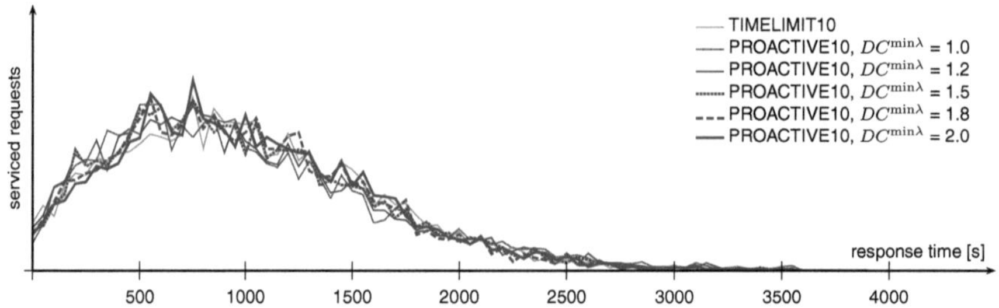

Fig. C.4 Request response times on S^{REAL} using 10 vehicles and $DC^{\min\lambda} = 1.0, 1.2, 1.5, 1.8$, and 2.0, quadratic customer inconvenience grouped by intervals of 50 seconds

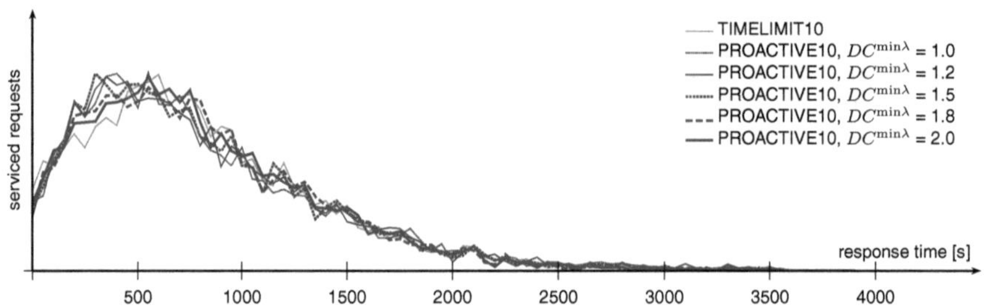

Fig. C.5 Request response times on S^{REAL} using 12 vehicles and $DC^{\min\lambda} = 1.0, 1.2, 1.5, 1.8$, and 2.0, linear2X customer inconvenience grouped by intervals of 50 seconds

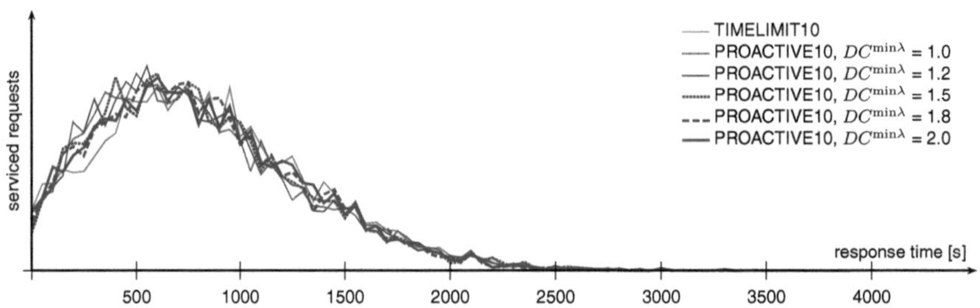

Fig. C.6 Request response times on S^{REAL} using 12 vehicles and $DC^{\min\lambda} = 1.0, 1.2, 1.5, 1.8$, and 2.0, quadratic customer inconvenience grouped by intervals of 50 seconds

C.2 Request Response Times of S^{GEN}

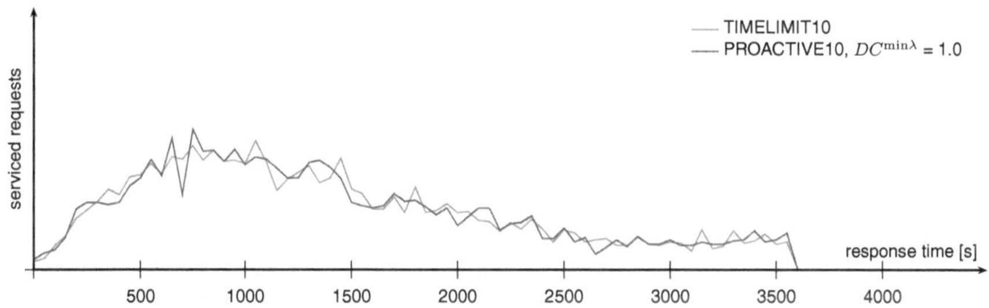

Fig. C.7 Request response times on S^{GEN} using 8 vehicles and $DC^{\min\lambda} = 1.0$, 1.2, 1.5, 1.8, and 2.0, linear2X customer inconvenience, $RD = 0.00$ and $TD = 0.00$ grouped by intervals of 50 seconds

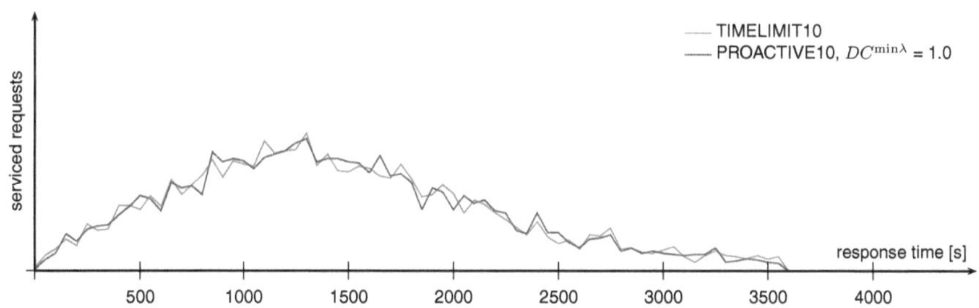

Fig. C.8 Request response times on S^{GEN} using 8 vehicles and $DC^{\min\lambda} = 1.0$, 1.2, 1.5, 1.8, and 2.0, quadratic customer inconvenience, $RD = 0.00$ and $TD = 0.00$ grouped by intervals of 50 seconds

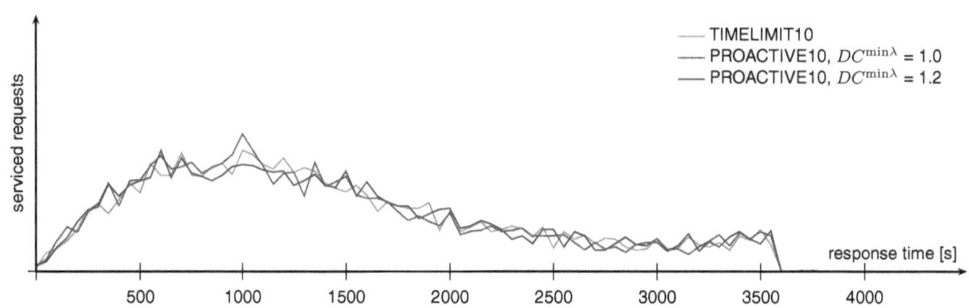

Fig. C.9 Request response times on S^{GEN} using 8 vehicles and $DC^{\min\lambda} = 1.0$, 1.2, 1.5, 1.8, and 2.0, linear2X customer inconvenience, $RD = 0.25$ and $TD = 0.25$ grouped by intervals of 50 seconds

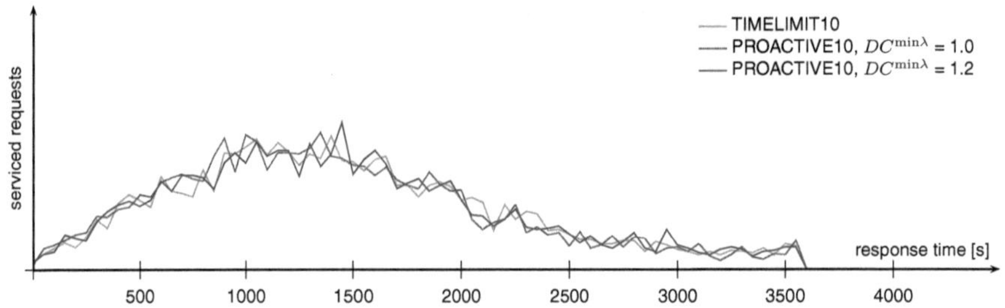

Fig. C.10 Request response times on S^{GEN} using 8 vehicles and $DC^{\min\lambda} = 1.0$, 1.2, 1.5, 1.8, and 2.0, quadratic customer inconvenience, $RD = 0.25$ and $TD = 0.25$ grouped by intervals of 50 seconds

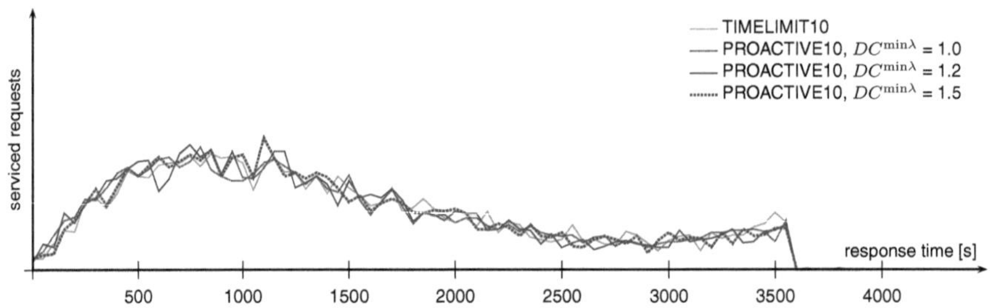

Fig. C.11 Request response times on S^{GEN} using 8 vehicles and $DC^{\min\lambda} = 1.0$, 1.2, 1.5, 1.8, and 2.0, linear2X customer inconvenience, $RD = 0.50$ and $TD = 0.50$ grouped by intervals of 50 seconds

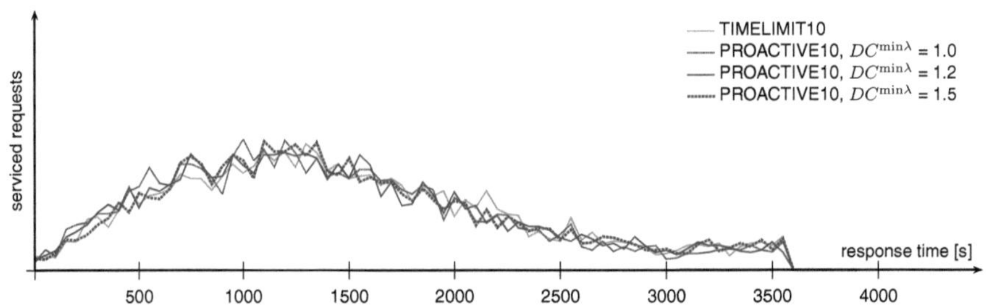

Fig. C.12 Request response times on S^{GEN} using 8 vehicles and $DC^{\min\lambda} = 1.0$, 1.2, 1.5, 1.8, and 2.0, quadratic customer inconvenience, $RD = 0.50$ and $TD = 0.50$ grouped by intervals of 50 seconds

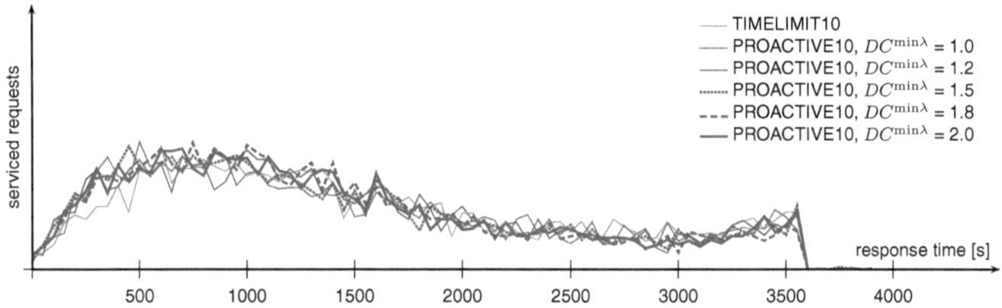

Fig. C.13 Request response times on S^{GEN} using 8 vehicles and $DC^{\min\lambda} = 1.0, 1.2, 1.5, 1.8$, and 2.0, linear2X customer inconvenience, $RD = 0.75$ and $TD = 0.75$ grouped by intervals of 50 seconds

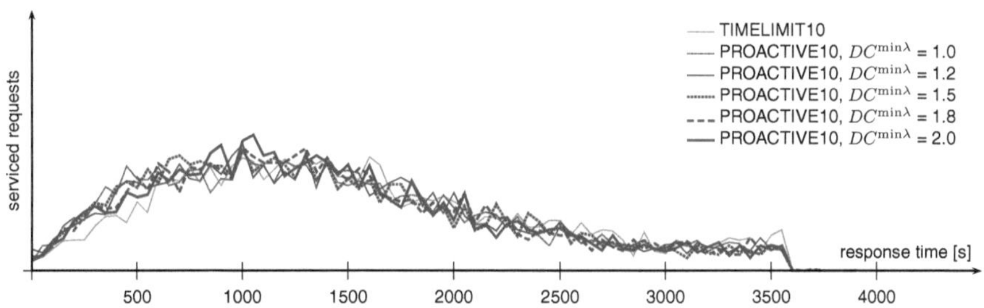

Fig. C.14 Request response times on S^{GEN} using 8 vehicles and $DC^{\min\lambda} = 1.0, 1.2, 1.5, 1.8$, and 2.0, quadratic customer inconvenience, $RD = 0.75$ and $TD = 0.75$ grouped by intervals of 50 seconds

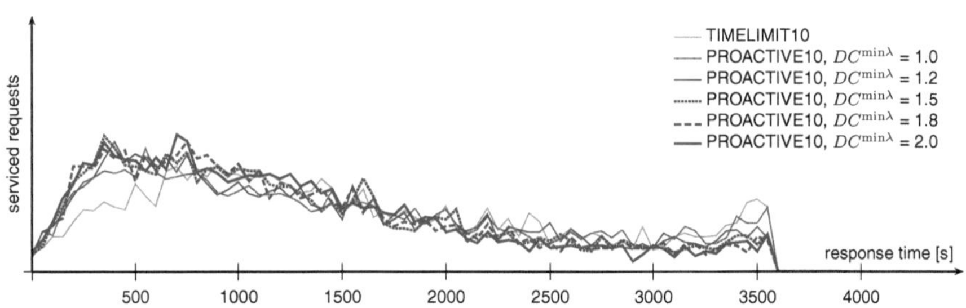

Fig. C.15 Request response times on S^{GEN} using 8 vehicles and $DC^{\min\lambda} = 1.0, 1.2, 1.5, 1.8$, and 2.0, linear2X customer inconvenience, $RD = 1.00$ and $TD = 1.00$ grouped by intervals of 50 seconds

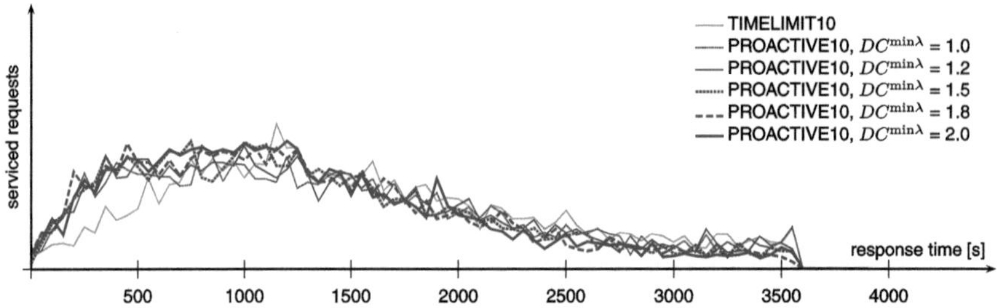

Fig. C.16 Request response times on S^{GEN} using 8 vehicles and $DC^{\min\lambda} = 1.0$, 1.2, 1.5, 1.8, and 2.0, quadratic customer inconvenience, $RD = 1.00$ and $TD = 1.00$ grouped by intervals of 50 seconds

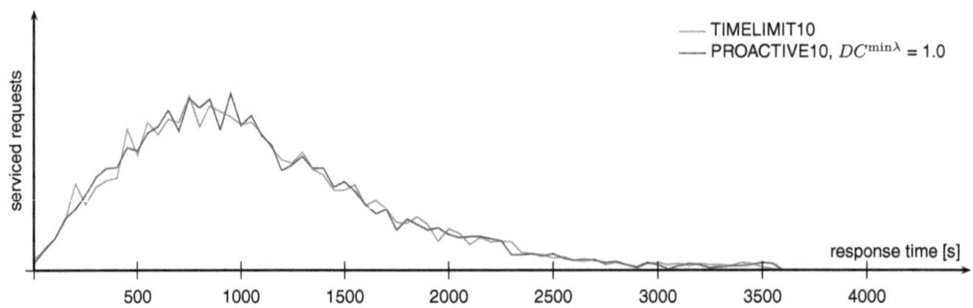

Fig. C.17 Request response times on S^{GEN} using 10 vehicles and $DC^{\min\lambda} = 1.0$, 1.2, 1.5, 1.8, and 2.0, linear2X customer inconvenience, $RD = 0.00$ and $TD = 0.00$ grouped by intervals of 50 seconds

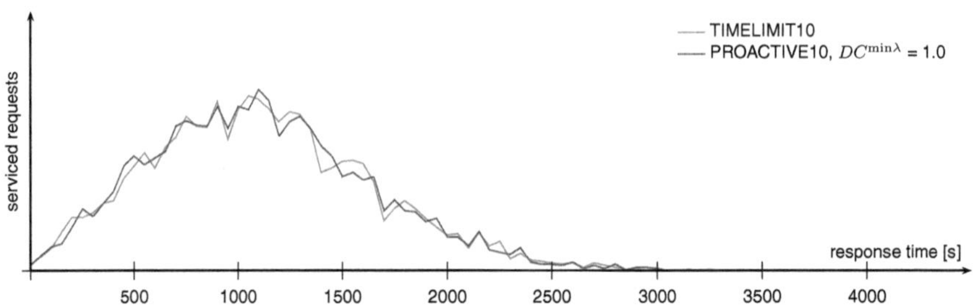

Fig. C.18 Request response times on S^{GEN} using 10 vehicles and $DC^{\min\lambda} = 1.0$, 1.2, 1.5, 1.8, and 2.0, quadratic customer inconvenience, $RD = 0.00$ and $TD = 0.00$ grouped by intervals of 50 seconds

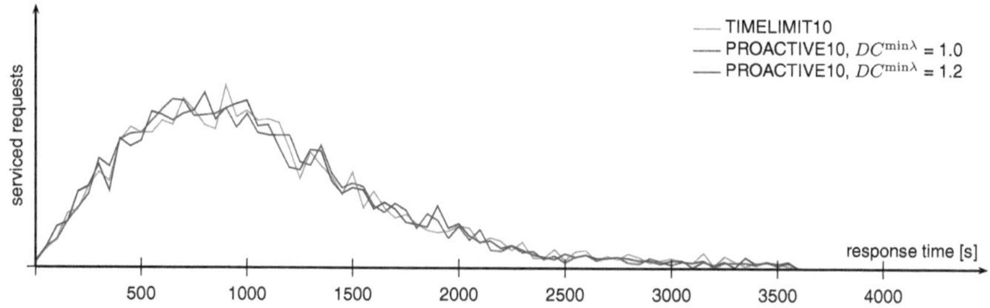

Fig. C.19 Request response times on S^{GEN} using 10 vehicles and $DC^{\min\lambda} = 1.0, 1.2, 1.5, 1.8,$ and 2.0, linear2X customer inconvenience, $RD = 0.25$ and $TD = 0.25$ grouped by intervals of 50 seconds

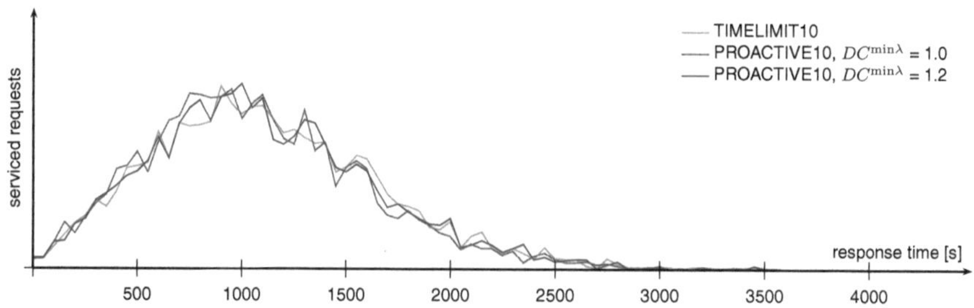

Fig. C.20 Request response times on S^{GEN} using 10 vehicles and $DC^{\min\lambda} = 1.0, 1.2, 1.5, 1.8,$ and 2.0, quadratic customer inconvenience, $RD = 0.25$ and $TD = 0.25$ grouped by intervals of 50 seconds

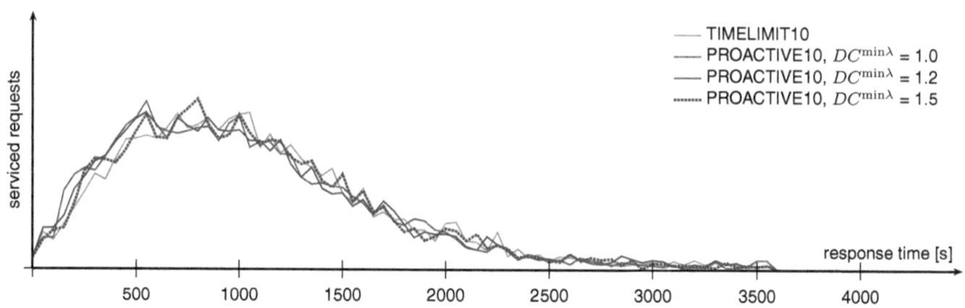

Fig. C.21 Request response times on S^{GEN} using 10 vehicles and $DC^{\min\lambda} = 1.0, 1.2, 1.5, 1.8,$ and 2.0, linear2X customer inconvenience, $RD = 0.50$ and $TD = 0.50$ grouped by intervals of 50 seconds

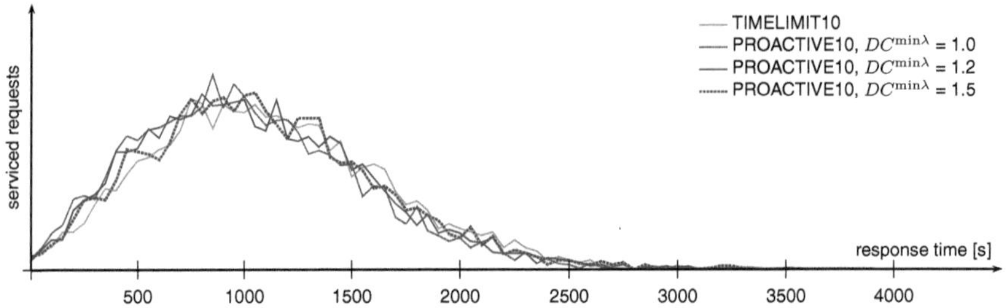

Fig. C.22 Request response times on S^{GEN} using 10 vehicles and $DC^{\min\lambda} = 1.0$, 1.2, 1.5, 1.8, and 2.0, quadratic customer inconvenience, $RD = 0.50$ and $TD = 0.50$ grouped by intervals of 50 seconds

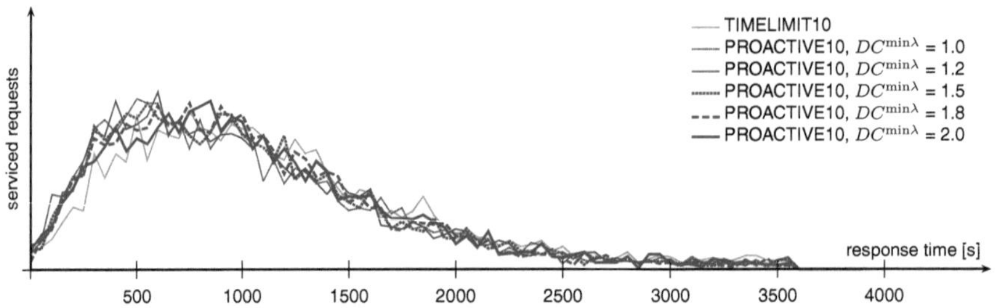

Fig. C.23 Request response times on S^{GEN} using 10 vehicles and $DC^{\min\lambda} = 1.0$, 1.2, 1.5, 1.8, and 2.0, linear2X customer inconvenience, $RD = 0.75$ and $TD = 0.75$ grouped by intervals of 50 seconds

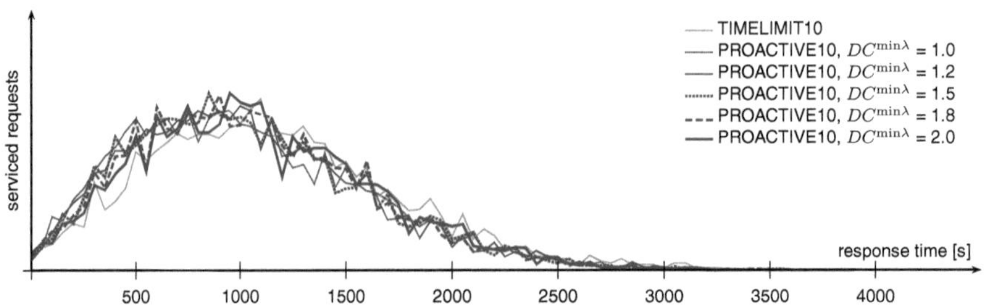

Fig. C.24 Request response times on S^{GEN} using 10 vehicles and $DC^{\min\lambda} = 1.0$, 1.2, 1.5, 1.8, and 2.0, quadratic customer inconvenience, $RD = 0.75$ and $TD = 0.75$ grouped by intervals of 50 seconds

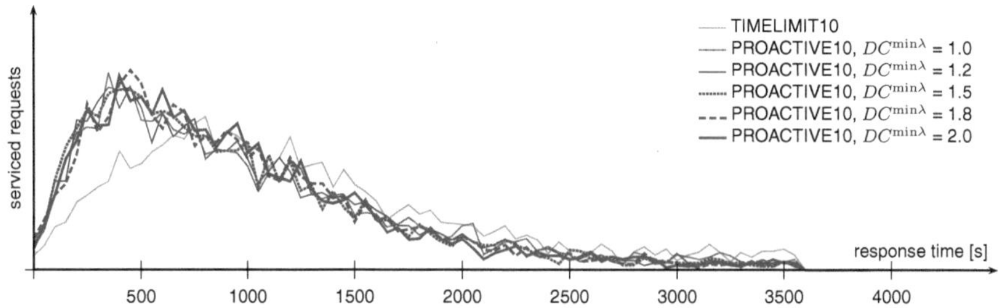

Fig. C.25 Request response times on S^{GEN} using 10 vehicles and $DC^{\min\lambda} = 1.0$, 1.2, 1.5, 1.8, and 2.0, linear2X customer inconvenience, $RD = 1.00$ and $TD = 1.00$ grouped by intervals of 50 seconds

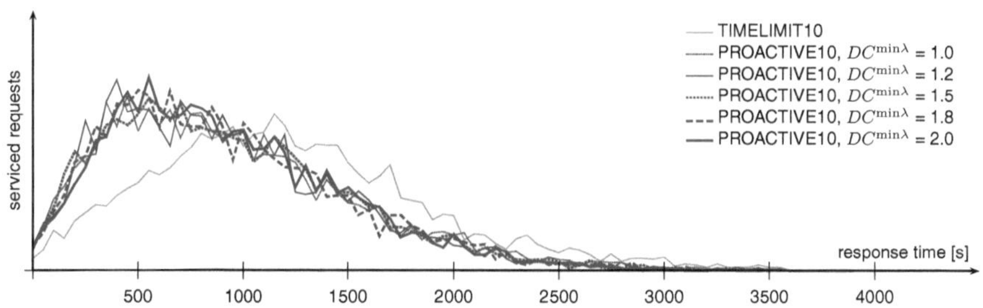

Fig. C.26 Request response times on S^{GEN} using 10 vehicles and $DC^{\min\lambda} = 1.0$, 1.2, 1.5, 1.8, and 2.0, quadratic customer inconvenience, $RD = 1.00$ and $TD = 1.00$ grouped by intervals of 50 seconds

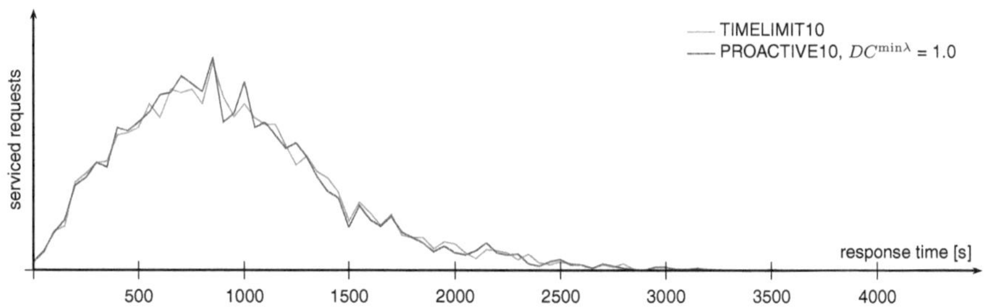

Fig. C.27 Request response times on S^{GEN} using 12 vehicles and $DC^{\min\lambda} = 1.0$, 1.2, 1.5, 1.8, and 2.0, linear2X customer inconvenience, $RD = 0.00$ and $TD = 0.00$ grouped by intervals of 50 seconds

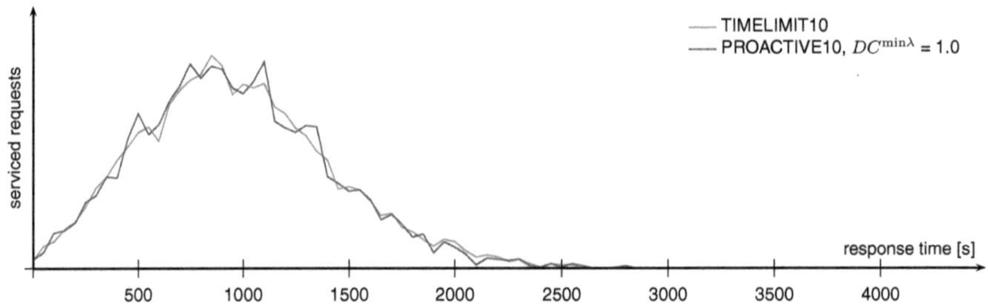

Fig. C.28 Request response times on S^{GEN} using 12 vehicles and $DC^{\min\lambda} = 1.0, 1.2, 1.5, 1.8,$ and 2.0, quadratic customer inconvenience, $RD = 0.00$ and $TD = 0.00$ grouped by intervals of 50 seconds

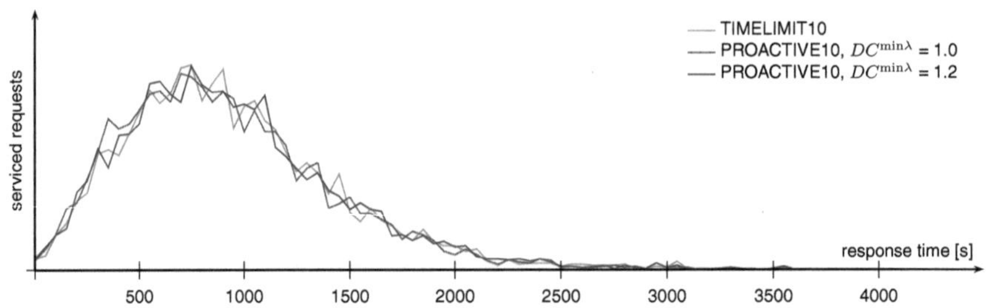

Fig. C.29 Request response times on S^{GEN} using 12 vehicles and $DC^{\min\lambda} = 1.0, 1.2, 1.5, 1.8,$ and 2.0, linear2X customer inconvenience, $RD = 0.25$ and $TD = 0.25$ grouped by intervals of 50 seconds

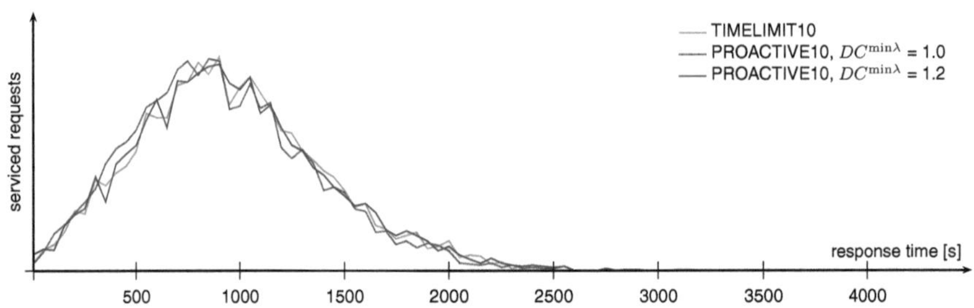

Fig. C.30 Request response times on S^{GEN} using 12 vehicles and $DC^{\min\lambda} = 1.0, 1.2, 1.5, 1.8,$ and 2.0, quadratic customer inconvenience, $RD = 0.25$ and $TD = 0.25$ grouped by intervals of 50 seconds

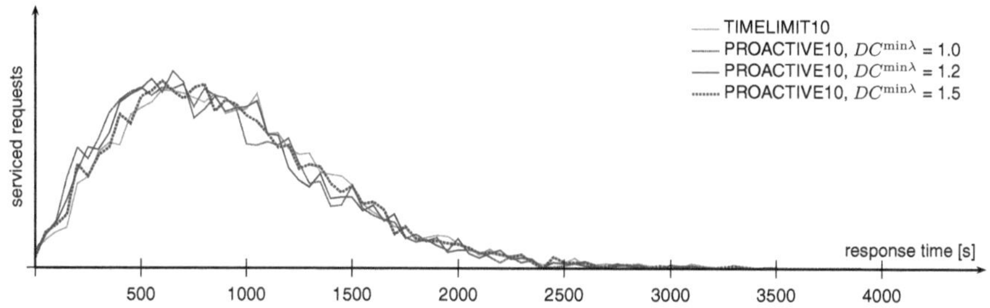

Fig. C.31 Request response times on S^{GEN} using 12 vehicles and $DC^{\min\lambda} = 1.0, 1.2, 1.5, 1.8,$ and 2.0, linear2X customer inconvenience, $RD = 0.50$ and $TD = 0.50$ grouped by intervals of 50 seconds

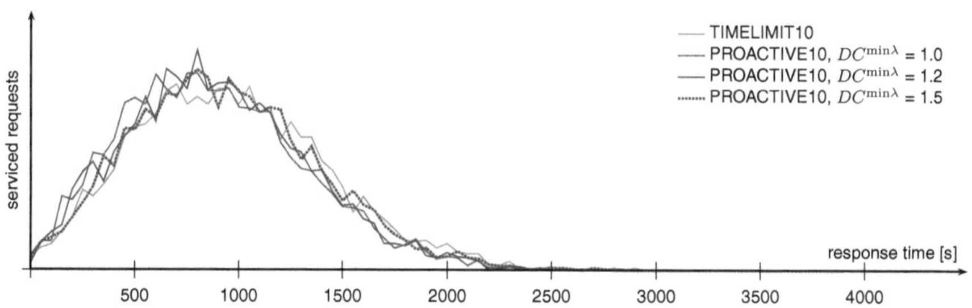

Fig. C.32 Request response times on S^{GEN} using 12 vehicles and $DC^{\min\lambda} = 1.0, 1.2, 1.5, 1.8,$ and 2.0, quadratic customer inconvenience, $RD = 0.50$ and $TD = 0.50$ grouped by intervals of 50 seconds

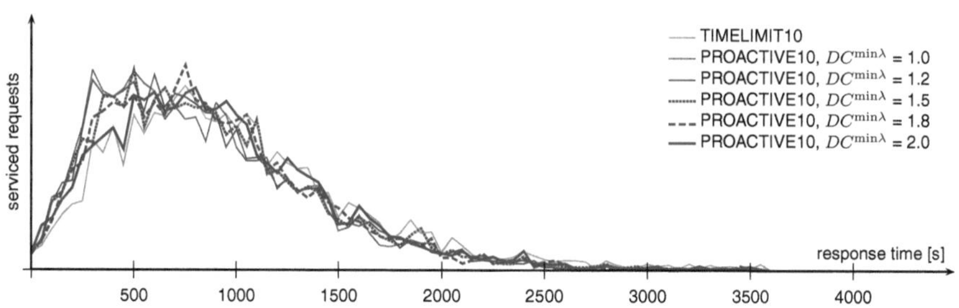

Fig. C.33 Request response times on S^{GEN} using 12 vehicles and $DC^{\min\lambda} = 1.0, 1.2, 1.5, 1.8,$ and 2.0, linear2X customer inconvenience, $RD = 0.75$ and $TD = 0.75$ grouped by intervals of 50 seconds

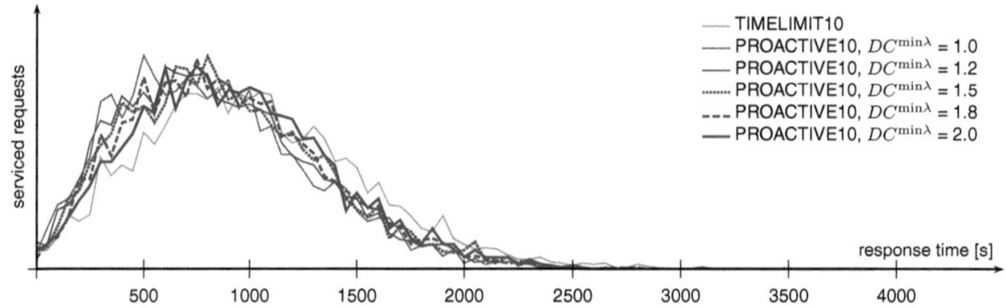

Fig. C.34 Request response times on S^{GEN} using 12 vehicles and $DC^{\min\lambda} = 1.0, 1.2, 1.5, 1.8,$ and 2.0, quadratic customer inconvenience, $RD = 0.75$ and $TD = 0.75$ grouped by intervals of 50 seconds

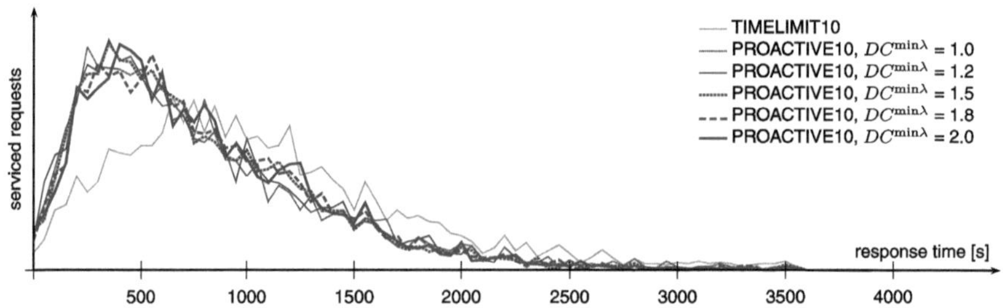

Fig. C.35 Request response times on S^{GEN} using 12 vehicles and $DC^{\min\lambda} = 1.0, 1.2, 1.5, 1.8,$ and 2.0, linear2X customer inconvenience, $RD = 1.00$ and $TD = 1.00$ grouped by intervals of 50 seconds

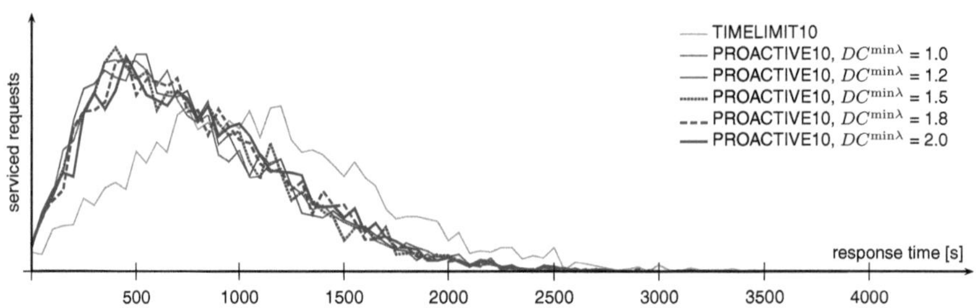

Fig. C.36 Request response times on S^{GEN} using 12 vehicles and $DC^{\min\lambda} = 1.0, 1.2, 1.5, 1.8,$ and 2.0, quadratic customer inconvenience, $RD = 1.00$ and $TD = 1.00$ grouped by intervals of 50 seconds

About the Author

Francesco Ferrucci was born in 1980 in Herford, Germany. He studied Business Computing at the University of Paderborn and received his Diploma Degree (with distinction) in 2006. During his time as a PhD student, he wrote different publications and visited the CIRRELT in Montreal for three months where he jointly worked with Professor Michel Gendreau. He received his PhD (summa cum laude) under the guidance of Professor Stefan Bock in 2012. His research interests include dynamic routing problems, especially real-time control approaches and forecasting approaches.

F. Ferrucci, *Pro-active Dynamic Vehicle Routing*, Contributions to Management Science,
DOI 10.1007/978-3-642-33472-6, © Springer-Verlag Berlin Heidelberg 2013

References

Aberle, G. (2009). *Transportwirtschaft: Einzelwirtschaftliche und gesamtwirtschaftliche Grundlagen (Wolls Lehr- und Handbücher der Wirtschafts- und Sozialwissenschaften)* (5th ed.). München: Oldenbourg Wissenschaftsverlag. ISBN 3486579517.

Ahuja, R. K., Magnanti, T. L., & Orlin, J. B. (1993). *Network flows: theory, algorithms, and applications*. Englewood Cliffs: Prentice-Hall.

Angelelli, E., Savelsbergh, M. W. P., & Grazia Speranza, M. (2007). Competitive analysis of a dispatch policy for a dynamic multi-period routing problem. *Operations Research Letters*, *35*(6), 713–721.

Ascheuer, N., Krumke, S. O., & Rambau, J. (2000). Online dial-a-ride problems: minimizing the completion time. In H. Reichel & S. Tison (Eds.), *Lecture notes in computer science: Vol. 1770. STACS 2000* (pp. 639–650). Berlin: Springer.

Attanasio, A., Bregman, J., Ghiani, G., & Manni, E. (2007). Real-time fleet management at Ecourier Ltd. In V. Zeimpekis, G. M. Giaglis, I. Minis & C. D. Tarantilis (Eds.), *Operations research/computer science interfaces series: Vol. 38. Dynamic fleet management* (pp. 219–238). Boston: Springer. ISBN 978-0-387-71722-7.

Badeau, P., Guertin, F., Gendreau, M., Potvin, J.-Y., & Taillard, É. D. (1997). A parallel tabu search heuristic for the vehicle routing problem with time windows. *Transportation Research. Part C, Emerging Technologies*, *5*(2), 109–122.

Baldacci, R., Toth, P., & Vigo, D. (2007). Recent advances in vehicle routing exact algorithms. *4OR: A Quarterly Journal of Operations Research*, *5*(4), 269–298.

Baldacci, R., Bartolini, E., & Laporte, G. (2010). Some applications of the generalized vehicle routing problem. *Journal of the Operational Research Society*, *61*(7), 1072–1077.

Barceló, J., Grzybowska, H., & Pardo, S. (2007). Vehicle routing and scheduling models, simulation and city logistics. In V. Zeimpekis, G. M. Giaglis, I. Minis & C. D. Tarantilis (Eds.), *Operations research/computer science interfaces series: Vol. 38. Dynamic fleet management* (pp. 163–195). Boston: Springer. ISBN 978-0-387-71722-7.

Barnhart, C., Johnson, E. L., Nemhauser, G. L., Savelsbergh, M. W. P., & Vance, P. H. (1998). Branch-and-price: column generation for solving huge integer programs. *Operations Research*, *46*(3), 316–329.

Bauer, R., Delling, D., Sanders, P., Schieferdecker, D., Schultes, D., & Wagner, D. (2008). Combining hierarchical and goal-directed speed-up techniques for Dijkstra's algorithm. In C. McGeoch (Ed.), *Lecture notes in computer science: Vol. 5038. Experimental algorithms* (pp. 303–318). Berlin: Springer.

Bektas, T. (2006). The multiple traveling salesman problem: an overview of formulations and solution procedures. *Omega*, *34*(3), 209–219.

Bell, W. J., Dalberto, L. M., Fisher, M. L., Greenfield, A. J., Jaikumar, R., Kedia, P., Mack, R. G., & Prutzman, P. J. (1983). Improving the distribution of industrial gases with an on-line comput-

DOI 10.1007/978-3-642-33472-6,

erized routing and scheduling optimizer. *Interfaces*, *13*(6), 4–23.

Bellman, R. (1954). The theory of dynamic programming. *Bulletin, New Series, of the American Mathematical Society*, *60*(6), 503–515.

Bellman, R. E. (2003). *Dover books on mathematics. Dynamic programming*. Mineola: Dover. ISBN 9780486428093.

Bent, R., & van Hentenryck, P. (2003). Dynamic vehicle routing with stochastic requests. In *Proceedings of the eighteenth international joint conference on artificial intelligence (IJCAI-2003)*, Acapulco, Mexico (pp. 1362–1363).

Bent, R., & van Hentenryck, P. (2004a). A two-stage hybrid local search for the vehicle routing problem with time windows. *Transportation Science*, *38*(4), 515–530.

Bent, R., & van Hentenryck, P. (2004b). Scenario-based planning for partially dynamic vehicle routing with stochastic customers. *Operations Research*, *52*(6), 977–987.

Bent, R., & van Hentenryck, P. (2004c). Regrets only! online stochastic optimization under time constraints. In *Proceedings of the nineteenth national conference on artificial intelligence (AAAI-2004)*, San Jose, California (pp. 501–506).

Bent, R., & van Hentenryck, P. (2005). Online stochastic optimization without distributions. In *Proceedings of the fifteenth international conference on automated planning and scheduling (ICAPS-2005)*, Monterey, California (pp. 171–180).

Bent, R., & van Hentenryck, P. (2007). Waiting and relocation strategies in online stochastic vehicle routing. In *Proceedings of the twentieth international joint conference on artificial intelligence (IJCAI 2007)*, Hyderabad, India (pp. 1816–1821).

Bent, R., Katriel, I., & van Hentenryck, P. (2005). Sub-optimality approximations. In P. van Beek (Ed.), *Lecture notes in computer science: Vol. 3709. Principles and practice of constraint programming—CP 2005* (pp. 122–136). Berlin: Springer.

Bent, R., van Hentenryck, P., & Upfal, E. (2010). Online stochastic optimization under time constraints. *Annals of Operations Research*, *177*(1), 151–183.

Berbeglia, G., Cordeau, J. F., & Laporte, G. (2010). Dynamic pickup and delivery problems. *European Journal of Operational Research*, *202*(1), 8–15.

Bertsimas, D. J., & van Ryzin, G. V. (1991). A stochastic and dynamic vehicle routing problem in the Euclidean plane. *Operations Research*, *39*(4), 601–615.

Bertsimas, D. J., & van Ryzin, G. V. (1993a). Stochastic and dynamic vehicle routing in the Euclidean plane with multiple capacitated vehicles. *Operations Research*, *41*(1), 60–76.

Bertsimas, D. J., & van Ryzin, G. V. (1993b). Stochastic and dynamic vehicle routing with general demand and interarrival time distributions. *Advances in Applied Probability*, *25*(4), 947–978.

BGL e.V. (2011). Marktanteile deutscher und ausländischer Lkw in Deutschland. http://www.bgl-ev.de/web/daten/grenz_lkwind.htm. Accessed on 2011-04-25.

Bianchi, L. (2000). *Notes on dynamic vehicle routing—the state of the art* (Technical report IDSIA-05-01). Manno-Lugano, Switzerland.

Bieding, T., Görtz, S., & Klose, A. (2009). On-line routing per mobile phone: a case on subsequent deliveries of newspapers. In J. A. E. E. van Nunen, M. G. Speranza & L. Bertazzi (Eds.), *Lecture notes in economics and mathematical systems: Vol. 619. Innovations in distribution logistics* (pp. 29–51). Berlin: Springer. ISBN 978-3-540-92944-4.

Biethahn, J., Hummeltenberg, W., Schmidt, B., Stähly, P., & Witte, T. (1999). *Simulation als betriebliche Entscheidungshilfe: State of the Art und neuere Entwicklungen*. Heidelberg: Physica-Verlag. ISBN 3790811785.

Bloos, M., Schönberger, J., & Kopfer, H. (2009). Supporting cooperative demand fulfillment in supply networks using autonomous control and multi-agent-systems. In S. Fischer, E. Maehle, & R. Reischuk (Eds.), *Informatik 2009: Im Focus das Leben. Proceedings of the 39th annual conference of the Gesellschaft für Informatik e.V. (GI)*, 28.9.–2.10.2009, Lübeck (pp. 3590–3604). ISBN 978-3-88579-248-2.

BMW Group (2011). Annual Report 2010. http://annual-report.bmwgroup.com/2010/gb/files/pdf/en/BMW_Group_AR2010.pdf. Accessed on 2011-05-12.

Bock, S. (2004). *Echtzeitfähige Steuerung von Speditionsnetzwerken: Nutzung moderner Informations- und Kommunikationstechnologien zur effizienten Durchführung von Transporten:*

Habilitationsschrift Universität Paderborn, 2004 (1st ed.). Wiesbaden: Deutscher Universitäts-Verlag. Gabler Edition Wissenschaft: Schriften zur quantitativen Betriebswirtschaftlehre. ISBN 3-8244-8214-2.

Bock, S. (2010). Real-time control of freight forwarder transportation networks by integrating multimodal transport chains. *European Journal of Operational Research*, *200*(3), 733–746.

Bodin, L., & Golden, B. (1981). Classification in vehicle routing and scheduling. *Networks*, *11*(2), 97–108.

Borndörfer, R., Grötschel, M., Klostermeier, F., & Küttner, C. (1997). *Telebus Berlin: vehicle scheduling in a dial-a-ride system* (Technical Report SC 97-23). Konrad-Zuse-Zentrum für Informationstechnik Berlin.

Brady, M. K., & Cronin Jr., J. J. (2001). Some new thoughts on conceptualizing perceived service quality: a hierarchical approach. *Journal of Marketing*, *65*(3), 34–49.

Branchini, R. M., Vinícius, A. A., & Løkketangen, A. (2009). Adaptive granular local search heuristic for a dynamic vehicle routing problem. *Computers & Operations Research*, *36*(11), 2955–2968.

Brandão, J., & Mercer, A. (1997). A tabu search algorithm for the multi-trip vehicle routing and scheduling problem. *European Journal of Operational Research*, *100*(1), 180–191.

Branke, J., Middendorf, M., Noeth, G., & Dessouky, M. (2005). Waiting strategies for dynamic vehicle routing. *Transportation Science*, *39*(3), 298–312.

Bräysy, O., & Gendreau, M. (2005a). Vehicle routing problem with time windows, Part I: route construction and local search algorithms. *Transportation Science*, *39*(1), 104–118.

Bräysy, O., & Gendreau, M. (2005b). Vehicle routing problem with time windows, Part II: metaheuristics. *Transportation Science*, *39*(1), 119–139.

Brotcorne, L., Laporte, G., & Semet, F. (2003). Ambulance location and relocation models. *European Journal of Operational Research*, *147*(3), 451–463.

Cathey, F. W., & Dailey, D. J. (2003). A prescription for transit arrival/departure prediction using automatic vehicle location data. *Transportation Research. Part C, Emerging Technologies*, *11*(3-4), 241–264.

Chang, H., Givan, R., & Chong, E. (2000). On-line scheduling via sampling. In *Artificial intelligence planning and scheduling (AIPS)*, Breckenridge, Colorado (pp. 62–71).

Christiansen, C. H., & Lysgaard, J. (2007). A branch-and-price algorithm for the capacitated vehicle routing problem with stochastic demands. *Operations Research Letters*, *35*(6), 773–781.

Cordeau, J. F., & Laporte, G. (2003). The dial-a-ride problem (DARP): variants, modeling issues and algorithms. *4OR: A Quarterly Journal of Operations Research*, *1*(2), 89–101.

Cordeau, J. F., & Laporte, G. (2007). The dial-a-ride problem: models and algorithms. *Annals of Operations Research*, *153*(1), 29–46.

Cordeau, J. F., Gendreau, M., Hertz, A., Laporte, G., & Sormany, J. S. (2005). New heuristics for the vehicle routing problem. In A. Langevin & D. Riopel (Eds.), *Logistics systems: design and optimization* (pp. 279–297). New York: Springer. ISBN 978-0-387-24977-3.

Cordeau, J. F., Laporte, G., Potvin, J.-Y., & Savelsbergh, M. W. P. (2007a). Transportation on demand. In C. Barnhart (Ed.), *Handbooks in operations research and management science: Vol. 14. Transportation* (pp. 429–466). Amsterdam: Elsevier/North-Holland. ISBN 0444513469.

Cordeau, J. F., Laporte, G., Savelsbergh, M. W. P., & Vigo, D. (2007b). Vehicle routing. In C. Barnhart (Ed.), *Handbooks in operations research and management science: Vol. 14. Transportation* (pp. 367–428). Amsterdam: Elsevier/North-Holland. ISBN 0444513469.

Cortés, C. E., Núñez, A., & Sáez, D. (2008). Hybrid adaptive predictive control for a dynamic pickup and delivery problem including traffic congestion. *International Journal of Adaptive Control and Signal Processing*, *22*(2), 103–123.

Cortés, C. E., Sáez, D., Núñez, A., & Muðoz-Carpintero, D. (2009). Hybrid adaptive predictive control for a dynamic pickup and delivery problem. *Transportation Science*, *43*(1), 27–42.

Crainic, T. G., & Kim, K. H. (2007). Intermodal transportation. In C. Barnhart (Ed.), *Handbooks in operations research and management science: Vol. 14. Transportation* (pp. 467–537). Amsterdam: Elsevier/North-Holland. ISBN 0444513469.

Cubillos, C., Crawford, B., & Rodríguez, N. (2007). Distributed planning for the on-line dial-a-ride problem. In F. Preparata & Q. Fang (Eds.), *Lecture notes in computer science: Vol. 4613. Frontiers in algorithmics* (pp. 124–135). Berlin: Springer.

Daimler Group (2011). Annual Report 2010. http://www.daimler.com/Projects/c2c/channel/documents/1985489_Daimler_Annual_Report_2010.pdf. Accessed on 2011-05-12.

Dakin, R. J. (1965). A tree-search algorithm for mixed integer programming problems. *Computer Journal*, *8*(3), 250–255.

Dantzig, G., Fulkerson, R., & Johnson, S. (1954). Solution of a large-scale Traveling-Salesman Problem. *Journal of the Operations Research Society of America*, *2*(4), 393–410.

Dantzig, G. B., & Ramser, J. H. (1959). The truck dispatching problem. *Management Science*, *6*(1), 80–91.

Davis, M. M., & Maggard, M. J. (1990). An analysis of customer satisfaction with waiting times in a two-stage service process. *Journal of Operations Management*, *9*(3), 324–334.

de Oliveira, H. C. B., Rocha, G. M., de Souza, M. M., Ciscon, L. A., Borges, V. R., & Mateus, G. R. (2008). A vehicular waiting time heuristic for dynamic vehicle routing problem. In *Proceedings of the 2008 ACM symposium on applied computing (SAC '08)* (pp. 13–17). New York: ACM.

de Paepe, W. E. (2002). *Complexity results and competitive analysis for vehicle routing problems*. PhD Thesis, Eindhoven: Technische University of Eindhoven.

Desaulniers, G., Desrosiers, J., & Solomon, M. M. (2005). *GERAD 25th anniversary series. Column generation*. Boston: Springer. ISBN 0387254854.

Desrochers, M., & Laporte, G. (1991). Improvements and extensions to the Miller-Tucker-Zemlin subtour elimination constraints. *Operations Research Letters*, *10*(1), 27–36.

Desrochers, M., Lenstra, J. K., Savelsbergh, M. W. P., & Soumis, F. (1988). Vehicle routing with time windows: optimization and approximation. In *Vehicle routing: Methods and studies* (pp. 65–84). Amsterdam: North-Holland.

Desrosiers, J., Dumas, Y., Solomon, M. M., & Soumis, F. (1995). Time constrained routing and scheduling. In M. O. Ball, T. L. Magnanti, C. L. Monma & G. L. Nemhauser (Eds.), *Handbooks in operations research and management science: Vol. 8. Network routing* (pp. 35–139). Amsterdam: North-Holland. ISBN 978-0444821416.

Deutsch, P., & Gailly, J. L. (1996). RFC1950: ZLIB compressed data format specification version 3.3. http://www.ietf.org/rfc/rfc1950.txt. Accessed on 2009-03-04.

DHL Express Germany GmbH (2010). Export Services. http://www.dhl.de/en/express/export-services.html. Accessed on 2010-12-06.

Dijkstra, E. W. (1959). A note on two problems in connexion with graphs. *Numerische Mathematik*, *1*(1), 269–271.

Domschke, W., & Drexl, A. (2011). *Einführung in operations research* (8th ed.). Berlin: Springer. ISBN 3642181112.

Domschke, W., & Scholl, A. (2010). *Oldenbourgs Lehr- und Handbücher der Wirtschafts- und Sozialwissenschaften: Vol. 2. Rundreisen und Touren* (5th revised and extended ed.). München: Oldenbourg. ISBN 3486590936.

Dror, M. (1989). Vehicle routing with stochastic demands: properties and solution frameworks. *Transportation Science*, *23*(3), 166–176.

Dror, M. (2005). Vehicle routing with stochastic demands: models & computational methods. In M. Dror, P. L'Ecuyer & F. Szidarovszky (Eds.), *International series in operations research & management science: Vol. 46. Modeling uncertainty* (pp. 625–649). Boston: Springer. ISBN 978-0-306-48102-4.

Dror, M., Laporte, G., & Louveaux, F. V. (1993). Vehicle routing with stochastic demands and restricted failures. *Mathematical Methods of Operations Research*, *37*(3), 273–283.

Dumas, Y., Soumis, F., & Desrosiers, J. (1990). Technical note—optimizing the schedule for a fixed vehicle path with convex inconvenience costs. *Transportation Science*, *24*(2), 145–152.

Dumas, Y., Desrosiers, J., Gelinas, E., & Solomon, M. M. (1995). An optimal algorithm for the Traveling Salesman Problem with time windows. *Operations Research*, *43*(2), 367–371.

Edmonds, R. (2010). NAA Report: newspaper inserts "Under Siege". http://www.poynter.org/latest-news/buiness-news/the-biz-blog/100805/naa-report-newspaper-inserts-under-siege/. Accessed on 2011-05-18.

Ehmke, J. F., Meisel, S., Engelmann, S., & Mattfeld, D. C. (2009). Data chain management for planning in city logistics. *International Journal of Data Mining, Modelling and Management, 1*(4), 335–356.

Ehmke, J. F., Meisel, S., & Mattfeld, D. C. (2010). Floating car data based analysis of urban travel times for the provision of traffic quality. In J. Barceló & M. Kuwahara (Eds.), *International series in operations research & management science: Vol. 144. Traffic data collection and its standardization* (pp. 129–149). New York: Springer. ISBN 978-1-4419-6070-2.

Eiselt, H. A., Gendreau, M., & Laporte, G. (1995). Arc routing problems, part I: the Chinese postman problem. *Operations Research, 43*(2), 231–242.

Eksioglu, B., Vural, A. V., & Reisman, A. (2009). The vehicle routing problem: a taxonomic review. *Computers & Industrial Engineering, 57*(4), 1472–1483.

El-Sherbeny, N. A. (2010). Vehicle routing with time windows: an overview of exact, heuristic and metaheuristic methods. *Journal of King Saud University: Computer and Information Sciences, 22*(3), 123–131.

European Commission (Ed.) (2010). *European union energy & transport in figures: Statistical pocketbook 2010*. Luxembourg: Publications office of the European communities. ISBN 978-92-79-13815-7.

Fagerholt, K. (2000). Evaluating the trade-off between the level of customer service and transportation costs in a ship scheduling problem. *Maritime Policy and Management, 27*(2), 145–153.

Fagerholt, K. (2001). Ship scheduling with soft time windows: an optimisation based approach. *European Journal of Operational Research, 131*(3), 559–571.

Faria, J. A., Nunes, E., & Matos, M. A. (2010). Cost and quality of service analysis of production systems based on the cumulative downtime. *International Journal of Production Research, 48*(6), 1653–1684.

Ferrucci, F. (2006). *Entwicklung und Validierung einer Echtzeitsteuerung für realitätsnahe Pickup-and-Delivery-Probleme*. Diploma Thesis, University of Paderborn.

Ferrucci, F., & Bock, S. (2008). Statische Analyse des Optimierungspotenzials einer Nachlieferungssteuerung für Tageszeitungen. In D. C. Mattfeld, S. Voß, L. Suhl & H.-O. Günther (Eds.), *DSOR Beiträge zur Wirtschaftsinformatik: Vol. 5. Informations- und Kommunikationssysteme in Supply Chain Management, Logistik und Transport*. Norderstedt: Books on Demand. ISBN 978-3-8334-8931-0.

Ferrucci, F., Bock, S., & Gendreau, M. (2013). A pro-active real-time control approach for dynamic vehicle routing problems dealing with the delivery of urgent goods. *European Journal of Operational Research, 225*(1), 130–141.

Fischer, K., Müller, J. P., & Pischel, M. (1996). Cooperative transportation scheduling: an application domain for DAI. *Applied Artificial Intelligence, 10*(1), 1–34.

Fisher, M. L. (2004). The Lagrangian relaxation method for solving integer programming problems. *Management Science, 50*(12), 1861–1871.

Flatberg, T., Hasle, G., Kloster, O., Nilssen, E. J., & Riise, A. (2005). *Dynamic and stochastic aspects in vehicle routing: a literature survey* (Technical report STF90A05413 SINTEF). Oslo, Norway.

Fleischmann, B., Gietz, M., & Gnutzmann, S. (2004a). Time-varying travel times in vehicle routing. *Transportation Science, 38*(2), 160–173.

Fleischmann, B., Gnutzmann, S., & Sandvoß, E. (2004b). Dynamic vehicle routing based on online traffic information. *Transportation Science, 38*(4), 420–433.

Fukasawa, R., Longo, H., Lysgaard, J., Aragão, M. P., Reis, M., Uchoa, E., & Werneck, R. F. (2006). Robust branch-and-cut-and-price for the capacitated vehicle routing problem. *Mathematical Programming, 106*(3), 491–511.

Garey, M. R., & Johnson, D. S. (1978). "Strong" NP-completeness results: motivation, examples, and implications. *Journal of the ACM, 25*(3), 499–508.

Garey, M. R., & Johnson, D. S. (1979). *A series of books in the mathematical sciences. Computers and intractability: a guide to the theory of NP-completeness* (1st ed.). New York: Freeman. ISBN 0716710455.

Gass, S. I. (2003). *Linear programming: methods and applications* (5th ed.). New York: Dover. ISBN 048643284X.

Gendreau, M., & Potvin, J.-Y. (1998). Dynamic vehicle routing and dispatching. In T. G. Crainic (Ed.), *Center for research on transportation 25th anniversary series, 1971–1996. Fleet management and logistics* (pp. 115–126). Heidelberg: Springer. ISBN 0792381610.

Gendreau, M., Laporte, G., & Séguin, R. (1996). Stochastic vehicle routing. *European Journal of Operational Research, 88*(1), 3–12.

Gendreau, M., Guertin, F., Potvin, J.-Y., & Séguin, R. (1998). *Neighborhood search heuristics for a dynamic vehicle dispatching problem with pick-ups and deliveries* (Technical Report CRT-98-10). Centre de Recherche sur les Transport (CRT), Universite de Montreal.

Gendreau, M., Guertin, F., Potvin, J.-Y., & Taillard, É. D. (1999a). Parallel tabu search for real-time vehicle routing and dispatching. *Transportation Science, 33*(4), 381–390.

Gendreau, M., Laporte, G., Musaraganyi, C., & Taillard, É. D. (1999b). A tabu search heuristic for the heterogeneous fleet vehicle routing problem. *Computers & Operations Research, 26*(12), 1153–1173.

Gendreau, M., Laporte, G., & Semet, F. (2001). A dynamic model and parallel tabu search heuristic for real-time ambulance relocation. *Parallel Computing, 27*(12), 1641–1653.

Gendreau, M., Guertin, F., Potvin, J.-Y., & Séguin, R. (2006). Neighborhood search heuristics for a dynamic vehicle dispatching problem with pick-ups and deliveries. *Transportation Research. Part C, Emerging Technologies, 14*(3), 157–174.

Gendreau, M., Potvin, J.-Y., Bräumlaysy, O., Hasle, G., & Løkketangen, A. (2008). Metaheuristics for the vehicle routing problem and its extensions: a categorized bibliography. In B. L. Golden, S. Raghavan & E. A. Wasil (Eds.), *Operations research/computer science interfaces series: Vol. 43. The vehicle routing problem: latest advances and new challenges* (pp. 143–169). Heidelberg: Springer. ISBN 0387777776.

Ghiani, G., Guerriero, F., Laporte, G., & Musmanno, R. (2003). Real-time vehicle routing: solution concepts, algorithms and parallel computing strategies. *European Journal of Operational Research, 151*(1), 1–11.

Ghiani, G., Laporte, G., & Musmanno, R. (2004). *Wiley-Interscience series in systems and optimization. Introduction to logistics systems planning and control.* Chichester: Wiley. ISBN 9780470849170.

Ghiani, G., Manni, E., Quaranta, A., & Triki, C. (2009). Anticipatory algorithms for same-day courier dispatching. *Transportation Research. Part E, Logistics and Transportation Review, 45*(1), 96–106.

Ghiani, G., Manni, E., & Thomas, B. W. (2012). A comparison of anticipatory algorithms for the dynamic and stochastic Traveling Salesman Problem. *Transportation Science, 46*(3), 374–387.

Giaglis, G. M., Minis, I., Tatarakis, A., & Zeimpekis, V. (2004). Minimizing logistics risk through real-time vehicle routing and mobile technologies. *International Journal of Physical Distribution & Logistics Management, 34*(9), 749–764.

Glover, F. (1989). Tabu search—part I. *ORSA Journal on Computing, 1*(3), 190–206.

Glover, F., & Laguna, M. (1998). *Tabu search* (4th ed.). Heidelberg: Springer. ISBN 0792381874.

Godfrey, G. A., & Powell, W. B. (2002). An adaptive dynamic programming algorithm for dynamic fleet management, I: single period travel times. *Transportation Science, 36*(1), 21–39.

Golden, B. L., & Stewart Jr., W. (1978). Vehicle routing with probabilistic demands. In *Computer science and statistics: tenth annual symposium on the interface: proceedings of the 10th annual symposium held at the National Bureau of Standards*, Gaithersburg, Maryland (pp. 252–259).

Golden, B. L., Raghavan, S., & Wasil, E. A. (Eds.) (2008). *Operations research/computer science interfaces series: Vol. 43. The vehicle routing problem: latest advances and new challenges.* Heidelberg: Springer. ISBN 0387777776.

Hadjiconstantinou, E., & Roberts, D. (2002). Routing under uncertainty: an application in the scheduling of field service engineers. In P. Toth & D. Vigo (Eds.), *SIAM monographs on dis-*

crete mathematics and applications. The vehicle routing problem (pp. 331–352). Philadelphia: Society for Industrial and Applied Mathematics. ISBN 0898714982.

Haghani, A., & Yang, S. (2007). Real-time emergency response fleet deployment: concepts, systems, simulation & case studies. In V. Zeimpekis, G. M. Giaglis, I. Minis & C. D. Tarantilis (Eds.), *Operations research/computer science interfaces series: Vol. 38. Dynamic fleet management* (pp. 133–162). Boston: Springer. ISBN 978-0-387-71722-7.

Haklay, M., & Weber, P. (2008). OpenStreetMap: user-generated street maps. *IEEE Pervasive Computing*, *7*(4), 12–18.

Halonen, T. (2003). *GSM, GPRS, and edge performance: evolution towards 3G/UMTS* (2nd ed.). Chichester: Wiley. ISBN 0470866942.

Hansen, P., & Mladenovic, N. (2001). Variable neighborhood search: principles and applications. *European Journal of Operational Research*, *130*(3), 449–467.

Held, M., & Karp, R. M. (1970). The Traveling-Salesman Problem and minimum spanning trees. *Operations Research*, *18*(6), 1138–1162.

Held, M., & Karp, R. M. (1971). The Traveling-Salesman Problem and minimum spanning trees: Part II. *Mathematical Programming*, *1*(1), 6–25.

Hillier, F. S., & Lieberman, G. J. (2008). *Introduction to operations research* (8th ed.). New York: McGraw-Hill. ISBN 007123828X.

Homberger, J., & Gehring, H. (2005). A two-phase hybrid metaheuristic for the vehicle routing problem with time windows. *European Journal of Operational Research*, *162*(1), 220–238.

Hoos, H. H., & Stützle, T. (2005). *Stochastic local search: foundations and applications*. San Francisco: Morgan Kaufmann. ISBN 1558608729.

Horn, M. E. T. (2002a). Fleet scheduling and dispatching for demand-responsive passenger services. *Transportation Research. Part C, Emerging Technologies*, *10*(1), 35–63.

Horn, M. E. T. (2002b). Multi-modal and demand-responsive passenger transport systems: a modelling framework with embedded control systems. *Transportation Research. Part A, Policy and Practice*, *36*(2), 167–188.

Hülsmann, M., Kopfer, H., Cordes, P., & Bloos, M. (2009). Collaborative transportation planning in complex adaptive logistics systems: a complexity science-based analysis of decision-making problems of "groupage systems". In J. Zhou (Ed.), *Lecture notes of the institute for computer sciences, social informatics and telecommunications engineering: Vol. 4. Complex sciences* (pp. 1160–1166). Berlin: Springer. ISBN 978-3-642-02466-5.

Hvattum, L. M., Løkketangen, A., & Laporte, G. (2006). Solving a dynamic and stochastic vehicle routing problem with a sample scenario hedging heuristic. *Transportation Science*, *40*(4), 421–438.

Hvattum, L. M., Løkketangen, A., & Laporte, G. (2007). A branch-and-regret heuristic for stochastic and dynamic vehicle routing problems. *Networks*, *49*(4), 330–340.

Ichoua, S. (2001). *Problèmes de gestion de flottes de véhicules en temps réel*. PhD thesis, Université de Montréal, Montréal, Canada.

Ichoua, S., Gendreau, M., & Potvin, J.-Y. (2000). Diversion issues in real-time vehicle dispatching. *Transportation Science*, *34*(4), 426–438.

Ichoua, S., Gendreau, M., & Potvin, J.-Y. (2003). Vehicle dispatching with time-dependent travel times. *European Journal of Operational Research*, *144*(2), 379–396.

Ichoua, S., Gendreau, M., & Potvin, J.-Y. (2006). Exploiting knowledge about future demands for real-time vehicle dispatching. *Transportation Science*, *40*(2), 211–225.

Ichoua, S., Gendreau, M., & Potvin, J.-Y. (2007). Planned route optimization for real-time vehicle routing. In V. Zeimpekis, G. M. Giaglis, I. Minis & C. D. Tarantilis (Eds.), *Operations research/computer science interfaces series: Vol. 38. Dynamic fleet management* (pp. 1–18). Boston: Springer. ISBN 978-0-387-71722-7.

Irani, S., Lu, X., & Regan, A. (2004). On-line algorithms for the dynamic traveling repair problem. *Journal of Scheduling*, *7*(3), 243–258.

Israel, D. (2008). *Data analysis in business research: a step-by-step nonparametric approach*. Los Angeles: SAGE. ISBN 9788178298757.

Jaillet, P. (1987). Stochastic routing problems. In G. Andreatta, F. Mason, & P. Serafini (Eds.), *Advanced school on stochastics in combinatorial optimization* (pp. 197–213). Singapore: World Scientific.

Jézéquel, A. (1985). *Probabilistic vehicle routing problems*. M.Sc. Dissertation, Massachusetts Institute of Technology, Cambridge, MA.

Johnston, R., & Clark, G. (2008). *Service operations management: improving service delivery* (3rd ed.). Harlow: Prentice Hall. ISBN 9781405847322.

Jula, H., Dessouky, M., & Ioannou, P. A. (2008). Real-time estimation of travel times along the arcs and arrival times at the nodes of dynamic stochastic networks. *IEEE Transactions on Intelligent Transportation Systems*, *9*(1), 97–110.

Kaplan, E. D., & Hegarty, C. J. (2006). *Understanding GPS: principles and applications* (2nd ed.). Boston: Artech House. ISBN 1580538940.

Kendall, D. G. (1953). Stochastic processes occurring in the theory of queues and their analysis by the method of the imbedded Markov chain. *The Annals of Mathematical Statistics*, *24*(3), 338–354.

Kenyon, A. S., & Morton, D. P. (2003). Stochastic vehicle routing with random travel times. *Transportation Science*, *37*(1), 69–82.

Kilby, P., Prosser, P., & Shaw, P. (1998). *Dynamic VRPs: a study of scenarios* (Technical Report APES-06-1998). University of Strathclyde.

Klein Haneveld, W. K., & van der Vlerk, M. H. (1999). Stochastic integer programming: general models and algorithms. *Annals of Operations Research*, *85*, 39–57.

Kleywegt, A. J., Savelsbergh, M., & Uyar, E. (2009). *A dynamic stochastic routing problem* (Technical Report). Georgia Institute of Technology, Atlanta, GA.

Kohout, R., & Erol, K. (1999). In-time agent-based vehicle routing with a stochastic improvement heuristic. In *Proceedings of the 16th national conference on artificial intelligence and the 11th on innovative applications of artificial intelligence*, Orlando, USA (pp. 864–869).

Korte, B., & Vygen, J. (2012). *Algorithms and combinatorics: Vol. 21. Combinatorial optimization: theory and algorithms* (5th ed.). Berlin: Springer. ISBN 3642244874.

Kristensen, K., Kanji, G. K., & Dahlgaard, J. J. (1992). On measurement of customer satisfaction. *Total Quality Management & Business Excellence*, *3*(2), 123–128.

Krumke, S. O., Rambau, J., & Torres, L. M. (2002). Real-time dispatching of guided and unguided automobile service units with soft time windows. In R. Möhring & R. Raman (Eds.), *Lecture notes in computer science: Vol. 2461. Algorithms—ESA 2002* (pp. 417–424). Heidelberg: Springer.

Kvam, P. H., & Vidakovic, B. (2007). *Nonparametric statistics with applications to science and engineering. Wiley series in probability and statistics*. Hoboken: Wiley. ISBN 0470081473.

Kwan, M. K. (1962). Graphic programming using odd or even points. *Chinese Mathematics*, *1*, 273–277.

Lackner, A. (2004). *Göttinger Wirtschaftsinformatik: Vol. 47. Dynamische Tourenplanung mit ausgewählten Metaheuristiken: Eine Untersuchung am Beispiel des kapazitätsrestriktiven dynamischen Tourenplanungsproblems mit Zeitfenstern: Univ., Diss.–Göttingen, 2004* (1st ed.). Göttingen: Cuvillier. ISBN 3865370845.

Lambert, V., Laporte, G., & Louveaux, F. V. (1993). Designing collection routes through bank branches. *Computers & Operations Research*, *20*(7), 783–791.

Land, A. H., & Doig, A. G. (1960). An automatic method of solving discrete programming problems. *Econometrica*, *28*(3), 497–520.

Laporte, G. (1992a). The vehicle routing problem: an overview of exact and approximate algorithms. *European Journal of Operational Research*, *59*(3), 345–358.

Laporte, G. (1992b). The traveling salesman problem: an overview of exact and approximate algorithms. *European Journal of Operational Research*, *59*(2), 231–247.

Laporte, G., & Louveaux, F. (1990). Formulations and bounds for the stochastic capacitated vehicle routing problem with uncertain supplies. In J. Gabszewicz, J.-F. Richard & L. A. Wolsey (Eds.), *Economic decision-making* (pp. 443–455). Amsterdam: North-Holland. ISBN 044488422X.

Laporte, G., & Nobert, Y. (1980). A cutting planes algorithm for the m-Salesmen Problem. *Journal of the Operational Research Society*, *31*(11), 1017–1023.
Laporte, G., Louveaux, F., & Mercure, H. (1992). The vehicle routing problem with stochastic travel times. *Transportation Science*, *26*(3), 161–170.
Laporte, G., Louveaux, F. V., & van Hamme, L. (2002). An integer L-shaped algorithm for the capacitated vehicle routing problem with stochastic demands. *Operations Research*, *50*(3), 415–423.
Larsen, A. (2000). *The dynamic vehicle routing problem*. PhD Thesis, Technical University of Denmark, Lyngby, Denmark.
Larsen, A., Madsen, O. B. G., & Solomon, M. M. (2002). Partially dynamic vehicle routing—models and algorithms. *Journal of the Operational Research Society*, *53*(6), 637–646.
Larsen, A., Madsen, O. B. G., & Solomon, M. M. (2004). The a priori dynamic Traveling Salesman Problem with time windows. *Transportation Science*, *38*(4), 459–472.
Larsen, A., Madsen, O. B. G., & Solomon, M. M. (2007). Classification of dynamic vehicle routing systems. In V. Zeimpekis, G. M. Giaglis, I. Minis & C. D. Tarantilis (Eds.), *Operations research/computer science interfaces series: Vol. 38. Dynamic fleet management* (pp. 19–40). Boston: Springer. ISBN 978-0-387-71722-7.
Larsen, A., Madsen, O. B. G., & Solomon, M. M. (2008). Recent developments in dynamic vehicle routing systems. In B. L. Golden, S. Raghavan & E. A. Wasil (Eds.), *Operations research/computer science interfaces series: Vol. 43. The vehicle routing problem: latest advances and new challenges* (pp. 199–218). Heidelberg: Springer. ISBN 0387777776.
Larson, R. C., & Odoni, A. R. (1981). *Urban operations research*. Englewood Cliffs: Prentice-Hall. http://web.mit.edu/urban_or_book/www/book/.
Law, A. M., & Kelton, W. D. (2000). *McGraw-Hill series in industrial engineering and management science. Simulation modeling and analysis* (3rd ed.). Boston: McGraw-Hill. ISBN 0070592926.
Lenstra, J. K., & Rinnooy Kan, A. H. G. (1979). Computational complexity of discrete optimization problems. In P. L. Hammer, E. L. Johnson & B. H. Korte (Eds.), *Proceedings of the advanced research institute on discrete optimization and systems applications of the systems science panel of NATO and of the discrete optimization symposium: Vol. 4. Annals of discrete mathematics: discrete optimization I* (pp. 121–140). Amsterdam: Elsevier. ISBN 978-0-444-85322-6.
Lenstra, J. K., & Rinnooy Kan, A. H. G. (1981). Complexity of vehicle routing and scheduling problems. *Networks*, *11*(2), 221–227.
Lew, A., & Mauch, H. (2007). *Dynamic programming: a computational tool* (1st ed.). Berlin: Springer. ISBN 3642072003.
Li, J. Q., Mirchandani, P. B., & Borenstein, D. (2009a). Real-time vehicle rerouting problems with time windows. *European Journal of Operational Research*, *194*(3), 711–727.
Li, J. Q., Mirchandani, P. B., & Borenstein, D. (2009b). A Lagrangian heuristic for the real-time vehicle rescheduling problem. *Transportation Research. Part E, Logistics and Transportation Review*, *45*(3), 419–433.
Lin, S. (1965). Computer solutions of the traveling salesman problem. *The Bell System Technical Journal*, *44*(10), 2245–2269.
Little, J. D. C., Murty, K. G., Sweeney, D. W., & Karel, C. (1963). An algorithm for the Traveling Salesman Problem. *Operations Research*, *11*(6), 972–989.
Lund, K., Madsen, O. B. G., & Rygaard, J. M. (1996). *Vehicle routing with varying degrees of dynamism* (Technical report IMM-REP-1996-1). Institute of Mathematical Modelling, Technical University of Denmark, Lyngby, Denmark.
MacQueen, J. (1967). Some methods for classification and analysis of multivariate observations. In L. M. Le Cam & J. Neyman (Eds.), *Proceedings of the fifth Berkeley symposium on mathematical statistics and probability* (Vol. 1, pp. 281–297). Berkeley: University of California Press.
Malandraki, C., & Daskin, M. S. (1992). Time dependent vehicle routing problems: formulations, properties and heuristic algorithms. *Transportation Science*, *26*(3), 185–200.

Markus, K. G. (2009). *Online stochastic vehicle routing*. Master's Thesis, Department of Mathematics and Computer Science (IMADA), University of Southern Denmark.

Mendoza, J. E., Castanier, B., Guéret, C., Medaglia, A. L., & Velasco, N. (2009). *Constructive heuristics for the multi-compartment vehicle routing problem with stochastic demands* (Technical report). Universidad de los Andes.

Mihram, G. A. (1972). *Mathematics in science and engineering: Vol. 92. Simulation: statistical foundations and methodology*. New York: Academic Press. ISBN 0124959504.

Miller, C. E., Tucker, A. W., & Zemlin, R. A. (1960). Integer programming formulation of traveling salesman problems. *Journal of the ACM*, *7*(4), 326–329.

Mintsis, G., Basbas, S., Papaioannou, P., Taxiltaris, C., & Tziavos, I. N. (2004). Applications of GPS technology in the land transportation system: new technologies in transportation systems. *European Journal of Operational Research*, *152*(2), 399–409.

Mitchell, M. (1998). *An introduction to genetic algorithms*. Cambridge: MIT Press. ISBN 0262631857.

Mitrovic-Minic, S., & Laporte, G. (2004). Waiting strategies for the dynamic pickup and delivery problem with time windows. *Transportation Research. Part B: Methodological*, *38*(7), 635–655.

Mitrovic-Minic, S., Krishnamurti, R., & Laporte, G. (2004). Double-horizon based heuristics for the dynamic pickup and delivery problem with time windows. *Transportation Research. Part B: Methodological*, *38*(8), 669–685.

Moon, T. K. (2005). *Error correction coding: mathematical methods and algorithms*. Hoboken: Wiley-Interscience. ISBN 0471648000.

Mu, Q., Fu, Z., Lysgaard, J., & Eglese, R. (2010). Disruption management of the vehicle routing problem with vehicle breakdown. *Journal of the Operational Research Society*, *62*(4), 742–749.

Muyldermans, L., & Pang, G. (2010). On the benefits of co-collection: experiments with a multi-compartment vehicle routing algorithm. *European Journal of Operational Research*, *206*(1), 93–103.

Nanry, W. P., & Wesley Barnes, J. (2000). Solving the pickup and delivery problem with time windows using reactive tabu search. *Transportation Research. Part B: Methodological*, *34*(2), 107–121.

Newspaper Association of America (2011). Annual Advertising Expenditures. http://www.naa.org/Trends-and-Numbers/Advertising-Expenditures/Annual-All-Categories.aspx. Accessed on 2011-06-05.

Novoa, C., & Storer, R. (2009). An approximate dynamic programming approach for the vehicle routing problem with stochastic demands. *European Journal of Operational Research*, *196*(2), 509–515.

Novoa, C., Berger, R., Linderoth, J., & Storer, R. (2006). *A set-partitioning-based model for the stochastic vehicle routing problem* (Technical Report 06T-008). Industrial and Systems Engineering, Lehigh University, USA.

Núñez, A., Sáez, D., & Cortés, C. E. (2008). Hybrid predictive control for the vehicle dynamic routing problem based on Evolutionary Multiobjective Optimization (EMO). In *17th IFAC world congress*, Seoul, Korea.

OpenStreetMap (2011). The OpenStreetMap Project. http://www.openstreetmap.org. Accessed on 2011-06-20.

Or, I. (1976). *Traveling salesman-type combinatorial problems and their relation to the logistics of regional blood banking*. PhD Thesis, Northwestern University, Evanston.

Pankratz, G. (2002). *Speditionelle Transportdisposition: Modell- und Verfahrensentwicklung unter Berücksichtigung von Dynamik und Fremdvergabe. Diss. FernUniv. Hagen*. Wiesbaden: Deutscher Universitätsverlag. ISBN 3824476681.

Papastavrou, J. D. (1996). A stochastic and dynamic routing policy using branching processes with state dependent immigration. *European Journal of Operational Research*, *95*(1), 167–177.

Parasuraman, A. (1987). Customer-oriented corporate cultures are crucial to services marketing success. *The Journal of Services Marketing*, *1*(1), 39–46.

Parasuraman, A., Zeithaml, V. A., & Berry, L. L. (1985). A conceptual model of service quality and its implications for future research. *The Journal of Marketing*, *49*(4), 41–50.

Parasuraman, A., Berry, L. L., & Zeithaml, V. A. (1991). Understanding customer expectations of service. *Sloan Management Review*, *32*(3), 39–48.

Park, S. K., & Miller, K. W. (1988). Random number generators: good ones are hard to find. *Communications of the ACM*, *31*(10), 1192–1201.

Park, S. K., Miller, K. W., & Stockmeyer, P. K. (1993). Technical correspondence: remarks on choosing and implementing random number generators. *Communications of the ACM*, *36*(7), 105. doi:10.1145/159544.376068.

Parragh, S., Doerner, K., & Hartl, R. (2008). A survey on pickup and delivery problems: Part I: transportation between customers and depot. *Journal für Betriebswirtschaft*, *58*(1), 21–51.

Perugini, D., Lambert, D., Sterling, L., & Pearce, A. (2003). A distributed agent approach to global transportation scheduling. In *2003 IEEE/WIC international conference on intelligent agent technology (IAT '03)*, Halifax, Canada.

Pillac, V., Guéret, A., & Medaglia, A. (2010). *Vehicle routing problems: state of the art and prospects* (Technical report, 10/4/AUTO). Ecole des Mines de Nantes, France.

Pollack, B. L. (2009). Linking the hierarchical service quality model to customer satisfaction and loyalty. *The Journal of Services Marketing*, *23*(1), 42–50.

Potvin, J.-Y., & Rousseau, J. M. (1995). An exchange heuristic for routeing problems with time windows. *Journal of the Operational Research Society*, *46*(12), 1433–1446.

Powell, W. B. (1988). A comparative review of alternative algorithms for the dynamic vehicle allocation problem. In B. L. Golden (Ed.), *Studies in management science and systems. Vehicle routing* (pp. 249–291). Amsterdam: Elsevier. ISBN 0444704078.

Powell, W. B. (1996). A stochastic formulation of the dynamic assignment problem, with an application to truckload motor carriers. *Transportation Science*, *30*(3), 195–219.

Powell, W. B. (2007). *Approximate dynamic programming: solving the curses of dimensionality*. Hoboken: Wiley-Interscience. ISBN 0470171553.

Powell, W. B. (2009). What you should know about approximate dynamic programming. *Naval Research Logistics*, *56*(3), 239–249.

Powell, W. B., & Topaloglu, H. (2005). Fleet management. In S. W. Wallace & W. T. Ziemba (Eds.), *MPS-SIAM series on optimization: Vol. 5. Applications of stochastic programming* (pp. 185–215). Philadelphia: Society for Industrial and Applied Mathematics. ISBN 0898715555.

Powell, W. B., Jaillet, P., & Odoni, A. (1995). Stochastic and dynamic networks and routing. In *Handbooks in operations research and management science* (Vol. 8, pp. 141–295).

Psaraftis, H. N. (1980). Dynamic programming solution to the single vehicle many-to-many immediate request dial-a-ride problem. *Transportation Science*, *14*(2), 130–154.

Psaraftis, H. N. (1988). Dynamic vehicle routing problems. In B. L. Golden (Ed.), *Studies in management science and systems. Vehicle routing* (pp. 223–248). Amsterdam: Elsevier. ISBN 0444704078.

Psaraftis, H. N. (1995). Dynamic vehicle routing: status and prospects. *Annals of Operations Research*, *61*(1), 143–164.

Pureza, V., & Laporte, G. (2008). Waiting and buffering strategies for the dynamic pickup and delivery problem with time windows. *INFOR. Information Systems and Operational Research*, *46*(3), 165–176.

Puterman, M. L. (2005). *Wiley series in probability and statistics. Markov decision processes: discrete stochastic dynamic programming*. Hoboken: Wiley-Interscience. ISBN 9780470316887.

Regan, A. C., Mahmassani, H. S., & Jaillet, P. (1995). Improving efficiency of commercial vehicle operations using real-time information: potential uses and assignment strategies. *Transportation Research Record*, *1493*, 188–198.

Regan, A. C., Mahmassani, H. S., & Jaillet, P. (1998). Evaluation of dynamic fleet management systems: simulation framework. *Transportation Research Record*, *1645*, 176–184.

Rice, J., & Caniato, F. (2003). Building a secure and resilient supply network. *Supply Chain Management Review*, *7*(5), 22–30.

Richter, A. (2005). *Dynamische Tourenplanung: Modifikation von klassischen Heuristiken für das Dynamische Rundreiseproblem (DTSP) und das Dynamische Tourenplanungsproblem (DVRP) mit der Möglichkeit der Änderung des aktuellen Fahrzeugzuges*. PhD Thesis, University of Dresden, Germany.

Rodgers, J. L., & Nicewander, W. A. (1988). Thirteen ways to look at the correlation coefficient. *American Statistician*, *42*(1), 59–66.

Ropke, S., & Pisinger, D. (2006). An adaptive large neighborhood search heuristic for the pickup and delivery problem with time windows. *Transportation Science*, *40*(4), 455–472.

Ross, S. M. (2010). *Introduction to probability models* (10th ed.). Burlington: Academic Press/Elsevier. ISBN 0123756863.

Sáez, D., Cortés, C. E., & Núñez, A. (2008). Hybrid adaptive predictive control for the multi-vehicle dynamic pick-up and delivery problem based on genetic algorithms and fuzzy clustering. *Computers & Operations Research*, *35*(11), 3412–3438.

Sapsford, R., & Jupp, V. (2006). *Data collection and analysis* (2nd ed.). London: SAGE. ISBN 9780761943631.

Savelsbergh, M. W. P. (1992). The vehicle routing problem with time windows: minimizing route duration. *ORSA Journal on Computing*, *4*, 146–154.

Savelsbergh, M. W. P., & Sol, M. (1995). The general pickup and delivery problem. *Transportation Science*, *29*(1), 17–29.

Savelsbergh, M. W. P., & Sol, M. (1998). DRIVE: dynamic routing of independent vehicles. *Operations Research*, *46*(4), 474–490.

Schmid, V. (2012). Solving the dynamic ambulance relocation and dispatching problem using approximate dynamic programming. *European Journal of Operational Research*, *219*(3), 611–621.

Schmidt, J. W., & Taylor, R. E. (1970). *Simulation and analysis of industrial systems*. Homewood: Richard D. Irwin.

Schönberger, J. (2010). Adaptive demand peak management in online transport process planning. *OR Spektrum*, *32*(3), 831–859.

Schultes, D. (2008). *Route Planning in Road Networks*. PhD Thesis, Karlsruhe Institute of Technology.

Secomandi, N. (2000). Comparing neuro-dynamic programming algorithms for the vehicle routing problem with stochastic demands. *Computers & Operations Research*, *27*(11-12), 1201–1225.

Secomandi, N., & Margot, F. (2009). Reoptimization approaches for the vehicle-routing problem with stochastic demands. *Operations Research*, *57*(1), 214–230.

Sedgewick, R. (1983). *Addison-Wesley series in computer science. Algorithms*. Reading: Addison-Wesley. ISBN 0201066726.

Séguin, R., Potvin, J.-Y., Gendreau, M., Crainic, T. G., & Marcotte, P. (1997). Real-time decision problems: an operational research perspective. *Journal of the Operational Research Society*, *48*(2), 162–174.

Sexton, T., & Choi, Y. (1986). Pickup and delivery of partial loads with "soft" time windows. *American Journal of Mathematical and Management Sciences*, *6*(3-4), 369–398.

Sexton, T. R., & Bodin, L. D. (1985). Optimizing single vehicle many-to-many operations with desired delivery times: I. scheduling. *Transportation Science*, *19*(4), 378–410.

Sleator, D. D., & Tarjan, R. E. (1985). Amortized efficiency of list update and paging rules. *Communications of the ACM*, *28*(2), 202–208.

Smith, A. K., Bolton, R. N., & Wagner, J. (1999). A model of customer satisfaction with service encounters involving failure and recovery. *Journal of Marketing Research*, *36*(3), 356–372.

Solomon, M. M. (1987). Algorithms for the vehicle routing and scheduling problems with time window constraints. *Operations Research*, *35*(2), 254–265.

Stenfors, S., Tanner, L., Syrjänen, M., Seppälä, T., & Haapalinna, I. (2007). Executive views concerning decision support tools. *European Journal of Operational Research*, *181*(2), 929–938.

Stone, J., Stewart, R., & Otis, D. (2002). RFC3309: Stream Control Transmission Protocol (SCTP) checksum change. http://www.ietf.org/rfc/rfc3309.txt. Accessed on 2008-08-03.

Stone, J. R. (2001). *Checksums and the internet*. PhD diss, Department of Computer Science, Stanford University, Stanford, CA, USA. http://wwwlib.umi.com/dissertations/fullcit/3028177.

Stumpf, P. (1998). *GVB: Schriftenreihe: Vol. 39 Tourenplanung im speditionellen Güterfernverkehr: Univ., Diss.–Augsburg, 1998* (1st ed.). Nürnberg: GVB. ISBN 3922665349.

Swihart, M. R., & Papastavrou, J. D. (1999). A stochastic and dynamic model for the single-vehicle pick-up and delivery problem. *European Journal of Operational Research*, *114*(3), 447–464.

Tarantilis, C. D., & Kiranoudis, C. T. (2007). A flexible adaptive memory-based algorithm for real-life transportation operations: two case studies from dairy and construction sector. *European Journal of Operational Research*, *179*(3), 806–822.

Tarantilis, C. D., Diakoulaki, D., & Kiranoudis, C. T. (2004). Combination of geographical information system and efficient routing algorithms for real life distribution operations: new technologies in transportation systems. *European Journal of Operational Research*, *152*(2), 437–453.

Tax, S. S., Brown, S. W., & Chandrashekaran, M. (1998). Customer evaluations of service complaint experiences: implications for relationship marketing. *The Journal of Marketing*, *62*(2), 60–76.

Telefónica Germany GmbH & Co. OHG (2011). Tarif o2 Active Data Card. http://www.o2online.de/nw/tarife/datentarif/datentarif.html. Accessed on 2011-06-20.

Thangiah, S. R., Osman, I. H., Vinayagamoorthy, R., & Sun, T. (1993). Algorithms for the vehicle routing problems with time deadlines. *American Journal of Mathematical and Management Sciences*, *13*(3-4), 323–355.

The Media Audit (2011). Daily Newspaper Websites see 5.6 Percent Growth. http://www.themediaaudit.com/press/archived-newsletters/the-media-audit-fyi/march-2011/daily-newspaper-websites-see-56-percent-growth. Accessed on 2011-07-29.

Thomas, B. W. (2007). Waiting strategies for anticipating service requests from known customer locations. *Transportation Science*, *41*(3), 319–331.

Thomas, B. W., & White III, C. C. (2004). Anticipatory route selection. *Transportation Science*, *38*(4), 473–487.

Tijms, H. C. (2003). *A first course in stochastic models* (2nd ed.). Chichester: Wiley.

Tillman, F. A. (1969). The multiple terminal delivery problem with probabilistic demands. *Transportation Science*, *3*(3), 192–204.

Toth, P., & Vigo, D. (Eds.) (2002). *SIAM monographs on discrete mathematics and applications. The vehicle routing problem*. Philadelphia: Society for Industrial and Applied Mathematics. ISBN 0898714982.

Uyar, E. (2008). *Routing in stochastic environments*. PhD Thesis, Georgia Institute of Technology.

van de Klundert, J., & Wormer, L. (2010). ASAP: the After-Salesman Problem. *Manufacturing & Service Operations Management*, *12*, 627–641.

van Hemert, J. I., & La Poutré, J. A. (2004). Dynamic routing problems with fruitful regions: models and evolutionary computation. In X. Yao, E. Burke, J. A. Lozano, J. Smith, J. J. Merelo-Guervós, J. A. Bullinaria, J. Rowe, P. Tino, A. Kabán & H.-P. Schwefel (Eds.), *Lecture notes in computer science: Vol. 3242. Parallel Problem Solving from Nature—PPSN VIII* (pp. 692–701). Berlin: Springer.

van Hentenryck, P., & Bent, R. (2009). *Online stochastic combinatorial optimization*. Cambridge: MIT Press. ISBN 0262513471.

Verweij, B., Ahmed, S., Kleywegt, A. J., Nemhauser, G., & Shapiro, A. (2003). The sample average approximation method applied to stochastic routing problems: a computational study. *Computational Optimization and Applications*, *24*(2-3), 289–333.

Weiss, G. (2000). *Multiagent systems: a modern approach to distributed artificial intelligence* (2nd ed.). Cambridge: MIT Press. ISBN 0262731312.

Westphal, S., & Krumke, S. O. (2008). Pruning in column generation for service vehicle dispatching. *Annals of Operations Research*, *159*(1), 355–371.

Wilson, N. H. M., & Colvin, N. J. (1977). *Computer control of the Rochester dial-a-ride system* (Technical report R-77-30). Department of Civil Engineering, Massachusetts Institute of Technology, Center for Transportation Studies, Cambridge, MA.

Wu, C. W., Beck, J. C., & Brown, K. N. (2004). Dynamic vehicle routing with uncertain customer demand. In *Proceedings of Changes '04 (International workshop on constraint solving under change and uncertainty at CP2004)*, Toronto, Canada (pp. 75–76).

Xu, R. (2005). Survey of clustering algorithms. *IEEE Transactions on Neural Networks*, *16*(3), 645–678.

Yang, J., Jaillet, P., & Mahmassani, H. S. (2004). Real-time multivehicle truckload pickup and delivery problems. *Transportation Science*, *38*(2), 135–148.

Yang, W. H., Mathur, K., & Ballou, R. H. (2000). Stochastic vehicle routing problem with restocking. *Transportation Science*, *34*(1), 99–112.

Yates, D., Moore, D., & McCabe, G. (1999). *The practice of statistics* (1st ed.). New York: Freeman. ISBN 0716737744.

Zeimpekis, V., & Giaglis, G. M. (2005). A dynamic real-time vehicle routing system for distribution operations. In G. J. Doukidis & A. P. Vrechopoulos (Eds.), *Consumer driven electronic transformation* (pp. 23–37). Berlin: Springer. ISBN 978-3-540-27059-1.

Zeimpekis, V., Giaglis, G. M., & Minis, I. (2005). A dynamic real-time fleet management system for incident handling in city logistics. In *IEEE 61st vehicular technology conference, VTC 2005-Spring*, Stockholm, Sweden (Vol. 5, pp. 2900–2904).

Zeithaml, V. A., Berry, L. L., & Parasuraman, A. (1996). The behavioral consequences of service quality. *The Journal of Marketing*, *60*(2), 31–46.

Zhan, F. B. (1997). Three fastest shortest path algorithms on real road networks: data structures and procedures. *Journal of Geographic Information and Decision Analysis*, *1*(1), 69–82.

Zhan, F. B., & Noon, C. E. (1998). Shortest path algorithms: an evaluation using real road networks. *Transportation Science*, *32*(1), 65–73.

Zhan, F. B., & Noon, C. E. (2000). A comparison between label-setting and label-correcting algorithms for computing one-to-one shortest paths. *Journal of Geographic Information and Decision Analysis*, *4*(2), 1–11.

Index

F. Ferrucci, *Pro-active Dynamic Vehicle Routing*, Contributions to Management Science,
DOI 10.1007/978-3-642-33472-6,

MIX
Papier aus verantwortungsvollen Quellen
Paper from responsible sources
FSC® C105338

If you have any concerns about our products,
you can contact us on
ProductSafety@springernature.com

In case Publisher is established outside the EU,
the EU authorized representative is:
Springer Nature Customer Service Center GmbH
Europaplatz 3, 69115 Heidelberg, Germany

Printed by Libri Plureos GmbH
in Hamburg, Germany